T0299357

Probing the Meaning of

Quantum Mechanics

Information, Contextuality, Relationalism
and Entanglement

Other Related Titles from World Scientific

Probing the Meaning of Quantum Mechanics: Superpositions, Dynamics, Semantics and Identity
edited by Diederik Aerts, Christian de Ronde, Hector Freytes and Roberto Giuntini
ISBN: 978-981-3146-27-3

Probing the Meaning of Quantum Mechanics: Physical, Philosophical and Logical Perspectives
edited by Diederik Aerts, Sven Aerts and Christian de Ronde
ISBN: 978-981-4596-28-2

Universal Measurements: How to Free Three Birds in One Move
by Diederik Aerts and Massimiliano Sassoli de Bianchi
ISBN: 978-981-3220-15-7
ISBN: 978-981-3220-16-4 (pbk)

Schrödinger's Mechanics: Interpretation
by David B Cook
ISBN: 978-1-78634-490-8

The World According to Quantum Mechanics: Why the Laws of Physics Make Perfect Sense After All
2nd Edition
by Ulrich Mohrhoff and Manu Jaiswal
ISBN: 978-981-3273-69-6

Probing the Meaning of

Quantum Mechanics

Information, Contextuality, Relationalism and Entanglement

Proceedings of the II International Workshop on Quantum Mechanics and Quantum Information. Physical, Philosophical and Logical Approaches

CLEA, Brussels Free University, Belgium   23 – 24 July 2015

Editors

Diederik Aerts
CLEA, Brussels Free University, Belgium

Maria Luisa Dalla Chiara
University of Florence, Italy

Christian de Ronde
CONICET, University of Buenos Aires, Argentina
National University Arturo Jauretche, Argentina
CLEA, Brussels Free University, Belgium

Décio Krause
Federal University of Santa Catarina, Brazil

World Scientific

NEW JERSEY · LONDON · SINGAPORE · BEIJING · SHANGHAI · HONG KONG · TAIPEI · CHENNAI

Published by

World Scientific Publishing Co. Pte. Ltd.

5 Toh Tuck Link, Singapore 596224

USA office: 27 Warren Street, Suite 401-402, Hackensack, NJ 07601

UK office: 57 Shelton Street, Covent Garden, London WC2H 9HE

Library of Congress Cataloging-in-Publication Data
Names: Quantum Mechanics and Quantum Information: Physical, Philosophical and
    Logical Approaches (Workshop) (2015 : CLEA, Brussels Free University) |
    Aerts, Diederik, 1953– editor. | Dalla Chiara, Maria Luisa, 1938– editor. |
    de Ronde, Christian, 1976– editor. | Krause, Décio, editor.
Title: Probing the meaning of quantum mechanics : information, contextuality, relationalism and
    entanglement : Quantum Mechanics and Quantum Information: Physical, Philosophical and
    Logical Approaches, CLEA, Brussels Free University, Belgium, 23–24 July 2015 / editors:
    Diederik Aerts (Brussels Free University, Belgium), Maria Luisa Dalla Chiara
    (University of Florence, Italy), Christian de Ronde (University of Buenos Aires, Argentina &
    National University Arturo Jauretche, Argentina & Brussels Free University, Belgium),
    Décio Krause (University of Santa Catarina, Brazil).
Description: Singapore ; Hackensack, NJ : World Scientific Publishing Co. Pte. Ltd., [2019] |
    Proceedings of the II International Workshop on Quantum Mechanics and Quantum Information.
    Physical, Philosophical and Logical Approaches. | Includes bibliographical references.
Identifiers: LCCN 2018049796| ISBN 9789813276888 (hardcover ; alk. paper) |
    ISBN 9813276886 (hardcover ; alk. paper)
Subjects: LCSH: Quantum theory--Congresses.
Classification: LCC QC173.96 .Q3632 2015 | DDC 530.12--dc23
LC record available at https://lccn.loc.gov/2018049796

British Library Cataloguing-in-Publication Data
A catalogue record for this book is available from the British Library.

Copyright © 2019 by World Scientific Publishing Co. Pte. Ltd.

*All rights reserved. This book, or parts thereof, may not be reproduced in any form or by any means,
electronic or mechanical, including photocopying, recording or any information storage and retrieval
system now known or to be invented, without written permission from the publisher.*

For photocopying of material in this volume, please pay a copying fee through the Copyright Clearance
Center, Inc., 222 Rosewood Drive, Danvers, MA 01923, USA. In this case permission to photocopy
is not required from the publisher.

For any available supplementary material, please visit
https://www.worldscientific.com/worldscibooks/10.1142/11172#t=suppl

Desk Editor: Ng Kah Fee

Typeset by Stallion Press
Email: enquiries@stallionpress.com

Printed in Singapore

PREFACE

After more than a century since its birth, quantum mechanics continues to be an intriguing theory. Its key elements — namely, quantum superpositions and entanglement — are being used with great success at the forefront of today's most powerful and challenging technological innovations, such as quantum computation, quantum information, quantum sensoring and quantum cryptography. Its formalism is systematically giving rise to new multidisciplinary theoretical domains like quantum biology, quantum cognition, quantum natural language processing and quantum economics. However, regardless of this amazing productive capacity, it is still considered by the majority of the scientific community as a theory which has not been yet properly understood. It is for this reason that till today, its foundational aspects continue to raise passionate debates and discussions not only within physical, but also within philosophical, mathematical and logical circles.

This book is part of a series which addresses foundational questions about quantum mechanics and its applications in an interdisciplinary manner considering simultaneously physical, philosophical, mathematical and logical perspectives and analysis. Going from philosophy of quantum mechanics to quantum logic, from categorical approaches to quantum information processing, the originality of this book resides in the multiplicity of approaches which contribute to the common aim to grasp the meaning of the theory of quanta. We believe that the advancement of such understanding will be possible not by fragmenting the theory into smaller and smaller separated topics of research, but on the very contrary, through the common sharing and discussion of multiple perspectives coming from different fields and disciplines.

As in the previous editions, the novelty of the book comes from the multiple perspectives put forward by top researchers in quantum mechanics, from Europe as well as North and South America, discussing the meaning and structure of the theory of quanta. The book comprises in a balanced manner physical, philosophical, logical and mathematical approaches to quantum mechanics and quantum information. From quantum superpositions and entanglement to dynamics and the problem of identity, from quantum logic, computation and quasi-set theory to categorical approaches and teleportation, from realism and empiricism to operationalism and instrumentalism, the book touches from different viewpoints some of the most intriguing questions about quanta. From Buenos Aires to Brussels and Cagliari, from Florence to Florianópolis, the interaction between different groups is reflected in the many articles approaching different questions and problems. This book is interesting not only to the specialists but also to the general public attempting to get a grasp on some of the most fundamental questions of present quantum physics.

D. Aerts, M. L. Dalla Chiara, C. de Ronde & D. Krause

CONTENTS

viii

A NOTE ON THE STATISTICAL SAMPLING ASPECT OF DELAYED CHOICE ENTANGLEMENT SWAPPING

KARL SVOZIL

Institute for Theoretical Physics, Vienna University of Technology
Wiedner Hauptstraße 8-10/136, A-1040 Vienna, Austria.
Department of Computer Science, University of Auckland
Private Bag 92019, Auckland 1142, New Zealand.
E-mail: svozil@tuwien.ac.at

Quantum and classical models for delayed choice entanglement swapping by postselection of measurements are discussed.

Keywords: Entanglement; entanglement swapping; quantum information, postselection.

1. Quantum case

This is a very brief reflection on the sampling aspect of a paper [1] on delayed choice for entanglement swapping [2]. The basic idea of entanglement swapping is as follows: at first two uncorrelated pairs of entangled two-state particles in a singlet state are produced independently. Then from each one of the two different pairs a single particle is taken. These two particles are subsequently subjected to a measurement of their relational (joint) properties. Depending on these properties the remaining two particles (of the two particle pairs) can be sorted into four groups in a manner which guarantees that within each group the pairs of remaining particles are entangled. That is, effectively, (within each sort group) the remaining particles, although initially produced independently, become entangled.

More explicitly, suppose the particles in the first pair are labelled by 1 and 2, and in the second pair by 3 and 4, respectively. In the following only pure states will be considered. The wave function is given by a product of two singlet state wave functions

$$|\Psi\rangle = |\Psi_{1,2}^-\rangle |\Psi_{3,4}^-\rangle, \tag{1}$$

where $|\Psi_{i,j}^\pm\rangle = 2^{-\frac{1}{2}} (|0_i 1_j\rangle \pm |1_i 0_j\rangle)$ and $|\Phi_{i,j}^\pm\rangle = 2^{-\frac{1}{2}} (|0_i 0_j\rangle \pm |1_i 1_j\rangle)$ are the states associated with the Bell basis $\mathfrak{B}_1 = \{|\Psi^-\rangle, |\Psi^+\rangle, |\Phi^-\rangle, |\Phi^+\rangle\}$ (or, equivalently, the associated context), "0" and "1" refers to the quantum numbers of the particles, and the subscripts indicate the particle number. In addition, consider the product states $|00\rangle$, $|01\rangle$, $|10\rangle$, and $|11\rangle$, forming another possible basis (among a continuum of bases) $\mathfrak{B}_2 = \{|--\rangle, |-+\rangle, |+-\rangle, |++\rangle\}$ of, or context in, four dimensional Hilbert space.

Associated with these eight unit vectors in $\mathfrak{B}_1$ and $\mathfrak{B}_2$ are the eight projection operators from the dyadic products $\mathbf{E}^\psi = |\psi\rangle\langle\psi|$, with ψ running over the entangled and product states, respectively.

Notice, for the sake of concreteness, that these states and projection operators can be represented by the vector components $|0\rangle = (1,0)^T$ and $|1\rangle = (0,1)^T$, respectively; but these representations will not be explicitly used here.

The product (1) is a sum of products of the states of the two "outer" particles (particle 1 from pair 1 & particle 4 from pair 2) and the two "inner" particles (particle 2 from pair 1 & particle 3 from pair 2); it can be recast in terms of the two bases in two ways:

$$|\Psi\rangle = \frac{1}{2}\left(|\Psi_{1,4}^+\rangle|\Psi_{2,3}^+\rangle - |\Psi_{1,4}^-\rangle|\Psi_{2,3}^-\rangle\right.$$
$$\left. +|\Phi_{1,4}^+\rangle|\Phi_{2,3}^+\rangle - |\Phi_{1,4}^+\rangle|\Phi_{2,3}^+\rangle\right) \tag{2}$$

in terms of the bell basis; and, in terms of the product basis by

$$|\Psi\rangle = |0_1 1_4\rangle|1_2 0_3\rangle - |0_1 0_4\rangle|1_2 1_3\rangle$$
$$-|1_1 1_4\rangle|0_2 0_3\rangle + |1_1 0_4\rangle|0_2 1_3\rangle. \tag{3}$$

Suppose an agent Alice is recording the "outer" particle 1, agent Bob is recording the "outer" particle 4, and agent Eve is recording the "inner" particles 2 and 3, respectively. Suppose further that Eve is free to choose her *type* of experiment – that is, either by observing the context $\mathbf{E}^{--}$, $\mathbf{E}^{-+}$, $\mathbf{E}^{+-}$, and $\mathbf{E}^{++}$ associated with the product basis, exclusive or observing the context $\mathbf{E}^{\Psi^-}$, $\mathbf{E}^{\Psi^+}$, $\mathbf{E}^{\Phi^-}$, and $\mathbf{E}^{\Phi^+}$, corresponding to the Bell basis states. As a consequence of Eve's choice the resulting state on Alice's and Bob's end is either a projection onto some (non-entangled) product state $|++\rangle$, $|+-\rangle$, $|-+\rangle$, and $|--\rangle$, exclusive or onto some entangled Bell basis state $|\Psi^-\rangle$, $|\Psi^+\rangle$, $|\Phi^-\rangle$, and $|\Phi^+\rangle$, respectively.

Peres' idea was to augment entanglement swapping with delayed choice; even to the point that Alice and Bob record their particles first; and let Eve later, by a delayed choice [3], decide the type of measurement she chooses to perform: Eve may measure propositions either corresponding to the elements of the Bell basis $\mathfrak{B}_1$, or of the product basis $\mathfrak{B}_2$. In the first case, in some quantum Hocus Pocus way, "entanglement is produced *a posteriori*, after the entangled particles have been measured and may even no longer exist [1]."

In order to obtain a clearer picture, let us observe that, while Eve can choose between the two contexts (or measurement bases) $\mathfrak{B}_1$ or $\mathfrak{B}_2$, she has no control of the particular outcome – that is, according to the axioms of quantum mechanics, the concrete state in which she finds the particles 2 & 3 occurs irreducibly random, with probability $1/4$ for each one of the terms in (2) and (3).

This can be interpreted as yet another instance of the peaceful coexistence [4,5] between relativity and quantum mechanics, mediated by parameter independence but outcome dependence of such events: Eve wilfully chooses the parameters – in this case the Bell basis $\mathfrak{B}_1$ *versus* the product basis $\mathfrak{B}_2$ – but quantum mechanics, and in particular, the recordings of Alice and Bob, are insensitive to that. Yet, Eve cannot in any way choose or stimulate the outcomes at her side, which quantum mechanics is sensitive to. (Actually, if Eve could somehow manipulate the outcome – maybe by stimulated emission [6] – this would be another instance of faster-than-light quantum communication, and possibly also the end of peaceful coexistence.)

Eve's task is twofold: (i) in communicating the *type* of measurement performed (Bell state *versus* product state observables), Eve tells Alice and Bob whether she samples an entangled or a product state; and (ii) in communicating her concrete measurement outcome Eve informs Alice and Bob about the concrete entangled state they are dealing with. For the sake of an example of a protocol sentence of Eve, consider this one: "*I decided to measure my ith set of two particles 2 & 3 in the Bell basis, and found the particles to be in the singlet state $|\Psi_{2,3}^-\rangle$ (so your state should have also been a singlet one, namely $|\Psi_{1,4}^-\rangle$); and your outcomes are the inverse of mine; that is, $i, j \to [(i+1) \mod 2], [(j+1) \mod 2])$.*"

Thereby Eve is not merely *sampling*, but also *partitioning* the table of Alice's and Bob's recordings – both according to her one choice of context, as well as through her measurement outcomes. Already Peres addressed this issue by stating "the point is that it is meaningless to assert that two particles are entangled without specifying in which state they are entangled, just as it is meaningless to assert that a quantum system is in a pure state without specifying that state [1]."

4

2. Classical analogue

For the sake of making explicitly what this means, consider a classical analogue, and study binary observables in one measurement direction only. Classical singlet states have been defined previously [7], but as long as effectively one-dimensional (with respect to the measurement direction) configurations are considered it suffices to consider pairs of outcomes "$0_i - 1_j$" or "$1_i - 0_j$," where the subscripts refer to the particle constituents. These product states satisfy the property that the observables of the particles constituting that singlet are always different. The associated observables are either joint observables, or separable ones.

Already at this point, it could quite justifiably be objected that this is an improper model for quantum singlets, as it implies that the two particles constituting the singlet have definite individual observable values. In contrast, a singlet quantum state is solely defined in terms of the *correlations* (joint probability distributions) [8–10], or, by another term, the *relational properties* [11,12] among the quanta; whereby (with some reasonable side assumptions such as non-contextuality) the supposition that the quanta carry additional information about their definite individual states leads to a complete contradiction [13,14].

Nevertheless, if one accepts this classical model with the aforementioned provisions, it is possible to explicitly study the partitioning of joint outcomes as follows. Consider a concrete list of possible outcomes of two uncorrelated singlets – note that, as per definition, the constituents forming each singlet are (intrinsically, that is within each singlet) correlated; but the two singlets are externally uncorrelated – as tabulated in Table 1. This is an enumeration of simulated empirical data – essentially binary observables – which are interpreted by *assigning* or *designating* some properties of a subensemble, thereby effectively *inducing* or *rendering* some other properties or features on the remaining subensemble.

What is important here is to realize that the data allow many views or interpretations. Consequently, what is a property of the data is purely conventionalized and means relative. The only ontology relates to the pairs of statistically independent singlets; how their constituents relate to each other is entirely epistemic. To emphasize this, Peres could be quoted a third time by repeating that "it is obvious that from the raw data collected by Alice and Bob it is possible to select in many different ways subsets that correspond to entangled pairs. The only role that Eve has in this experiment is to tell Alice and Bob how to select such a subset [1]." It is amusing to notice that Peres' entire abstract applies to the analogue situation just

Table 1. (Color online) Three partitions of, or views on, one and the same data set A1, E2, E3, B4 created through 30 simulated runs of an experiment. There are the two uncorrelated singlet sources A1–E2 and E3–B4, producing random $0-1$- or $1-0$-pairs of data. The difference between the three partitions lies in the choice of how the data E2 and E3 are interpreted: If E2–E3 is interpreted as coincidence measurements "revealing" their relational properties, indicated by c and a gray background, then A1 and B4 are characterized by their relational properties; in particular, by the even and odd parity, indicated by green and red backgrounds, respectively. If, on the other hand, E2 and E3 are interpreted as measurements of single events, indicated by p and a white background, then A1 and B4 are characterized by their separate pairs of outcomes, indicated by light yellow and blue backgrounds, respectively.

#	A1	E2	E3	B4	c/p	E	A1	E2	E3	B4	c/p	E	A1	E2	E3	B4	c/p	E
1	0	1	0	1	p	p3	0	1	0	1	p	p3	0	1	0	1	c	o2
2	0	1	1	0	c	e2	0	1	1	0	c	e2	0	1	1	0	c	e2
3	0	1	0	1	c	o2	0	1	0	1	c	o2	0	1	0	1	c	o2
4	0	1	1	0	c	e2	0	1	1	0	p	p4	0	1	1	0	c	e2
5	0	1	1	0	p	p4	0	1	1	0	c	e2	0	1	1	0	c	e2
6	0	1	0	1	c	o2	0	1	0	1	c	o2	0	1	0	1	p	p3
7	1	0	0	1	p	p1	1	0	0	1	c	e1	1	0	0	1	p	p1
8	1	0	1	0	c	o1	1	0	1	0	c	o1	1	0	1	0	p	p2
9	1	0	1	0	p	p2	1	0	1	0	p	p2	1	0	1	0	p	p2
10	0	1	1	0	c	e2	0	1	1	0	c	e2	0	1	1	0	p	p4
11	1	0	1	0	p	p2	1	0	1	0	c	o1	1	0	1	0	c	o1
12	0	1	0	1	c	o2	0	1	0	1	c	o2	0	1	0	1	c	o2
13	0	1	0	1	c	o2	0	1	0	1	c	o2	0	1	0	1	p	p3
14	1	0	0	1	p	p1	1	0	0	1	p	p1	1	0	0	1	p	p1
15	1	0	0	1	c	e1	1	0	0	1	p	p1	1	0	0	1	c	e1
16	0	1	0	1	p	p3	0	1	0	1	c	o2	0	1	0	1	c	o2
17	1	0	0	1	p	p1	1	0	0	1	p	p1	1	0	0	1	c	e1
18	0	1	0	1	p	p3	0	1	0	1	p	p3	0	1	0	1	c	o2
19	1	0	1	0	p	p2	1	0	1	0	c	o1	1	0	1	0	p	p2
20	1	0	1	0	c	o1	1	0	1	0	c	o1	1	0	1	0	p	p2
21	0	1	1	0	p	p4	0	1	1	0	p	p4	0	1	1	0	p	p4
22	0	1	1	0	c	e2	0	1	1	0	c	e2	0	1	1	0	p	p4
23	0	1	0	1	c	o2	0	1	0	1	p	p3	0	1	0	1	p	p3
24	1	0	0	1	c	e1	1	0	0	1	c	e1	1	0	0	1	c	e1
25	0	1	0	1	p	p3	0	1	0	1	p	p3	0	1	0	1	p	p3
26	0	1	1	0	c	e2	0	1	1	0	p	p4	0	1	1	0	p	p4
27	1	0	1	0	p	p2	1	0	1	0	p	p2	1	0	1	0	c	o1
28	1	0	0	1	c	e1	1	0	0	1	c	e1	1	0	0	1	c	e1
29	1	0	0	1	c	e1	1	0	0	1	c	e1	1	0	0	1	p	p1
30	0	1	1	0	c	e2	0	1	1	0	p	p4	0	1	1	0	c	e2
⋮	⋮	⋮	⋮	⋮	⋮	⋮	⋮	⋮	⋮	⋮	⋮	⋮	⋮	⋮	⋮	⋮	⋮	⋮

discussed (but we refrain from repeating it here because of fear of copyright infringement).

What are the differences between the classical analogue and the quantum original? In answering this question one can consult another paper by Peres [7] on the hypothetical (non-)existence of counterfactuals (or, in Specker's scholastic terminology [15], *Infuturabilien*). One of the most striking differences is the fact that classical configurations allow a truth table (that is, physical properties) of the constituents of the singlets, whereas hypothetical (counterfactual) truth tables associated with entangled quantum states, when viewed at different directions or contexts, in general do not; at least not statistically [7], but also not on a per particle pair basis [13,14].

That is, if we analyse the Bell states sampled according to Eve's directives by Alice and Bob, they will be not only correlated but also entangled; in particular, particles in a sampled singlet state will perform like a singlet state produced from a common source [16]. In particular, their correlations, involving more than one measurement directions, violate Bell-type inequalities.

3. Type of randomness

There is also another important difference in the perception of randomness involved. The randomness in the classical analogue resides in the (pseudo-)random creation of the two singlet pairs.

In quantum mechanics certain entangled states, such as the states in the Bell basis $\mathfrak{B}_1$, exclude the separate existence of single-particle observables. Formally this is easily seen, as tracing out one particle (i.e., taking the partial trace with respect to this particle) yields the identity density matrix for the other particle: for instance, $\mathrm{Tr}_i \left(|\Psi_{i,j}^{\pm}\rangle\langle\Psi_{i,j}^{\pm}| \right) = \mathrm{Tr}_i \left(|\Phi_{i,j}^{\pm}\rangle\langle\Phi_{i,j}^{\pm}| \right) = \mathrm{Tr}_j \left(|\Psi_{i,j}^{\pm}\rangle\langle\Psi_{i,j}^{\pm}| \right) = \mathrm{Tr}_j \left(|\Phi_{i,j}^{\pm}\rangle\langle\Phi_{i,j}^{\pm}| \right) = (1/2)\,\mathbb{I}_2$.

This is a consequence of the fact that (under certain mild side assumptions such as non-contextual value definiteness) a quantum state can only be value definite with respect to a single one proposition [14]– that is, the proposition corresponding to the state preparation, which in turn corresponds to a single direction, and a unit vector in the 2^n-dimensional Hilbert space of n 2-state particles. (A generalization to particles with k states per particles is straightforward.) Relative to this single value definite proposition, all other propositions corresponding to non-orthogonal vectors are indeterminate. Zeilinger's Foundational Principle [11,12] is a corollary of this fact, once an orthonormal basis system including the vector corresponding to this determinate property is fixed: it is always possible to define filters corresponding to equipartitions of basis states which are co-measurable and resolve states corresponding to single basis elements [17,18].

As has already been mentioned earlier, Schrödinger [8], was the first to notice that, as expressed by Everett [9], in general "a constituent subsystem cannot be said to be in any single well-defined state, independently of the remainder of the composite system." The entire state of multiple quanta can be expressed completely in terms of correlations (joint probability distributions) [10,19], or, by another term, relational properties [11], among observables belonging to the subsystems. There is "a complete knowledge of the whole without knowing the state of any one part. That a thing can be in a definite state, even though its parts were not [20]."

Some have thus suggested that, upon "forcing" the "measurement" of such indeterminate observables the "outcomes" allow one to obtain "irreducible randomness." In theological terms, this is a *creatio continua;* quasi *ex nihilo.* Indeed, this appears to be the canonical position at present.

I have argued [21,22] that in such cases a context translation takes place that is effectively mediated by the measurement apparatus. In many cases this apparatus may be considered quasi-classical; with many degrees of freedom which are, for all practical purposes (but not in principle), impossible to resolve. Therefore, the forced single outcome reflects both the microstate of the "measurement device" as well as the "object," whereby the cut between those two is purely conventional [23] and, in close analogy to statistical mechanics [24] means relative.

4. Concluding remarks

Pointedly stated any set of raw data from correlated sources, quantum or otherwise, can be combined and (re-)interpreted in many different ways. Any such way presents a particular view on, or interpretation of, these data. There is no unique way of representation; everything remains means relative and conventional.

Temporal considerations are not important here, because no causation, just correlations are involved. This is not entirely dissimilar to what has already been pointed out by Born, "there are deterministic relations which are not causal; for instance, any time table or programmatic statement [25]."

The difference between the sampling of quantum and classical system is the scarcity of information encoded in entangled quantum states, which carry relational information about joint properties of the particles involved (a property they share with their classical counterparts) but do not carry information about their single constituents, as classical states additionally do.

5. Acknowledgements

This work was supported in part by the European Union, Research Executive Agency (REA), Marie Curie FP7-PEOPLE-2010-IRSES-269151-RANPHYS grant. Responsibility for the information and views expressed in this article lies entirely with the author. The author declares no conflict of interest.

References

1. A. Peres, *Journal of Modern Optics* **47**, 139 (2000).
2. M. Zukowski, A. Zeilinger, M. A. Horne and A. K. Ekert, *Physical Review Letters* **71**, 4287(Dec 1993).
3. X.-S. Ma, J. Kofler and A. Zeilinger, *Reviews of Modern Physics* **88**, p. 015005(Mar 2016).
4. C. G. Timpson, Entanglement and relativity, in *Understanding Physical Knowledge*, eds. R. Lupacchini and V. Fano (University of Bologna, CLUEB, Bologna, 2002) pp. 147–166.
5. M. P. Seevinck, *Can quantum theory and special relativity peacefully coexist?*, tech. rep. (2010).
6. K. Svozil, What is wrong with SLASH?, eprint arXiv:quant-ph/0103166, (1989).
7. A. Peres, *American Journal of Physics* **46**, 745 (1978).
8. E. Schrödinger, *Mathematical Proceedings of the Cambridge Philosophical Society* **31**, 555 (1935).
9. H. Everett III, *Reviews of Modern Physics* **29**, 454 (1957).
10. D. N. Mermin, *American Journal of Physics* **66**, 753 (1998).
11. A. Zeilinger, *Foundations of Physics* **29**, 631 (1999).
12. C. G. Timpson, *Studies in History and Philosophy of Science Part B: Studies in History and Philosophy of Modern Physics* **34**, 441 (2003), Quantum Information and Computation.
13. I. Pitowsky, *Journal of Mathematical Physics* **39**, 218 (1998).
14. A. A. Abbott, C. S. Calude and K. Svozil, *Journal of Mathematical Physics* **56**, 102201(1 (2015).
15. E. Specker, *Dialectica* **14**, 239 (1960).
16. X.-S. Ma, S. Zotter, J. Kofler, R. Ursin, T. Jennewein, C. Brukner and A. Zeilinger, *Nature Physics* **8**, 479 (2012).
17. N. Donath and K. Svozil, *Physical Review A* **65**, p. 044302 (2002).
18. K. Svozil, *Physical Review A* **66**, p. 044306 (2002).
19. W. K. Wootters, Local accessibility of quantum states, in *Complexity, Entropy, and the Physics of Information*, ed. W. H. ZurekSFI Studies in the Sciences of Complexity, Vol. VIII (Addison-Wesley, Boston, 1990) pp. 39–46.

20. IBM, Charles Bennett – a founder of quantum information theory (2016), May 3rd, 2016, accessed July 16th, 2016.
21. K. Svozil, *Journal of Modern Optics* **51**, 811 (2004).
22. K. Svozil, *International Journal of Theoretical Physics* **53**, 3648 (2014).
23. K. Svozil, *Foundations of Physics* **32**, 479 (2002).
24. W. C. Myrvold, *Studies in History and Philosophy of Science Part B: Studies in History and Philosophy of Modern Physics* **42**, 237 (2011).
25. M. Born, *Science News* **17**, 93 (1949), Joule Memorial Lecture, 1950. Reprinted in Ref. [26, p. 78-83].
26. M. Born, *Physics in my generation* (Pergamon Press, London & New York, 1956).

THE EXTENDED BLOCH REPRESENTATION OF QUANTUM MECHANICS FOR INFINITE-DIMENSIONAL ENTITIES

DIEDERIK AERTS

Center Leo Apostel for Interdisciplinary Studies and Department of Mathematics,
Brussels Free University, 1050 Brussels, Belgium
E-mail: diraerts@vub.ac.be

MASSIMILIANO SASSOLI DE BIANCHI

Center Leo Apostel for Interdisciplinary Studies,
Brussels Free University, 1050 Brussels, Belgium
E-mail: msassoli@vub.ac.be

We show that the extended Bloch representation of quantum mechanics also applies to infinite-dimensional entities, to the extent that the number of (possibly infinitely degenerate) outcomes of a measurement remains finite, which is always the case in practical situations.

Keywords: Probability; hidden-measurements; degenerate measurements; hidden-variables; Born rule; Bloch sphere; extended Bloch representation.

1. Introduction

The so-called 'spin quantum machine', also known as the 'ϵ-model', or 'sphere model' [1–3], is an extension of the standard (3-dimensional) Bloch sphere representation that includes a description also of the measurements, as (weighted) symmetry breaking processes selecting (in a non-predictable way) the hidden-measurement interactions responsible for producing the transitions towards the outcome-states. Recently, the model has been extended, so that measurements having an arbitrary number N of (possibly degenerate) outcomes can also be described, in what has been called the 'extended Bloch representation' (EBR) of quantum mechanics [4–10].

So far, the EBR has been formulated only for finite-dimensional quantum entities, although of arbitrary dimension. It is thus natural to ask if the representation remains consistent when dealing with infinite-dimensional entities. Of course, certain quantum entities, like spin entities, are intrinsically finite-dimensional. For instance, the Hilbert state space of a spin-s entity is $(2s + 1)$-dimensional and can be taken to be isomorphic to $\mathcal{H} = \mathbb{C}^{2s+1}$. However, an entity as simple as an electron, when considered in relation to position or momentum measurements, requires an infinite-dimensional Hilbert space $\mathcal{H} = L^2(\mathbb{R}^3)$, to account for all its possible states.

As we emphasized in [4], the EBR being valid for an arbitrary finite number N of dimensions, it can be advocated that if the physics of infinite-dimensional entities can be recovered by taking the limit $N \to \infty$ of suitably defined finite-dimensional entities, then a hidden-measurement description of quantum measurements should also apply for infinite-dimensional entities. More precisely, assuming that it is always possible to express the transition probabilities of an infinite-dimensional entity as the limit of a sequence of transition probabilities of finite-dimensional entities, and considering that a hidden-measurement interpretation holds for the latter, one would expect it to also hold for the former.

However, the possibility of a hidden-measurement interpretation does not necessarily imply the existence, for infinite-dimensional entities, of an explicit representation. For instance, some years ago Coecke was able to apply the hidden-measurement approach to measurements having an infinite set of outcomes, but to do so he had to take the space describing the hidden-measurement interactions to be the fixed interval $[0, 1]$, independently of the number N of outcomes, and this precisely to avoid problems when taking the infinite limit $N \to \infty$ [11]. His construction is thus very different from the canonical EBR, as in the latter the dimension of the set of hidden-measurements depends on the number of outcomes.

More precisely, in the EBR the set of hidden-measurement interactions, for a measurement having N possible outcomes, is given by a $(N - 1)$-dimensional simplex $\triangle_{N-1}$, inscribed in a convex region of states which, in turn, is inscribed in a $(N^2 - 1)$-dimensional unit sphere $B_1(\mathbb{R}^{N^2-1})$ [4]. A measurement then consists first in a deterministic process, producing a decoherence of the pre-measurement state, represented by an abstract point at the surface of the sphere that plunges into it, to reach the measurement simplex $\triangle_{N-1}$, following a path orthogonal to the latter. In this way, N different disjoint subregions A_i of $\triangle_{N-1}$ are defined, $i = 1, \ldots, N$, whose measures[a] describe the number of measurement-interactions that are available to actualize the corresponding outcomes. This means that the relative

---

[a] One should say, more precisely, $(N - 1)$-dimensional volumes, or Lebesgue measures.

measures $\frac{\mu(A_i)}{\mu(\triangle_{N-1})}$ of the different subregions can be interpreted as the probabilities to obtain the associated outcomes, and the remarkable result of the model is that these probabilities are exactly those predicted by the Born rule [4].

However, when taking the $N \to \infty$ limit of the extended Bloch construction, there is the following problem. The M-dimensional volume of a M-ball of radius r, given by:

$$\mu[B_r(\mathbb{R}^M)] = \frac{\pi^{\frac{M}{2}} r^M}{\Gamma(\frac{M}{2} + 1)}, \tag{1}$$

tends to zero, as $M \to \infty$. Indeed, if M is even, we have $\Gamma(\frac{M}{2} + 1) = \frac{M}{2}!$, so that according to Stirling's approximation, $M! \sim \sqrt{2\pi M}(\frac{M}{e})^M$, we have the asymptotic behavior:

$$\mu[B_r(\mathbb{R}^M)] \sim \frac{1}{\sqrt{2e}\pi r}\left(\sqrt{\frac{2\pi e}{M}}\, r\right)^{M+1}, \tag{2}$$

as $M \to \infty$, and a similar asymptotic formula can be found when M is odd, using $\Gamma(\frac{M}{2} + 1) = \sqrt{\pi}\, 2^{-\frac{M+1}{2}} M!!$. In other words, the measure of a M-dimensional ball of fixed radius r goes to zero extremely fast when the dimension M increases.[b]

The same is necessarily true for all structures of same dimension that are contained in it, like for instance inscribed simplexes. More precisely, the measure of a M-dimensional simplex $\triangle_M$, inscribed in a sphere of radius r is [4]:

$$\mu(\triangle_M) = \frac{\sqrt{M}}{M!}\left(\frac{M+1}{M}\right)^{\frac{M+1}{2}} r^M. \tag{3}$$

Using again Stirling's approximation, we thus obtain the asymptotic form:

$$\mu(\triangle_M) \sim \frac{1}{\sqrt{2\pi}}\left(\frac{e\,r}{M}\right)^M, \tag{4}$$

which goes even faster to zero than (2). So, if we naively consider the infinite-dimensional limit of the EBR, we find that the measures of the

---

[b]The fact the measure of a M-ball (its M-dimensional volume) of fixed radius tends exponentially fast to zero as M increases is counter intuitive. Indeed, for a unit radius $r = 1$, we have $\mu[B_1(\mathbb{R}^1)] = 2$, $\mu[B_1(\mathbb{R}^2)] = \pi > 2$, $\mu[B_1(\mathbb{R}^3)] = \frac{4}{3}\pi > \pi$, $\mu[B_1(\mathbb{R}^4)] = \frac{\pi^2}{2} > \frac{4}{3}\pi$, $\mu[B_1(\mathbb{R}^5)] = \frac{8\pi^2}{15} > \frac{\pi^2}{2}$, but $\mu[B_1(\mathbb{R}^6)] = \frac{\pi^3}{6} < \frac{8\pi^2}{15}$. In other words, the measure increases from $M = 1$ to $M = 5$, then it starts decreasing as from $M = 6$.

structures involved in the model rapidly go to zero. Nevertheless, considering that in actual measurement situations the number of distinguishable outcomes is always finite, this will prove to be unproblematic, as we are going to show. Also, as we will suggest in the last section, one can even speculate that the measurement-interactions would precisely supervene because of the meeting between an entity that is possibly infinite-dimensional, and the constraints exercised by a measurement context only allowing for a finite number of possible outcomes.

2. The infinite-dimensional limit

According to the EBR of quantum mechanics, the transition probability $\mathcal{P}[D_N(\mathbf{r}) \to P_N(\mathbf{n})]$, from an initial state $D_N(\mathbf{r})$ to a final outcome-state $P_N(\mathbf{n}_i)$, is given by the formula [4]:

$$\mathcal{P}[D_N(\mathbf{r}) \to P_N(\mathbf{n}_i)] = \frac{1}{N}[1 + (N-1)\,\mathbf{r} \cdot \mathbf{n}_i] = \frac{1}{N}[1 + (N-1)\,\mathbf{r}^{\parallel} \cdot \mathbf{n}_i]. \quad (5)$$

Here $D_N(\mathbf{r})$ and $P_N(\mathbf{n}_i)$ are one-dimensional projection operators acting in $\mathcal{H}_N = \mathbb{C}^N$, which can be written as:

$$D_N(\mathbf{r}) = \frac{1}{N}(\mathbb{I}_N + c_N\,\mathbf{r} \cdot \boldsymbol{\Lambda}), \quad P_N(\mathbf{n}_i) = \frac{1}{N}(\mathbb{I} + c_N\,\mathbf{n}_i \cdot \boldsymbol{\Lambda}), \quad (6)$$

where $\mathbf{r}$ and $\mathbf{n}_i$ are unit vectors in the generalized Bloch sphere $B_1(\mathbb{R}^{N^2-1})$, with $\mathbf{n}_i$ being also one of the N vertices of a given $(N-1)$-dimensional measurement simplex $\triangle_{N-1}$, inscribed in $B_1(\mathbb{R}^{N^2-1})$, $c_N = [N(N-1)/2]^{\frac{1}{2}}$, $\boldsymbol{\Lambda}$ is a vector whose components Λ_i are a choice of the $N^2 - 1$ generators of the group $SU(N)$, and the 'dot' denotes the scalar product in $\mathbb{R}^{N^2-1}$. In (5) we have also introduced the vector $\mathbf{r}^{\parallel} = \mathbf{r} - \mathbf{r}^{\perp}$, where $\mathbf{r}^{\perp}$ is the component of $\mathbf{r}$ perpendicular to $\triangle_{N-1}$, i.e., $\mathbf{r}^{\perp} \cdot \mathbf{n}_i = 0$, for all $i = 1, \ldots, N$ (we refer the reader to [4] for a detailed exposition).

It is straightforward to take the $N \to \infty$ limit of (5). By doing so, one just needs to keep in mind that also the dimension of the Hilbert space increases, as N increases. In other words, one has to assume that, as $N \to \infty$, both $D_N(\mathbf{r})$ and $P_N(\mathbf{n}_i)$ converge (in the Hilbert-Schmidt sense) to well-defined projection operators $D(\mathbf{r})$ and $P(\mathbf{n}_i)$, respectively, acting in

$\mathcal{H}_\infty = \ell^2(\mathbb{C}).^c$ More precisely, we have to assume that:

$$|\mathcal{P}(D_N \to P_N) - \mathcal{P}(D \to P)| = |\operatorname{Tr} D_N P_N - \operatorname{Tr} DP|$$
$$= |\operatorname{Tr} D_N(P_N - P) + \operatorname{Tr}(D_N - D)P|$$
$$\leq |\operatorname{Tr}(P_N - P)| + |\operatorname{Tr}(D_N - D)| \to 0, \text{ as } N \to \infty. \qquad (7)$$

Then, the $N \to \infty$ limit of (5) is:

$$\mathcal{P}[D(\mathbf{r}) \to P(\mathbf{n}_i)] = \mathbf{r} \cdot \mathbf{n}_i = \mathbf{r}^{\|} \cdot \mathbf{n}_i, \qquad (8)$$

where $\mathbf{r}$ and $\mathbf{n}_i$ are now vectors belonging to $\ell^2(\mathbb{R})$, the Hilbert space of infinite sequences $\{r_1, r_1, \dots\}$ of real numbers satisfying $\sum_{i=1}^\infty r_i^2 < \infty$, with scalar product $\mathbf{r} \cdot \mathbf{n}_i = \sum_{j=1}^\infty r_j[\mathbf{n}_i]_j$.

It is worth remembering that one of the differences between the standard Bloch representation ($N = 2$) and the EBR, for $N > 2$, is that in the latter not all vectors in $B_1(\mathbb{R}^{N^2-1})$ are necessarily representative of *bona fide* states. However, all good states are represented by vectors belonging to a convex region inscribed in $B_1(\mathbb{R}^{N^2-1})$. The vectors living outside of such convex region of states (the shape of which depends on the choice of the generators Λ_i) can be characterized by the fact that for them (5) would give unphysical negative values. This possibility is even more manifest in the infinite-dimensional formula (8), as is clear that a scalar product can take both positive and negative values. For instance, the unit vector $\mathbf{r} = -\mathbf{n}_i$ cannot be representative of a state, as for it the transition probability (8) would be equal to -1.

Let us now investigate what is the $N \to \infty$ limit of the $(N-1)$-dimensional measurement simplex $\triangle_{N-1}$. By definition, we have:

$$\triangle_{N-1} = \{\mathbf{t} \in \mathbb{R}^N | \mathbf{t} = \sum_{i=1}^N t_i \mathbf{n}_i, \sum_{i=1}^N t_i = 1, 0 \leq t_i \leq 1\}, \qquad (9)$$

where the $\mathbf{n}_i$, $i = 1, \dots, N$, are the N vertex vectors of $\triangle_{N-1}$, describing the N outcome states and obeying:

$$\mathbf{n}_i \cdot \mathbf{n}_j = -\frac{1}{N-1} + \delta_{ij}\frac{N}{N-1}, \qquad (10)$$

so that we also have $\sum_{i=1}^N \mathbf{n}_i = \mathbf{0}$. Taking the $N \to \infty$ limit of (9)-(10), we thus obtain:

$$\triangle_\infty = \{\mathbf{t} \in \mathbb{R}^\infty | \mathbf{t} = \sum_{i=1}^\infty t_i \mathbf{n}_i, \sum_{i=1}^\infty t_i = 1, 0 \leq t_i \leq 1\}, \qquad (11)$$

$^c\ell^2(\mathbb{C})$ is the Hilbert space of infinite sequences $\{v_0, v_1, \dots\}$ of complex numbers satisfying $\sum_{i=0}^\infty |v_i|^2 < \infty$, with scalar product $\langle v|w \rangle = \sum_{i=0}^\infty v_i^* w_i$.

with the outcome states now obeying:

$$\mathbf{n}_i \cdot \mathbf{n}_j = \delta_{ij}, \tag{12}$$

i.e., they are all mutually orthogonal unit vectors in $\ell^2(\mathbb{R})$.

Clearly, $\mathbf{0} \in \triangle_{N-1}$, for all $N < \infty$, i.e., finite-dimensional simplexes contain the null vector $\mathbf{0}$, which describes their center, representative of the operator-state $\frac{1}{N}\mathbb{I}_N$. For instance, for the $N = 2$ case, taking $s_1 = s_2 = \frac{1}{2}$, and considering that $\mathbf{n}_1 = -\mathbf{n}_2$, we clearly have $\mathbf{0} = \frac{1}{2}\mathbf{n}_1 + \frac{1}{2}\mathbf{n}_2 \in \triangle_1$. On the other hand, vectors belonging to $\triangle_\infty$ are convex combinations of mutually orthogonal unit vectors, so that $\mathbf{0} \notin \triangle_\infty$. In other words, by taking the infinite limit we shift from a representation where the null vector is the center of the simplexes, to a standard (infinite) representation where the null vector describes a point external to the simplex (see Appendix 4).

So, given a (pure point spectrum) observable A, acting in $\mathcal{H}_\infty = \ell^2(\mathbb{C})$, with spectral family $\{P(\mathbf{n}_1), P(\mathbf{n}_2), \dots\}$, where the $P(\mathbf{n}_i)$ are mutually orthogonal one-dimensional projection operators, $\operatorname{Tr} P(\mathbf{n}_i)P(\mathbf{n}_j) = \delta_{ij}\,\mathbb{I}$, we can associate to it an infinite dimensional (standard) simplex $\triangle_\infty$, with vertices $\mathbf{n}_i \cdot \mathbf{n}_j = \delta_{ij}$, in such a way that the transition probabilities from an operator-state $D(\mathbf{r})$ to the vector-eigenstates $P(\mathbf{n}_i)$ are simply given by the (real) scalar products (8) [see also (36)]. Also, to each vector $\mathbf{n}_i$, we can associate a region $A_i \subset \triangle_\infty$, corresponding to the convex closure of $\{\mathbf{n}_1, \dots, \mathbf{n}_{i-1}, \mathbf{r}^\|, \mathbf{n}_{i+1}, \dots\}$. However, since $\mu(A_i) = \mu(\triangle_\infty) = 0$, we cannot anymore define the transition probabilities $\mathcal{P}[D(\mathbf{r}) \to P(\mathbf{n}_i)]$ as the ratios $\frac{\mu(A_i)}{\mu(\triangle_\infty)}$, as they are now undefined "zero over zero" ratios. So, different from the finite-dimensional situation, it seems not to be anymore possible to understand the scalar product (8) as resulting from the processes of actualization of the available potential measurement-interactions.

Of course, as we mentioned already in Sec. 1, it is always possible to understand $\frac{\mu(A_i)}{\mu(\triangle_\infty)}$ as the limit of well-defined ratios, associated with finite-dimensional systems, considering that the EBR works for all finite N. However, we would like to elucidate if a hidden-measurement mechanism can also be directly highlighted for infinite-dimensional entities. More precisely, can we maintain that, when the abstract point particle representative of the state, initially located in $\mathbf{r}$, orthogonally "falls" onto the measurement simplex $\triangle_\infty$, thus producing the deterministic (decoherence-like) transition $\mathbf{r} \to \mathbf{r}^\|$, a subsequent indeterministic process takes place, describable as a weighted symmetry breaking over the available hidden-measurement interactions, in accordance with the Born rule?

To answer this question, we start by considering the simple situation where $\mathbf{r}^\parallel$ can be written as the convex combination of only two vertex vectors, i.e., $\mathbf{r}^\parallel = r_i^\parallel \mathbf{n}_i + r_j^\parallel \mathbf{n}_j$, for some i and j, $i \neq j$. In other words, we assume that following the transition $\mathbf{r} \to \mathbf{r}^\parallel$, the on-simplex vector $\mathbf{r}^\parallel$ belongs to one of the edges of the infinite simplex, i.e., $\mathbf{r}^\parallel \in \tilde{\Delta}_1(\mathbf{n}_i, \mathbf{n}_j) = \{\mathbf{t} \in \mathbb{R}^2 | \mathbf{t} = t_i \mathbf{n}_i + t_j \mathbf{n}_j, t_i + t_j = 1, 0 \leq t_i \leq 1\}$, with $\mathbf{n}_i \cdot \mathbf{n}_j = 0$.[d] The measure of $\tilde{\Delta}_1(\mathbf{n}_i, \mathbf{n}_j)$ is of course finite and equal to the edge's length, i.e., $\mu[\tilde{\Delta}_1(\mathbf{n}_i, \mathbf{n}_j)] = \|\mathbf{n}_j - \mathbf{n}_i\| = \sqrt{2}$, and can be considered to be representative of the available measurement-interactions. Also, we have that A_i is the convex closure of $\{\mathbf{r}^\parallel, \mathbf{n}_j\}$, and A_j is the convex closure of $\{\mathbf{r}^\parallel, \mathbf{n}_i\}$, so that:

$$\mu(A_j) = \|\mathbf{r}^\parallel - \mathbf{n}_i\| = \|(r_i^\parallel - 1)\mathbf{n}_i + r_j^\parallel \mathbf{n}_j\| = \|r_j^\parallel (\mathbf{n}_j - \mathbf{n}_i)\| = r_j^\parallel \sqrt{2}, \quad (13)$$

and similarly: $\mu(A_i) = r_i^\parallel \sqrt{2}$. Thus, in accordance with the Born rule, we have the well-defined ratio:

$$\mathcal{P}[D(\mathbf{r}) \to P(\mathbf{n}_k)] = \frac{\mu(A_k)}{\mu[\tilde{\Delta}_1(\mathbf{n}_i, \mathbf{n}_j)]} = \frac{\|\mathbf{r}^\parallel - \mathbf{n}_k\|}{\|\mathbf{n}_j - \mathbf{n}_i\|} = \frac{r_k^\parallel \sqrt{2}}{\sqrt{2}} = r_k^\parallel, \quad k = i, j.$$

$$(14)$$

The above simple exercise was to emphasize that the logic of the EBR remains intact also when working with the infinite (standard) simplex Δ_∞, if the initial state vector is the convex combination $\mathbf{r}^\parallel = r_i^\parallel \mathbf{n}_i + r_j^\parallel \mathbf{n}_j$, and of course the same reasoning also applies, *mutatis mutandis*, when $\mathbf{r}^\parallel$ is the convex combination of a finite arbitrary number N of vertex vectors. However, these are very special circumstances, as in general $\mathbf{r}^\parallel$ will be written as a convex combination of an infinite number of vertex vectors, and in that case we face the previously mentioned difficulty that $\mu(\Delta_\infty) = 0$. But, is it really so?

It is worth making the distinction between the fact that an infinite-dimensional quantum entity can in principle produce an infinity of outcome-states and the fact that in actual experimental situations not all these *a priori* possible outcome-states will be truly available to be actualized. In other words, actual measurement contexts, like those we create in our laboratories, only allow for a finite number of possible outcomes, because the number of detectors, however large, is necessarily finite, and their resolving power is also limited. This means that, even though the dimension of a quantum entity, like an electron, can be infinite, its measurement contexts are always finite-dimensional, and therefore described by degenerate measurements. This means that actual measurements need to be associated with

---

[d]We have introduced the notation $\tilde{\Delta}$ (with a tilde) to indicate that, contrary to (9), the simplex is a standard one, defined in terms of orthonormal vertex vectors.

effective finite-dimensional simplexes, for which the hidden-measurement interpretation always applies in a consistent way. Let us show how this works.

We introduce N different disjoint subsets I_i of $\mathbb{N}^*$, $i = 1\ldots, N$ (which may have each a finite or infinite number of elements), such that $\cup_{i=1}^{N} I_i = \mathbb{N}^*$. Then, we define the N projection operators $P_i = \sum_{j \in I_i} P(\mathbf{n}_j)$. According to the Lüders-von Neumann projection formula, the possibly degenerate outcomes that are associated with them correspond to the outcome-states:

$$D(\mathbf{s}_i) = \frac{P_i D(\mathbf{r}) P_i}{\operatorname{Tr} P_i D(\mathbf{r}) P_i}, \quad i = 1\ldots, N. \tag{15}$$

If we introduce the vector-state notation $|\phi_i\rangle = P_i|\psi\rangle/\|P_i|\psi\rangle\|$, with $D(\mathbf{r}) = |\psi\rangle\langle\psi|$ and $D(\mathbf{s}_i) = |\phi_i\rangle\langle\phi_i|$, we can also write the pre-measurement vector-state $|\psi\rangle$ as the superposition:

$$|\psi\rangle = \sum_{i=1}^{N} \|P_i|\psi\rangle\| \, |\phi_i\rangle, \tag{16}$$

as is clear that $\sum_{i=1}^{N} P_i = \mathbb{I}$. Since $\operatorname{Tr} D(\mathbf{s}_i)D(\mathbf{s}_j) = \langle\phi_i|\phi_j\rangle = \delta_{ij}$, $i, j \in \{1,\ldots,N\}$, we have $\mathbf{s}_i \cdot \mathbf{s}_j = \delta_{ij}$, $i, j = 1,\ldots, N$. This means that the N (infinite-dimensional) unit vectors $\mathbf{s}_i$ define a standard $(N-1)$-dimensional sub-simplex of Δ_∞:

$$\tilde{\triangle}_{N-1}(\mathbf{s}_1,\ldots,\mathbf{s}_N) = \{\mathbf{t} \in \mathbb{R}^\infty | \mathbf{t} = \sum_{i=1}^{N} t_i \, \mathbf{s}_i, \sum_{i=1}^{N} t_i = 1, 0 \le t_i \le 1\}. \tag{17}$$

Clearly, being $\tilde{\triangle}_{N-1}(\mathbf{s}_1,\ldots,\mathbf{s}_N)$ finite-dimensional, we have that its measure $\mu[\tilde{\triangle}_{N-1}(\mathbf{s}_1,\ldots,\mathbf{s}_N)] \ne 0$, so that the EBR can be consistently applied to it.

More precisely, writing $\mathbf{r} = \mathbf{r}^\perp + \mathbf{r}^\|$, with $\mathbf{r}^\perp$ the component of $\mathbf{r}$ perpendicular to $\tilde{\triangle}_{N-1}(\mathbf{s}_1,\ldots,\mathbf{s}_N)$, i.e. $\mathbf{r}^\perp \cdot \mathbf{s}_i = 0$, for all $i = 1,\ldots, N$, we have $\mathbf{r}^\| = \sum_{i=1}^{N} r_i^\| \mathbf{s}_i$, so that:

$$\mathcal{P}[D(\mathbf{r}) \to D(\mathbf{s}_i)] = \mathbf{r}^\| \cdot \mathbf{s}_i = r_i^\|, \tag{18}$$

and from the general properties of a simplex, we also have [4]:

$$r_i^\| = \frac{\mu(A_i)}{\mu[\tilde{\triangle}_{N-1}(\mathbf{s}_1,\ldots,\mathbf{s}_N)]}, \tag{19}$$

where A_i denotes the convex closure of $\{\mathbf{s}_1,\ldots,\mathbf{s}_{i-1}, \mathbf{r}^\|, \mathbf{s}_{i+1},\ldots,\mathbf{s}_N\}$. In other words, for as long as the number of outcomes remains finite, even though the quantum entity is infinite-dimensional we can still describe the outcome probabilities as a condition of lack of knowledge about the measurement-interactions that are actualized at each run of the measurement.

3. Continuous spectrum

In the previous section, starting from an observable having a pure point spectrum, we have shown that the degenerate observables that can be built from it admit a hidden-measurement description for the transition probabilities, if the number of degenerate outcomes is finite. This seems to exclude observables also having some continuous spectrum. To show that this is not the case, in this section we present an alternative derivation, using a representation where the generators Λ_i are constructed using the outcome states.

More precisely, we consider a quantum entity with a possibly infinite-dimensional Hilbert space $\mathcal{H}$, and N mutually orthogonal projection operators P_i, $i = 1, \ldots, N$, such that $\sum_i^N P_i = \mathbb{I}$. For example, if $\mathcal{H} = L^2(\mathbb{R})$, and we consider the position observable $Q = \int_{-\infty}^{\infty} dx\, x\, |x\rangle\langle x|$, they could be given by the integrals: $P_i = \int_{I_i} dx\, |x\rangle\langle x|$, where the I_i are disjoint intervals covering the entire real line, i.e., $\mathbb{R} = \cup_{i=1}^N I_i$, so that $\sum_i^N P_i = \sum_i^N \int_{I_i} dx\, |x\rangle\langle x| = \int_{-\infty}^{\infty} dx\, |x\rangle\langle x| = \mathbb{I}$.

If the $D(\mathbf{s}_i) = |\phi_i\rangle\langle\phi_i|$ are the outcomes defined in (15), we can use the N orthonormal vector-states $|\phi_i\rangle$ to construct the first $N^2 - 1$ generators of $SU(\infty)$ [12]: $\{\Lambda_i\}_{i=1}^{N^2-1} = \{U_{jk}, V_{jk}, W_l\}$, with:

$$U_{jk} = |\phi_j\rangle\langle\phi_k| + |\phi_k\rangle\langle\phi_j|, \quad V_{jk} = -i(|\phi_j\rangle\langle\phi_k| - |\phi_k\rangle\langle\phi_j|),$$

$$W_l = \sqrt{\frac{2}{l(l+1)}} \left(\sum_{j=1}^{l} |\phi_j\rangle\langle\phi_j| - l|\phi_{l+1}\rangle\langle\phi_{l+1}| \right),$$

$$1 \le j < k \le N, \quad 1 \le l \le N - 1. \tag{20}$$

We also define the operator $\mathbb{I}_N = \sum_{i=1}^N |\phi_i\rangle\langle\phi_i|$, acting as an indentity operator in the N-dimensional subspace $\mathrm{Span}\{|\phi_1\rangle, \ldots, |\phi_N\rangle\}$. Since $|\psi\rangle \in \mathrm{Span}\{|\phi_1\rangle, \ldots, |\phi_N\rangle\}$, the associated projection operator $|\psi\rangle\langle\psi|$ can be expanded on the basis $\{\mathbb{I}_N, \Lambda_1, \ldots, \Lambda_{N^2-1}\}$, and we can write:

$$|\psi\rangle\langle\psi| = D(\mathbf{r}) = \frac{1}{N}(\mathbb{I}_N + c_N\, \mathbf{r} \cdot \boldsymbol{\Lambda}) = \frac{1}{N}\left(\mathbb{I}_N + c_N \sum_{i=1}^{N^2-1} r_i \Lambda_i \right). \tag{21}$$

Note that despite the similarity with (6), in (21) the operator-state $D(\mathbf{r})$ is not finite-dimensional (all operators in (21) act in an infinite-dimensional Hilbert space $\mathcal{H}$).

Considering for instance the $N = 2$ case, we have the following three Pauli generators:

$$\Lambda_1 = |\phi_1\rangle\langle\phi_2| + |\phi_2\rangle\langle\phi_1|, \quad \Lambda_2 = -i(|\phi_1\rangle\langle\phi_2| - |\phi_2\rangle\langle\phi_1|),$$
$$\Lambda_3 = |\phi_1\rangle\langle\phi_1| - |\phi_2\rangle\langle\phi_2|, \tag{22}$$

and the indentity operator $\mathbb{I}_2 = |\phi_1\rangle\langle\phi_1| + |\phi_2\rangle\langle\phi_2|$, so that we can write:

$$|\psi\rangle\langle\psi| = D(\mathbf{r}) = \frac{1}{2}(\mathbb{I}_2 + \mathbf{r} \cdot \mathbf{\Lambda}) = \frac{1}{2}\left(\mathbb{I}_2 + \sum_{i=1}^{3} r_i \Lambda_i\right). \tag{23}$$

For the two outcome-states we have:

$$|\phi_1\rangle\langle\phi_1| = D(\mathbf{n}_1) = \frac{1}{2}(\mathbb{I}_2 + \Lambda_3), \quad |\phi_2\rangle\langle\phi_2| = D(\mathbf{n}_2) = \frac{1}{2}(\mathbb{I}_2 - \Lambda_3), \tag{24}$$

which means that $\mathbf{n}_1 = (0, 0, 1)$ and $\mathbf{n}_2 = (0, 0, -1)$. Note that the representation is that of a (non-standard) simplex Δ_1 of measure $\mu(\Delta_1) = 2$, as is clear that the two vertex vectors $\mathbf{n}_1$ and $\mathbf{n}_2$, are not orthogonal, but opposite: $\mathbf{n}_1 = -\mathbf{n}_2$.

Let us also consider, for sake of clarity, the $N = 3$ case. We have then the eight Gell-Mann operators:

$$\Lambda_1 = |\phi_1\rangle\langle\phi_2| + |\phi_2\rangle\langle\phi_1|, \quad \Lambda_2 = -i(|\phi_1\rangle\langle\phi_2| - |\phi_2\rangle\langle\phi_1|),$$
$$\Lambda_3 = |\phi_1\rangle\langle\phi_1| - |\phi_2\rangle\langle\phi_2|, \quad \Lambda_4 = |\phi_1\rangle\langle\phi_3| + |\phi_3\rangle\langle\phi_1|, \tag{25}$$
$$\Lambda_5 = -i(|\phi_1\rangle\langle\phi_3| - |\phi_3\rangle\langle\phi_1|), \quad \Lambda_6 = |\phi_2\rangle\langle\phi_3| - |\phi_3\rangle\langle\phi_2|$$
$$\Lambda_7 = -i(|\phi_2\rangle\langle\phi_3| - |\phi_3\rangle\langle\phi_2|), \quad \Lambda_8 = \frac{1}{\sqrt{3}}(|\phi_1\rangle\langle\phi_1| + |\phi_2\rangle\langle\phi_2| - 2|\phi_3\rangle\langle\phi_3|),$$

and the indentity operator $\mathbb{I}_3 = |\phi_1\rangle\langle\phi_1| + |\phi_2\rangle\langle\phi_2| + |\phi_3\rangle\langle\phi_3|$, so that we can write:

$$|\psi\rangle\langle\psi| = D(\mathbf{r}) = \frac{1}{3}(\mathbb{I}_3 + \sqrt{3}\,\mathbf{r} \cdot \mathbf{\Lambda}) = \frac{1}{3}\left(\mathbb{I}_2 + \sqrt{3}\sum_{i=1}^{8} r_i \Lambda_i\right). \tag{26}$$

For the three outcome-states we have:

$$|\phi_1\rangle\langle\phi_1| = D(\mathbf{n}_1) = \frac{1}{3}[\mathbb{I}_3 + \sqrt{3}\,(\frac{\sqrt{3}}{2}\Lambda_3 + \frac{1}{2}\Lambda_8)],$$

$$|\phi_2\rangle\langle\phi_2| = D(\mathbf{n}_2) = \frac{1}{3}[\mathbb{I}_3 + \sqrt{3}\,(-\frac{\sqrt{3}}{2}\Lambda_3 + \frac{1}{2}\Lambda_8)],$$

$$|\phi_3\rangle\langle\phi_3| = D(\mathbf{n}_3) = \frac{1}{3}[\mathbb{I}_3 + \sqrt{3}\,(-1)\Lambda_8], \tag{27}$$

and the associated 8-dimensional unit vectors are:

$$\mathbf{n}_1 = (0, 0, \frac{\sqrt{3}}{2}, 0, 0, 0, 0, \frac{1}{2}), \quad \mathbf{n}_2 = (0, 0, -\frac{\sqrt{3}}{2}, 0, 0, 0, 0, \frac{1}{2}),$$
$$\mathbf{n}_3 = (0, 0, 0, 0, 0, 0, 0, -1), \tag{28}$$

which clearly form an equilateral triangle, that is, a 2-simplex Δ_2.

We thus see that an EBR of the measurement context is still possible, if the latter only involves a finite number of outcomes, which can also correspond to operators projecting onto some continuous spectrum of the observable under consideration. For this, the outcome-states have to be used to construct the first N^2-1 generators Λ_i, which means that we have now to renounce using a same representation to describe different measurements, unless they would all produce outcomes belonging to the same subspace $\mathrm{Span}\{|\phi_1\rangle, \ldots, |\phi_N\rangle\}$.

Let us illustrate this last observation in the simple $N = 2$ case. We consider a measurement whose outcome-states are $|\phi_1'\rangle$ and $|\phi_2'\rangle$, which we assume also form a basis of $\mathrm{Span}\{|\phi_1\rangle, |\phi_2\rangle\}$. We can then generally write: $|\phi_1'\rangle = u_{11}|\phi_1\rangle + u_{12}|\phi_2\rangle$, and $|\phi_2'\rangle = u_{21}|\phi_1\rangle + u_{22}|\phi_2\rangle$. The condition $\langle\phi_1'|\phi_1'\rangle = 1$ implies: $|u_{11}|^2 + |u_{12}|^2 = 1$, and the condition $\langle\phi_2'|\phi_2'\rangle = 1$ implies: $|u_{21}|^2 + |u_{22}|^2 = 1$. Also, condition $\langle\phi_1'|\phi_2'\rangle = 0$ implies: $(u_{11}^*\langle\phi_1| + u_{12}^*\langle\phi_2|)(u_{21}|\phi_1\rangle + u_{22}|\phi_2\rangle) = 0$, i.e., $u_{11}^*u_{21} + u_{12}^*u_{22} = 0$. Thus, the 2×2 matrix U, with elements $[U]_{ij} = u_{ij}$, obeys:

$$UU^\dagger = \begin{bmatrix} u_{11} & u_{12} \\ u_{21} & u_{22} \end{bmatrix} \begin{bmatrix} u_{11}^* & u_{21}^* \\ u_{12}^* & u_{22}^* \end{bmatrix} = \begin{bmatrix} 1 & 0 \\ 0 & 1 \end{bmatrix}. \tag{29}$$

We have:

$$|\phi_1'\rangle\langle\phi_1'| = D(\mathbf{n}_1') = (u_{11}|\phi_1\rangle + u_{12}|\phi_2\rangle)(u_{11}^*\langle\phi_1| + u_{12}^*\langle\phi_2|)$$
$$= |u_{11}|^2 D(\mathbf{n}_1) + |u_{12}|^2 D(\mathbf{n}_2) + u_{11}u_{12}^*|\phi_1\rangle\langle\phi_2| + u_{12}u_{11}^*|\phi_2\rangle\langle\phi_1| \tag{30}$$
$$= \frac{1}{2}[\mathbb{I}_2 + (|u_{11}|^2 - |u_{12}|^2)\Lambda_3 + 2u_{11}u_{12}^*|\phi_1\rangle\langle\phi_2| + 2u_{12}u_{11}^*|\phi_2\rangle\langle\phi_1|]$$
$$= \frac{1}{2}[\mathbb{I}_2 + (|u_{11}|^2 - |u_{12}|^2)\Lambda_3 + (u_{11}u_{12}^* + u_{12}u_{11}^*)\Lambda_1 - i(u_{11}u_{12}^* - u_{12}u_{11}^*)\Lambda_2].$$

In other words, the components of $\mathbf{n}_1'$ are:

$$\mathbf{n}_1' = \left(2\Re(u_{11}u_{12}^*), 2\Im(u_{11}u_{12}^*), |u_{11}|^2 - |u_{12}|^2\right), \tag{31}$$

and we can check that $\mathbf{n}_1' \cdot \mathbf{n}_1' = 4|u_{11}|^2|u_{12}|^2 + (|u_{11}|^2 - |u_{12}|^2)^2 = (|u_{11}|^2 + |u_{12}|^2)^2 = 1$. Of course, a similar calculation can be done to determine the coordinates of $\mathbf{n}_2'$, associated with $|\phi_2'\rangle\langle\phi_2'|$. So, it is possible to describe, within the same 3-dimensional effective Bloch sphere, all two-outcome measurements with outcome-states $|\phi_1'\rangle$ and $|\phi_2'\rangle$ belonging to Span $\{|\phi_1\rangle, |\phi_2\rangle\}$, i.e., of the form:

$$\begin{bmatrix} |\phi_1'\rangle \\ |\phi_2'\rangle \end{bmatrix} = \begin{bmatrix} u_{11} & u_{12} \\ u_{21} & u_{22} \end{bmatrix} \begin{bmatrix} |\phi_1\rangle \\ |\phi_2\rangle \end{bmatrix}. \tag{32}$$

4. Conclusion

In this article, we emphasized that when we take the infinite-dimensional limit of the EBR we face the problem that the Lebesgue measures of the simplexes tend to zero, so preventing the direct use of the infinite EBR to express the outcome probabilities as relative measures of the simplexes' sub-regions. However, we have also shown that the problem can be overcome by observing that measurements are operations that in practice always present a finite number of outcome-states (possibly associated with an infinite dimension of degeneracy), so that their representation only requires finite-dimensional simplexes.

In other words, we have proposed to distinguish the dimension of a quantum entity *per se*, expressing its 'intrinsic potentiality', which can either be finite or infinite, from the dimension of a measurement (the number of outcomes that are available in our spatiotemporal theater, in a given experimental situation), which determines the 'effective potentiality' that can be manifested by a quantum entity, when submitted to the former. In practice, the dimension of an actual interrogative context is always finite, as the number of macroscopic entities that can play the role of detectors is finite and their resolving powers are limited. In that respect, one could even go as far as saying, albeit only speculatively, that the measurement-interactions responsible for the transitions to the different possible outcome-states in fact supervene and produce their effects only when the (possibly infinite) potentiality level associated with the quantum entity gets constrained by a finite number of possible outcomes, during the practical execution of a measurement.

To conclude, let us offer an analogy taken from the domain of human cognition. Consider a person submitted to an interrogative context, forced to choose one among a finite number of distinct answers. For this, the person's mind has to immerse into the semantic context created by the question and the available answers, gradually building up a tension with each one of them; a tension that in the end will have to be released, thus producing an outcome.[e] However, if the number of possibilities to be taken

---

[e]In quantum cognition, a more general version of the EBR, not limited to the Hilbert

simultaneously into account in providing an answer increases, it will become more and more difficult for the person's mind to maintain a sufficient cognitive interaction with each one of them. In other words, in the limit where the number of possible answers tends to infinity, the cognitive interactions associated with each one of them will either tend to zero, and then the process of actualization of an answer cannot take place (note that measurements might as well not produce an outcome), or the person's mind will start focusing on a finite subset of possibilities, with respect to which a tension-reduction process, yielding an outcome, can again take place.

Appendix: The standard simplex representation

To better understand what happens when one takes the infinite limit, one can adopt from the beginning a representation where the scalar product between the different vertex vectors is independent of the dimension N, and such that $\mathbf{0} \notin \triangle_{N-1}$. Indeed, in the representation used in [4], which naturally emerges from the Hilbert geometry, the scalar product between the different vertex vectors is given by (10), thus it depends on the dimension N. To eliminate this dependency, one can first introduce N mutually orthogonal vectors $\mathbf{m}_i$, $i = 1, \ldots, N$, of length $\sqrt{N}$, such that (see Fig. 1, for the $N = 2$ case):

$$\mathbf{n}_i = \sqrt{\frac{1}{N-1}}(\mathbf{m}_i - \mathbf{R}), \quad \mathbf{R} = \frac{1}{N}\sum_{i=1}^{N}\mathbf{m}_i, \quad \mathbf{m}_i \cdot \mathbf{m}_j = N\delta_{ij}. \tag{33}$$

We then have $\|\mathbf{R}\|^2 = 1$ and $\mathbf{m}_i \cdot \mathbf{R} = 1$, for all i, and one can check that, in accordance with (10), $\mathbf{n}_i \cdot \mathbf{n}_j = \frac{1}{N-1}(\mathbf{m}_i \cdot \mathbf{m}_j - \mathbf{m}_i \cdot \mathbf{R} - \mathbf{m}_j \cdot \mathbf{R} + \|\mathbf{R}\|^2) = \frac{1}{N-1}(N\delta_{ij} - 1 - 1 + 1) = \delta_{ij}\frac{N}{N-1} - \frac{1}{N-1}$. Introducing also the vector $\mathbf{s}^{\|}$, defined by: $\mathbf{r}^{\|} = \sqrt{\frac{1}{N-1}}(\mathbf{s}^{\|} - \mathbf{R})$, one finds for the transition probability (5):

$$\mathcal{P}[D_N(\mathbf{r}) \to P_N(\mathbf{n}_i)] = \frac{1}{N}[1 + (\mathbf{s}^{\|} - \mathbf{R}) \cdot (\mathbf{m}_i - \mathbf{R})] \tag{34}$$

$$= \frac{1}{N}[1 + (\mathbf{s}^{\|} \cdot \mathbf{m}_i - \mathbf{s}^{\|} \cdot \mathbf{R} - \mathbf{m}_i \cdot \mathbf{R} + \|\mathbf{R}\|^2)]$$

$$= \frac{1}{N}[1 + (\mathbf{s}^{\|} \cdot \mathbf{m}_i - 1 - 1 + 1)] = \frac{1}{N}\mathbf{s}^{\|} \cdot \mathbf{m}_i,$$

where we have used the fact that $\mathbf{R} \cdot \mathbf{n}_i = 0$, for all $i = 1, \ldots, N$, so that $\mathbf{R} \cdot \mathbf{r}^{\|} = 0$, and consequently $\mathbf{s}^{\|} \cdot \mathbf{R} = (\sqrt{N-1}\,\mathbf{r}^{\|} + \mathbf{R}) \cdot \mathbf{R} = 1$. Then, one

geometry for the state space, called the general tension-reduction (GTR) model, has been proposed to model human decision processes [7,15].

24

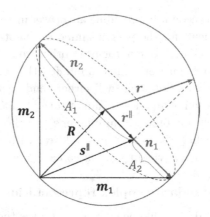

Fig. 1. For the $N = 2$ case, the following vectors are represented in the figure: the unit vector $\mathbf{r}$, describing the initial state of the entity, the orthogonally projected vector $\mathbf{r}^{\parallel}$, describing the on-simplex state, the two unit vectors $\mathbf{n}_1$ and $\mathbf{n}_2$, describing the two vertices of the one-dimensional measurement simplex $\triangle_1$ (corresponding to a diameter of the 3-dimensional Bloch sphere), the two orthogonal vectors $\mathbf{m}_1$ and $\mathbf{m}_2$, of length $\sqrt{2}$, the unit vector $\mathbf{R} = \frac{1}{2}(\mathbf{m}_1 + \mathbf{m}_2)$, corresponding to the center of the simplex and of the Bloch sphere, and the translated vector $\mathbf{s}^{\parallel} = \mathbf{r}^{\parallel} + \mathbf{R}$.

can introduce the unit vectors $\tilde{\mathbf{m}}_i = \sqrt{\frac{1}{N}}\,\mathbf{m}_i$, associated with the standard simplex:

$$\tilde{\triangle}_{N-1} = \{\mathbf{t} \in \mathbb{R}^N | \mathbf{t} = \sum_{i=1}^{N} t_i\, \tilde{\mathbf{m}}_i, \sum_{i=1}^{N} t_i = 1, 0 \leq t_i \leq 1\}, \qquad (35)$$

to which belongs the renormalized vectors $\tilde{\mathbf{R}} = \sqrt{\frac{1}{N}}\,\mathbf{R} = \frac{1}{N}\sum_{i=1}^{N}\tilde{\mathbf{m}}_i$ and $\tilde{\mathbf{s}}^{\parallel} = \sqrt{\frac{1}{N}}\,\mathbf{s}^{\parallel}$, so that (35) simply becomes:

$$\mathcal{P}[D_N(\mathbf{r}) \to P_N(\mathbf{n}_i)] = \tilde{\mathbf{s}}^{\parallel} \cdot \tilde{\mathbf{m}}_i, \qquad (36)$$

i.e., does not depend anymore explicitly on the dimension N, so that the $N \to \infty$ limit becomes trivial. Clearly, $\triangle_{\infty}$ is the limit of both $\tilde{\triangle}_{N-1}$ and $\triangle_{N-1}$, considering that $\tilde{\mathbf{R}} \to \mathbf{0}$, as $N \to \infty$, so that in this limit one also has: $\mathbf{n}_i = \sqrt{\frac{N}{N-1}}(\tilde{\mathbf{m}}_i - \tilde{\mathbf{R}}) \to \tilde{\mathbf{m}}_i$ and $\mathbf{r}^{\parallel} = \sqrt{\frac{N}{N-1}}(\tilde{\mathbf{s}}^{\parallel} - \tilde{\mathbf{R}}) \to \tilde{\mathbf{s}}^{\parallel}$.

References

1. D. Aerts, S. Aerts, B. Coecke, B. D. Hooghe, T. Durt and F. Valckenborgh, "A model with varying fluctuations in the measurement context." In: *New Developments on Fundamental Problems in Quantum Physics*, M. Ferrero and A. van der Merwe (Eds.), Kluwer Academic, Dordrecht, 7 (1997).

2. D. Aerts, "The hidden measurement formalism: what can be explained and where paradoxes remain," *International Journal of Theoretical Physics*, **37**, 291 (1998).

3. D. Aerts, "The Stuff the World is Made of: Physics and Reality," In: *The White Book of 'Einstein Meets Magritte'*, Diederik Aerts, Jan Broekaert and Ernest Mathijs (Eds.), pp. 129–183, Kluwer Academic Publishers, Dordrecht, (1999).

4. D. Aerts and M. Sassoli de Bianchi, "The Extended Bloch Representation of Quantum Mechanics and the Hidden-Measurement Solution to the Measurement Problem", *Annals of Physics*, **351**, 975-1025 (2014). Erratum: Annals of Physics 366, 197–198 (2016).

5. D. Aerts and M. Sassoli de Bianchi, "Many-Measurements or Many-Worlds? A Dialogue", *Foundations of Science*, **20**, 399-427 (2015).

6. D. Aerts and M. Sassoli de Bianchi, "Do spins have directions?", *Soft Computing*, **21**, 1483-1504 (2017).

7. D. Aerts and M. Sassoli de Bianchi, "The GTR-model: a universal framework for quantum-like measurements." In: *Probing the Meaning of Quantum Mechanics: Superpositions, Dynamics, Semantics and Identity*, pp. 91–140, D. Aerts, C. de Ronde, H. Freytes and R. Giuntini (Eds.), World Scientific Publishing Company, Singapore (2016).

8. D. Aerts and M. Sassoli de Bianchi, "A possible solution to the second entanglement paradox." In: *Superpositions, Dynamics, Semantics and Identity*, pp. 351-359, D. Aerts, C. de Ronde, H. Freytes and R. Giuntini (Eds.), World Scientific Publishing Company, Singapore (2016).

9. D. Aerts and M. Sassoli de Bianchi, "The Extended Bloch Representation of Quantum Mechanics. Explaining Superposition, Interference and Entanglement." *Journal Mathematical Physics*, **57**, 122110 (2016).

10. D. Aerts and M. Sassoli de Bianchi, "Quantum measurements as weighted symmetry breaking processes: the hidden measurement perspective.", *International Journal of Quantum Foundations*, **3**, 1–16 (2017).

11. B. Coecke, "Generalization of the proof on the existence of hidden measurements to experiments with an infinite set of outcomes", *Foundations of Physics Letters*, **8**, 437 (1995).

12. F. T. Hioe, J. H. Eberly, "N-level coherence vector and higher conservation laws in quantum optics and quantum mechanics", *Physical Review Letters*, **47**, 838–841 (1981).

13. R. Alicki, K. Lendi, *Quantum Dynamical Semigroups and Application*, Lecture Notes in Physics Vol. 286, Springer-Verlag, Berlin (1987).

14. T. Durt, B.-G. Englert, I. Bengtsson, K. Ayczkowski, "On mutually unbiased bases", *International Journal of Quantum Information*, **8**, 535-640 (2010).

15. D. Aerts and M. Sassoli de Bianchi, "The unreasonable success of quantum probability I. Quantum measurements as uniform fluctuations", *Journal Mathematical Psychology*, **67**, 51–75 (2015).

FREDKIN AND TOFFOLI QUANTUM GATES: FUZZY REPRESENTATIONS AND COMPARISON

Ranjith Venkatrama

Dipartimento di Filosofia Universitá di Cagliari
Via Is Mirrionis 1, 09123, Cagliari-Italia
E-mail: ranquest@gmail.com

Giuseppe Sergioli

Dipartimento di Filosofia Universitá di Cagliari
Via Is Mirrionis 1, 09123, Cagliari-Italia
E-mail: giuseppe.sergioli@gmail.com

Hector Freytes

Dipartimento di Filosofia Universitá di Cagliari
Via Is Mirrionis 1, 09123, Cagliari-Italia
Departmento de Matemática (FCEIA), Universidad Nacional de Rosario-CONICET,
Av. Pellegrini 250, C.P.2000 Rosario, Argentina
E-mail: hfreytes@gmail.com

Roberto Leporini

Dipartimento di Ingengneria gestionale, dell'informazione e della produzione,
Universitá di Bergamo
Viale Marconi 5 - 24044 Dalmine, Bergamo-Italia
E-mail: roberto.leporini@unibg.it

In the framework of quantum computation with mixed states, fuzzy representations based on continuous t-norms for Toffoli and Fredkin quantum gates are introduced. A comparison between both gates is also studied.

PACS numbers: 03.67.Lx, 02.10.-v

Keywords: Fredkin gate; Toffoli gate; continuous t-norms; density operators.

Introduction

Standard quantum computing is based on quantum systems described by finite dimensional Hilbert spaces, specially $\mathbb{C}^2$, that is the two-dimensional space where qbits live. A qubit (the quantum counterpart of the classical bit) is represented by a unit vector in $\mathbb{C}^2$ and, generalizing for a positive integer n, n-qubits are represented by unit vectors in $\otimes^n \mathbb{C}^2$. Similarly to the classical case, it is possible to study the behavior of a number of quantum logical gates (hereafter quantum gates, for short) operating on qbits. These quantum gates are represented by unitary operators.

In [3,5] a quantum gate system based on Toffoli gate is studied. This system is interesting for two main reasons: (i) it is related to continuous t-norms [14], i.e., continuous binary operations on the interval $[0,1]$ that are commutative, associative, non-decreasing and with 1 as the unit element. They are naturally proposed in fuzzy logic as interpretations of the conjunction [13]. (ii) A generalization of the mentioned system to mixed states allows us to connect it with sequential effect algebras [10], introduced to study the sequential action of quantum effects which are unsharp versions of quantum events [11,12]. However there exists another quantum gate, the Fredkin gate, whose behavior is similar to the Toffoli gate. Moreover, under particular conditions, it allows us to represent the same continuous t-norms that Toffoli gate. It suggests to introduce a comparison between Toffoli and Fredkin gates.

The aim of this paper is to study a probabilistic type representation of Toffoli and Fredkin gates based on Łukasiewicz negation $\neg x = 1 - x$, Łukasiewicz sum $x \oplus y = \min\{x + y, 1\}$ and Product t-norms $x \cdot y$ in the framework of quantum computation with mixed states and to establish a comparison between both. Note that, the interval $[0,1]$ equipped with the operations $\langle \oplus, \cdot, \neg \rangle$, defines an algebraic structure called *product MV-algebra* (*PMV*-algebra for short) [17]. In our representation, circuits made from assemblies of Toffoli and Fredkin gates, can be probabilistically represented as $\langle \oplus, \cdot, \neg \rangle$-polynomial expressions in a *PMV*-algebra. In this way, *PMV*-algebra structure related to Toffoli and Fredkin gates, plays a similar role than Boolean algebras describing digital circuits.

The paper is organized as follows: in Section 1 we introduce basic notions of quantum computational logic and we fix some mathematical notation. In Section 2 we briefly describe the Controlled Unitary Operations, that turn out to be very useful in the rest of the paper. In Section 3 and in Section 4 fuzzy representations related to Toffoli and Fredkin gates are respectively provided. In Section 5 we make a comparison between Toffoli and Fredkin gate.

1. Basic notions

In quantum computation, information is elaborated and processed by means of quantum systems. Pure states of a quantum system are described by unit vectors in a Hilbert space. A *quantum bit* or *qbit*, the fundamental concept of quantum computation, is a pure state in the Hilbert space $\mathbb{C}^2$. The *standard orthonormal basis* $\{|0\rangle, |1\rangle\}$ of $\mathbb{C}^2$ is generally called *quantum computational basis*. Intuitively, $|1\rangle$ is related to the truth logical value and $|0\rangle$ to the falsity. Thus, pure states $|\psi\rangle$ in $\mathbb{C}^2$ are superpositions of the basis vectors with complex coefficients: $|\psi\rangle = c_0|0\rangle + c_1|1\rangle$, where $|c_0|^2 + |c_1|^2 = 1$.

In the usual representation of quantum computational processes, a quantum circuit is identified with an appropriate composition of *quantum gates*, mathematically represented by *unitary operators* acting on pure states of a convenient (n-fold tensor product) Hilbert space $\otimes^n \mathbb{C}^2$ [18]. A special basis, called the 2^n-*standard orthonormal basis*, is chosen for $\otimes^n \mathbb{C}^2$. More precisely, it consists of the 2^n-orthogonal states $|\iota\rangle$, $0 \le \iota \le 2^n$ where ι is in binary representation and $|\iota\rangle$ can be seen as the tensor product of states $|\iota\rangle = |\iota_1\rangle \otimes |\iota_2\rangle \otimes \ldots \otimes |\iota_n\rangle$, where $\iota_j \in \{0,1\}$. It provides the standard quantum computational model, based on qbits and unitary operators.

In general, a quantum system is not in a pure state. This may be caused, for example, by the non-complete efficiency in the preparation procedure or by the fact that systems can not be completely isolated from the environment, undergoing decoherence of their states. On the other hand, there are interesting processes that can not be encoded in unitary evolutions. For example, at the end of the computation a non-unitary operation - a measurement - is applied, and the state becomes a probability distribution over pure states, or what is called a *mixed state*. In view of these facts, several authors [1,5,7,8,10] have paid attention to a more general model of quantum computational processes, where pure states are replaced by mixed states. In what follows we give a short description of this mathematical model.

To each vector of the quantum computational basis of $\mathbb{C}^2$ we may associate two density operators $P_0 = |0\rangle\langle 0|$ and $P_1 = |1\rangle\langle 1|$ that represent the standard basis in this framework. Let $P_1^{(n)}$ be the operator $P_1^{(n)} = (\otimes^{n-1} I) \otimes P_1$ on $\otimes^n \mathbb{C}^2$, where I is the 2×2 identity matrix. Clearly, $P_1^{(n)}$ is a 2^n-square matrix. By applying the Born rule, we consider the probability of a density operator ρ as follows:

$$p(\rho) = tr(P_1^{(n)} \rho) \tag{1}$$

We focus our attention in this probability values since it allows us to establish a link between Toffoli gate and fuzzy connectives. Note that, in the particular case in which $\rho = |\psi\rangle\langle\psi|$ where $|\psi\rangle = c_0|0\rangle + c_1|1\rangle$, we obtain $p(\rho) = |c_1|^2$. Thus, this probability value associated to ρ is the generalization, in this model, of the probability that a measurement over $|\psi\rangle$ yields $|1\rangle$ as output. A *quantum operation* [15] is a linear operator $\mathcal{E} : \mathcal{L}(H_1) \to \mathcal{L}(H_2)$ where $\mathcal{L}(H_i)$ is the space of linear operators in the complex Hilbert space H_i ($i = 1, 2$), representable as $\mathcal{E}(\rho) = \sum_i A_i \rho A_i^\dagger$, where A_i are operators satisfying $\sum_i A_i^\dagger A_i = I$ (Kraus representation [15]). It can be seen that a quantum operation maps density operators into density operators. Each unitary operator U gives rise to a quantum operation $\mathcal{O}_U$ such that $\mathcal{O}_U(\rho) = U\rho U^\dagger$ for any density operator ρ. In the case in which U is a real unitary operator, then probability of $\mathcal{O}_U(\rho)$ is simply given by

$$p(\mathcal{O}_U) = tr(P_1^{(n)} \cdot U\rho U) = tr((UP_1^{(n)}U) \cdot \rho). \tag{2}$$

The model based on density operators and quantum operations is called *"quantum computation with mixed states"*. It allows us to also represent irreversible processes as measurements in the middle of the computation.

The connection between a quantum operation $\mathcal{E}$ and continuous t-norms arises when the generic probability values $p(\mathcal{E}(-\otimes \ldots \otimes -))$ can be described in terms of the operations $\langle \oplus, \cdot, \neg \rangle$ defined in the introduction.

Let us define a $\langle \oplus, \cdot, \neg \rangle_n$-*polynomial expression* as a function $f : [0, 1]^n \to [0, 1]$ built only using the three operations $\langle \oplus, \cdot, \neg \rangle$ and n variables.

Now we can formally introduce the connection between quantum operations and continuous t-norms.

Definition 1.1. Let $\mathcal{E} : \mathcal{L}(\otimes^m \mathbb{C}^2) \to \mathcal{L}(\otimes^r \mathbb{C}^2)$ be a quantum operation. Then $\mathcal{E}$ is said to be $\langle \oplus, \cdot, \neg \rangle_n$-representable if and only if there exists a $\langle \oplus, \cdot, \neg \rangle_n$-*polynomial expression* $f : [0, 1]^n \to [0, 1]$ and natural numbers $k_1, \ldots k_n$ satisfying $k_1 + \ldots + k_n = m$, such that:

$$p(\mathcal{E}(\rho_1 \otimes \ldots \otimes \rho_n)) = f(p(\rho_1), \ldots, p(\rho_n))$$

where ρ_i is a density operator in $\otimes^{k_i} \mathbb{C}^2$.

This definition turns out to be crucial in the fuzzy representations of Toffoli and Fredkin gates provided in Section 3 and Section 4, respectively.

2. Controlled unitary operators

By following the standard construction of controlled operators (see, *e.g.*, section 4.3 in [18]), if $U^{(l)}$ is a unitary l-qubit gate, then the controlled-U gate operating on $l + 1$ qubits assumes the following block-representation:

$$CU^{(1,l)} = \left[\begin{array}{c|c} I^{(l)} & 0 \\ \hline 0 & U^{(l)} \end{array}\right].$$

This block representation allows us to end up with the following operational form of an arbitrary $CU^{(1,l)}$ gate:

$$CU^{(1,l)} = \left[\begin{array}{c|c} I^{(l)} & 0 \\ \hline 0 & U^{(l)} \end{array}\right] = \left[\begin{array}{c|c} I^{(l)} & 0 \\ \hline 0 & 0 \end{array}\right] + \left[\begin{array}{c|c} 0 & 0 \\ \hline 0 & U^{(l)} \end{array}\right]$$

$$= P_0 \otimes I^{(l)} + P_1 \otimes U^{(l)}.$$

Further, the generalized control unitary $CU^{(m,l)}$ gate is given by:

$$CU^{(m,l)} = I^{(m-1)} \otimes \left[\begin{array}{c|c} I^{(l)} & 0 \\ \hline 0 & U^{(l)} \end{array}\right] = P_0^{(m)} \otimes I^{(l)} + P_1^{(m)} \otimes U^{(l)} \quad (3)$$

$$= I^{(m-1)} \otimes \left(P_0 \otimes I^{(l)} + P_1 \otimes U^{(l)}\right) = I^{(m-1)} \otimes CU^{(1,l)}. \quad (4)$$

As a useful example, in the special case where the unitary operator U is the well known *Not* gate defined as $Not = \begin{bmatrix} 0 & 1 \\ 1 & 0 \end{bmatrix}$ and whose extension to higher dimensions is given by: $Not^{(l)} = I^{(l-1)} \otimes Not$, then the notion of *Control − Not* gate $CNot^{(m,l)}$ is given by :

$$CNot^{(m,l)} = P_0^{(m)} \otimes I^{(l)} + P_1^{(m)} \otimes Not^{(l)}. \quad (5)$$

3. Fuzzy representation of Toffoli gate

The Toffoli gate, introduced by Tommaso Toffoli [21], is a universal reversible logical gate, which means that any classical reversible circuit can be built from an ensemble of Toffoli gates. This gate has three input bits (x, y, z) and three output bits. Two of the bits, x and y, are control bits that are unaffected by the action of the gate. The third bit z is the target bit that is flipped if both control bits are set to 1, and otherwise is left unchanged. The application of the Toffoli gate to a set of three bits is dictated by:

$$T(x, y, z) = (x, y, xy\widehat{+}z)$$

where $\widehat{+}$ is the sum modulo 2. The Toffoli gate can be used to reproduce the classical AND gate when $z = 0$ and the $NAND$ gate when $z = 1$.

The classical definition of the Toffoli gate can extended as a quantum gate in the following way.

Definition 3.1. For any natural numbers $n, m, l \geq 1$ and for any vectors of the standard orthonormal basis $|x\rangle = |x_1 \ldots x_n\rangle \in \otimes^n \mathbb{C}^2$, $|y\rangle = |y_1 \ldots y_m\rangle \in \otimes^m \mathbb{C}^2$ and $|z_1 \ldots z_l\rangle \in \otimes^l \mathbb{C}^2$, the Toffoli quantum gate $T^{(m,n,l)}$ (from now on, shortly, Toffoli gate) on $\otimes^{n+m+l} \mathbb{C}^2$ is defined as follows:

$$T^{(n,m,l)}(|x\rangle \otimes |y\rangle \otimes |z\rangle) = |x\rangle \otimes |y\rangle \otimes |x_n y_m \widehat{+} z_l\rangle.$$

Taking into account that the Toffoli gate can be interpreted as a *Control-Control-Not gate* [9], we have that:

$$
\begin{aligned}
T^{(n,m,l)} &= CCNot^{(n,m,l)} = I^{(n-1)} \otimes \left[\begin{array}{c|c} I^{(m+l)} & 0 \\ \hline 0 & CNot^{(m,l)} \end{array} \right] \\
&= P_0^{(n)} \otimes I^{(m+l)} + P_1^{(n)} \otimes CNot^{(m,l)} \\
&= (I^{(n)} - P_1^{(n)}) \otimes I^{(m+l)} + P_1^{(n)} \otimes \\
&\quad \left((I^{(m)} - P_1^{(m)}) \otimes I^{(l)} + P_1^{(m)} \otimes Not^{(l)} \right) \\
&= (I^{(n+m)} - P_1^{(n)} \otimes P_1^{(m)}) \otimes I^{(l)} + P_1^{(n)} \otimes P_1^{(m)} \otimes Not^{(l)} = \\
&= I^{(n+m+l)} + P_1^{(m)} \otimes P_1^{(m)} \otimes (Not^{(l)} - I(l)) .
\end{aligned}
$$

The following Theorem provides a fuzzy representation founded of the probability value of the Toffoli gate.

Theorem 3.1. *Let ρ, σ, τ be density operators such that $\rho \in \otimes^n \mathbb{C}^2$, $\sigma \in \otimes^m \mathbb{C}^2$ and $\tau \in \otimes^l \mathbb{C}^2$. Then*

$$p(T^{(n,m,l)}(\rho \otimes \sigma \otimes \tau)T^{(n,m,l)}) = (1 - p(\tau))p(\rho)p(\sigma) + p(\tau)(1 - p(\rho)p(\sigma))$$

and the quantum operation associated to $T^{(m,n,l)}$ is $\langle \oplus, \cdot, \neg \rangle_3$-representable by $\neg z \cdot x \cdot y \ \oplus \ z \cdot \neg(x \cdot y)$.

Proof.

$$p(T^{(n,m,l)}(\rho \otimes \sigma \otimes \tau)T^{(n,m,l)}) =$$
$$= tr(P_1^{(n+m+l)}T^{(n,m,l)}(\rho \otimes \sigma \otimes \tau)T^{(n,m,l)}) =$$
$$= tr(P_1^{(n,m,l)}((I^{n+m} - I^{(n)} \otimes P_1^{(m)}) \otimes I^{(l)} +$$
$$+ P_1^{(n)} \otimes P_1^{(m)} \otimes Not^{(l)})(\rho \otimes \sigma \otimes \tau) \cdot$$
$$((I^{(n+m)} - P_1^{(n)} \otimes P_1^{(m)}) \otimes I^{(l)} + P_1^{(n)} \otimes P_1^{(m)} \otimes Not^{(l)})) =$$
$$= tr(((I^{(n+m)} - I^{(n)} \otimes P_1^{(m)}) \otimes P_1^{(l)} +$$
$$+ P_1^{(n)} \otimes P_1^{(m)} \otimes P_1^{(l)} Not^{(l)})(\rho \otimes \sigma \otimes \tau) \cdot$$
$$\cdot ((I^{(n+m)} - P_1^{(n)} \otimes P_1^{(m)}) \otimes I^{(l)} + P_1^{(n)} \otimes P_1^{(m)} \otimes Not^{(l)})) =$$
$$= tr(((I^{(n+m)} - P_1^{(n)} \otimes P_1^{(m)}) \otimes I^{(l)} + P_1^{(n)} \otimes P_1^{(m)} \otimes Not^{(l)})$$
$$((I^{(n+m)} - I^{(n)} \otimes P_1^{(m)}) \otimes P_1^{(l)} +$$
$$+ P_1^{(n)} \otimes P_1^{(m)} \otimes P_1^{(l)} Not^{(l)}(\rho \otimes \sigma \otimes \tau)) =$$
$$= tr(((I^{(n+m)} - P_1^{(n)} \otimes P_1^{(m)}) \otimes P_1^{(l)} +$$
$$+ P_1^{(n)} \otimes P_1^{(m)} \otimes Not^{(l)} P_1^{(l)} Not^{(l)}(\rho \otimes \sigma \otimes \tau)) =$$
$$= tr(((I^{(n+m)} - P_1^{(n)} \otimes P_1^{(m)})(\rho \otimes \sigma) \otimes P_1^{(l)} \tau) +$$
$$+ tr(P_1^{(m)} \rho \otimes P_1^{(m)} \rho \otimes P_0^{(l)} \tau) =$$
$$= tr(((I^{(n+m)} - P_1^{(n)} \otimes P_1^{(m)})(\rho \otimes \sigma))tr(P_1^{(l)} \tau) +$$
$$+ tr(P_1^{(n)} \rho)tr(P_1^{(m)} \sigma)tr(P_0^{(l)} \tau) =$$
$$= (1 - p(\rho)p(\sigma))p(\tau) + p(\rho)p(\sigma)(1 - p(\tau)).$$

Since $p(T^{(n,m,l)}(\rho \otimes \sigma \otimes \tau)T^{(n,m,l)}) \leq 1$, *then the expression* $(1 - p(\rho)p(\sigma))p(\tau) + p(\rho)p(\sigma)(1 - p(\tau)) = (1 - p(\rho)p(\sigma))p(\tau) \oplus p(\rho)p(\sigma)(1 - p(\tau))$. *In this way, we simply obtain that the quantum operation associated to* $T^{(m,n,l)}$ *is* $\langle \oplus, \cdot, \neg \rangle_3$-*representable by* $\neg z \cdot x \cdot y \oplus z \cdot \neg(x \cdot y)$. $\square$

4. Fredkin gate and its fuzzy representation

The Fredkin gate, introduced by Edward Fredkin [6], is another example of *universal reversible classical logic gate.*

Also Fredkin is a ternary gate, implementing a *Controlled-Swap* operation. More precisely, let (x, y, z) be a 3-bits input state. The first bit, say x, is taken to be the control bit, remaining unaffected by the action of the gate. The second and the third bits, say y and z, are the target bits that are swapped if the control bit x is set to 1; they remain unchanged otherwise. Formally:

$$F(x, y, z) = (x,\ y \mathbin{\widehat{+}} x(y \mathbin{\widehat{+}} z),\ z \mathbin{\widehat{+}} x(y \mathbin{\widehat{+}} z))\,, \tag{6}$$

where, once again, $\widehat{+}$ is the *addition modulo* 2 (equivalent to the XOR operation of the classical sharp logic).

Let us notice that the Fredkin can reproduce the classical AND gate (*i.e.*, when $z_{in} = 0$, $z_{out} = x_{in} \cdot y_{in}$), the classical NOT gate (*i.e.*, when $y_{in} = 0, z_{in} = 1$ $z_{out} = x_{in} \mathbin{\widehat{+}} 1$) and the classical OR gate (*i.e.*, when $z_{in} = 1$ then y_{out} is the OR between x_{in} and y_{in}).

Also the Fredkin gates can also be naturally extended as a quantum gate in the following way.

Definition 4.1. Let $|x\rangle = |x_1, x_2, \ldots, x_n\rangle$, $|y\rangle = |y_1, y_2, \ldots, y_m\rangle$ and $|z\rangle = |z_1, z_2 \ldots, z_l\rangle$ be vectors of the standard orthonormal basis in $\otimes^n \mathbb{C}^2$, $\otimes^m \mathbb{C}^2$ and $\otimes^l \mathbb{C}^2$, respectively. Then, the quantum Fredkin gate is defined by the following equation:

$$F^{(n,m,l)} |x, y, z\rangle =$$

$$= |x\rangle |y_1 \ldots y_{m-1},\ y_m \mathbin{\widehat{+}} x_n (y_m \mathbin{\widehat{+}} z_l)\rangle |z_1 \ldots z_{l-1},\ z_l \mathbin{\widehat{+}} x_n (y_m \mathbin{\widehat{+}} z_l)\rangle.$$

We also notice that, similarly to the Toffoli gate, also the Fredkin gate is a control unitary gate. Hence, it can be represented by using the argument given in Section 2. This unitary gate is the quantum $SWAP^{(m,l)}$ gate.

Note that $SWAP^{(m,l)}$ is a linear operator that swaps the last qubit (*i.e.*, m^{th} qubit) of the its first input with the last qubit (i.e., l^{th} bit) of its second input [18,22]. Formally, for every state $|y_1, \ldots, y_m, z_1, \ldots, z_l\rangle$ of the computational basis:

$$SWAP^{(m,l)} |y_1, \ldots, y_m\rangle |z_1, \ldots, z_l\rangle = |y_1, \ldots, y_{m-1}, z_l\rangle |z_1, \ldots, z_{l-1}, y_m\rangle\,. \tag{7}$$

In order to introduce a matrix representation of the $F^{(n,m,l)}$ gate, we first need to provide a matrix form of the $SWAP^{(m,l)}$ gate.

$$SWAP^{(1,1)} = \begin{bmatrix} 1 & 0 & 0 & 0 \\ 0 & 0 & 1 & 0 \\ 0 & 1 & 0 & 0 \\ 0 & 0 & 0 & 1 \end{bmatrix} = \left[\begin{array}{c|c} P_0 & L_1 \\ \hline L_0 & P_1 \end{array} \right], \tag{8}$$

where L_1 and L_0 are given by $L_1 \equiv |1\rangle\langle 0|$ and $L_0 \equiv |0\rangle\langle 1|$, respectively [a]. These operators can be extended to higher dimensions as $L_1^{(l)} = I^{(l-1)} \otimes L_1$ and $L_0^{(l)} = I^{(l-1)} \otimes L_0$, respectively. Hence, we end up with the following generalization of the *Swap gate* $SWAP^{(m,l)}$:

$$SWAP^{(m,l)} = I^{(m-1)} \otimes SWAP^{(1,l)} = I^{(m-1)} \otimes \begin{bmatrix} P_0^{(l)} & L_1^{(l)} \\ \hline L_0^{(l)} & P_1^{(l)} \end{bmatrix}. \quad (9)$$

By referring to Eq.(3) we easily obtain the generalized quantum Fredkin gate $F^{(n,m,l)}$ as follows.

$$\begin{aligned} F^{(n,m,l)} &= CSwap^{(n,m,l)} = I^{(n-1)} \otimes \begin{bmatrix} I^{(m+l)} & 0 \\ \hline 0 & SWAP^{(m,l)} \end{bmatrix} & (10) \\ &= P_0^{(n)} \otimes I^{(m+l)} + P_1^{(n)} \otimes SWAP^{(m,l)} & (11) \\ &= P_0^{(n)} \otimes I^{(m+l)} + P_1^{(n)} \otimes I^{(m-1)} \otimes \begin{bmatrix} P_0^{(l)} & L_1^{(l)} \\ \hline L_0^{(l)} & P_1^{(l)} \end{bmatrix} & (12) \\ &= I^{(n+m+l)} + P_1^{(n)} \otimes \left(SWAP^{(m,l)} - I^{(m+l)} \right). & (13) \end{aligned}$$

The following Theorem provides a fuzzy representation founded of the probability value of the Fredkin gate.

Theorem 4.1. *Let ρ, σ, τ be density operators such that $\rho \in \otimes^n \mathbb{C}^2$, $\sigma \in \otimes^m \mathbb{C}^2$ and $\tau \in \otimes^l \mathbb{C}^2$. Then*

$$p(F^{(n,m,l)}(\rho \otimes \sigma \otimes \tau)F^{(n,m,l)}) = (1 - p(\rho))\, p(\tau) + p(\rho)\, p(\sigma)$$

and the quantum operation associated to $F^{(m,n,l)}$ is $\langle \oplus, \cdot, \neg \rangle_3$-representable by $\neg x \cdot z \oplus x \cdot y$.

Proof. By using the matrix representation of $F^{(n,m,l)}$, we obtain:

---

[a] L_1 and L_0 are well known in atomic physics as *Ladder-raising* and the *Ladder-lowering*, respectively.

$$F^{(n,m,l)} \cdot P_1^{(n+m+l)} \cdot F^{(n,m,l)} =$$

$$= I^{(n-1)} \otimes$$
$$[(P_0 \otimes I^{(m+l)} + P_1 \otimes SWAP^{(m,l)}) \cdot (I^{(m+l)} \otimes P_1) \cdot$$
$$(P_0 \otimes I^{(m+l)} + P_1 \otimes SWAP^{(m,l)})] =$$

$$= I^{(n-1)} \otimes$$
$$\begin{bmatrix} ((P_0 \otimes I^{(m+l)}) \cdot (I^{(m+l)} \otimes P_1) \cdot (P_0 \otimes I^{(m+l)})) \\ + (P_0 \cdot I \cdot P_1) \otimes (\ldots) \\ + (P_1 \cdot I \cdot P_0) \otimes (\ldots) \\ + (P_1 \cdot I \cdot P_1) \otimes \\ (I^{(m-1)} \cdot I^{(m-1)} \cdot I^{(m-1)}) \otimes \\ (SWAP^{(1,l)} \cdot (I \otimes P_1^{(l)}) \cdot SWAP^{(1,l)}) \end{bmatrix}.$$

Let us recall that,

$P_0 \cdot I \cdot P_1 = P_1 \cdot I \cdot P_0$ that correspond to the null matrix 0.

Further, $SWAP^{(m,l)} = I^{(m-1)} \otimes SWAP^{(1,l)}$.

$$= I^{(n-1)} \otimes \begin{bmatrix} (P_0 \cdot I \cdot P_0) \otimes (I^{(m+l-1)} \cdot P_1^{(m+l-1)} \cdot I^{(m+l-1)}) \\ + 0 \\ + 0 \\ + (P_1 \cdot I \cdot P_1) \otimes (I^{(m-1)} \cdot I^{(m-1)} \cdot I^{(m-1)}) \otimes \\ (SWAP^{(1,l)} \cdot (I \otimes P_1^{(l)}) \cdot SWAP^{(1,l)}) \end{bmatrix}.$$

Let us recall that,

for density matrices A, B, C, D of appropriate dimension, is

$$SWAP^{(m,l)} \cdot \left(A^{(m-1)} \otimes B \otimes C^{(l-1)} \otimes D \right) \cdot SWAP^{(m,l)} =$$
$$A^{(m-1)} \otimes D \otimes C^{(l-1)} \otimes B . \text{ Hence,}$$

$$= I^{(n-1)} \otimes \left[P_0 \otimes P_1^{(m+l)} + P_1 \otimes I^{(m-1)} \otimes P_1 \otimes I^{(l)} \right]$$
$$= P_0^{(n)} \otimes I^{(m)} \otimes P_1^{(l)} + P_1^{(n)} \otimes P_1^{(m)} \otimes I^{(l)}$$
$$= \left(I^{(n)} - P_1^{(n)} \right) \otimes I^{(m)} \otimes P_1^{(l)} + P_1^{(n)} \otimes P_1^{(m)} \otimes I^{(l)}.$$

Therefore, by Eq.(2), the probability value of $F^{(n,m,l)}(\rho \otimes \sigma \otimes \tau)F^{(n,m,l)}$ is given by:

$$p(F^{(n,m,l)}(\rho \otimes \sigma \otimes \tau)F^{(n,m,l)}) =$$

$$= Tr\left[\left(\left(I^{(n)} - P_1^{(n)} \right) \otimes I^{(m)} \otimes P_1^{(l)} + P_1^{(n)} \otimes P_1^{(m)} \otimes I^{(l)} \right) \cdot (\rho \otimes \sigma \otimes \tau) \right]$$

which can be reduced in a straightforward manner to $(1 - p(\rho))\, p(\tau) + p(\rho)\, p(\sigma)$.

Since $p(F^{(n,m,l)}(\rho \otimes \sigma \otimes \tau)F^{(n,m,l)}) \leq 1$, then the expression $(1 - p(\rho))\, p(\tau) + p(\rho)\, p(\sigma) = (1 - p(\rho))\, p(\tau) \oplus p(\rho)\, p(\sigma)$. In this way, we have that the quantum operation associated to $F^{(m,n,l)}$ is $\langle \oplus, \cdot, \neg \rangle_3$-representable by $\neg x \cdot z \oplus x \cdot y$. $\qquad\qquad\square$

5. Comparing the Toffoli and Fredkin quantum gates

In this Section we show are both Toffoli and Fredkin gate are able to represent the product t-norm. However, from a physical point of view, Fredking gate turns out to be more efficient.

An immediate consequence of the Theorem 3.1 and the Theorem 4.1 is that in the special case where $\tau = P_0$ then, for any $\rho \in \otimes^m \mathbb{C}^2, \sigma \in \otimes^n \mathbb{C}^2, \tau \in \mathbb{C}^2$ is $p(T^{(n,m,1)}(\rho \otimes \sigma \otimes P_0)T^{(n,m,1)}) = p(F^{(n,m,1)}(\rho \otimes \sigma \otimes P_0)F^{(n,m,1)}) = p(\rho) \cdot p(\sigma)$. It shows that both quantum gates represent the product t-norm.

A crucial feature of the classical Fredkin gate, in contrast to the Toffoli gate, is that the Fredkin gate is *logically conservative*. This means to say that the number of 1's present in the output of the gate is the same as the number of 1's in its input. In other words, the *parity of bits* remains unchanged during the operation of logically-conservative gates like the Fredkin Gate [4,19–21].

This aspect turns out to be advantageous for building computational circuits that could dissipate less energy (in comparison to the circuits of non-conservative gates) during their operational cycles [4,19]. This is in the light of the well known *Landauer's principle*, according to which, *there is an unavoidable heat-dissipation-cost associated with every bit of information that gets erased*. The theoretical lower bound to the heat-generation of this type is argued to be $K_B T \log 2$. This link between the thermodynamical reversibility and the logical conservativity is due to the well known *statistical-inviolability* of the second law of thermodynamics [2,4,16,19] following a deconstruction of the much debated *Maxwell's demon*.

This crucial aspect of the classical conservative gates gets extended to the quantum gates as follows. Firstly, at a design level, if a gate-module in a given circuit is logically irreversible (meaning that the information encoded by the input states is not entirely recoverable by using the output states alone), then it must be the case that some information about the input states is lost from the gate-module in question. But, the problem of weather this information is irreversibly lost or not, depends on the details of the physical implementation of the gate: this information may either be irreversibly lost –resulting in heat-dissipation, or be just *hidden away* (in a *deterministically retrievable* manner) in some other module of the physical circuit. In a similar case, it may not result in a heat generation, but perhaps costing a memory-resource overhead. The *logically reversible gates* would naturally avoid this type of dissipation at the very design-level itself by leaving a one-to-one correspondence between the output and the input, thereby keeping all the information about the inputs within the same specific gate-module.

However, at the level of physical implementation, there is a further possibility that the operational cycles of even a reversible gate would involve an erasure of some bits of information. There are several factors which could contribute to this information-erasure resulting in dissipation. The main possible reason for this is as follows: though, in theory, all the memory-states are ideally expected to be equally probable, at any non-zero temperature, the memory states of a physical device would be unequally populated following a Boltzman-distribution. This is especially the case of those quantum systems in which the encoding is done onto the energy states of a quantum system. Therein, in ambient temperatures, the ground state is highly populated and the exited states are less populated, following a Boltzman distribution. Further, there is also a natural loss of population from the exited states (also called as the spontaneous emission). This would make it necessary that a standard repumping mechanism be incorporated to retain the memory-states that are encoded using the excited states. These factors would summarily result in an asymmetry in the operational (thermo-economical) cost of different memory-states belonging to the same physical system.

However, there is a possible way of circumventing the above type of dissipation by using those family gates which are not only *logically-reversible* but also *logically-conservative*, like *e.g.,* the Fredkin gate. The strategy is to use an encoding of information such that the most recurring bits of an input is mapped to the most-stable states respectively. Then, the conservativity of the Fredkin gate would guarantee that the number of excited states remains unaltered throughout the operational cycles of the gate: the output would have the same number of excited states as the input was, and hence no extra stabilization cost is required.

Thusly, following the above arguments, even though all the quantum gates by construction are reversible, it becomes desirable to design the circuits based on the logically conservative gates like the Fredkin gate that has been characterized in the present work.

Acknowledgments

This work is partially supported by Regione Autonoma della Sardegna within the project *"Time-logical evolution of correlated microscopic systems"*; CRP 55, L.R. 7/2007 (2015), CUP: F72F16002910002 and by Fondazione Sardegna within the project *"Strategies and Technologies for Scientific Education and Dissemination"* (2017).

References

1. Aharanov D., Kitaev A., Nisan N.: *"Quantum circuits with mixed states"*, Proc. 13th Annual ACM Symp. on Theory of Computation, STOC, 20-30 (1997).
2. C.H. Bennett, "Notes on Landauer's principle, reversible computation, and Maxwell's Demon", *Studies In History and Philosophy of Science Part B: Studies In History and Philosophy of Modern Physics*, **34,3**, 501-510 (2003).
3. Cattaneo G., Dalla Chiara M., Giuntini R., and Leporini R., *An unsharp logic from quantum computation*, Int. J. Theor. Phys. 43, 1803-1817 (2001).
4. G. Cattaneo, A. Leporati, R. Leporini, "Fredkin Gates for Finite-valued Reversible and Conservative Logics", *Journal of Physics A* **35,46** (2002).
5. Dalla Chiara M.L., Giuntini R., Greechie R.: Reasoning in Quantum Theory, Sharp and Unsharp Quantum Logics, Kluwer, Dordrecht-Boston-London (2004).
6. Fredkin, Edward, and Tommaso Toffoli. "Conservative logic." Collision-based computing. Springer London, 47-81 (2002).
7. Freytes H., Sergioli G., Aricó A.: *"Representing continuous t-norms in quantum computation with mixed states "*, Journal of Physics A **43,46** (2010).
8. Freytes H., Domenech G.: *"Quantum computational logic with mixed states"*, Mathematical Logic Quarterly 59, 27-50 (2013).
9. H. Freytes, G. Sergioli : Fuzzy approach for Toffoli gate in quantum computation with mixed states, *Reports on Mathematical Physics*, **74,2**, 159–180 (2014).
10. Gudder S.: *"Quantum computational logic"*, International Journal of Theoretical Physics 42, 39-47 (2003).
11. Gudder S., Greechie R.: *Sequential products on effect algebras*, Rep. Math. Phys. **49**, 87-111 (2002).
12. Gudder S., Greechie R.: *Uniqueness and Order in Sequential Effect Algebras*, Int. J. Theor. Phys. 44, 755-770 (2005).
13. Hàjek P.: Metamathematics of Fuzzy Logic, Trends in Logic vol 4, Dordrecht, Kluwer (1998).

14. Klement E. P., Mesiar R., Pap E., *Triangular norms*, Trends in Logic Vol. 8 (Kluwer, Dordrecht, 2000).

15. Kraus K.: States, effects and operations, Springer-Verlag, Berlin (1983).

16. R. Landauer, "Irreversibility and heat generation in the computing process", *IBM journal of research and development* **5,3**, 183-191 (1961).

17. Montagna F.: *"Functorial Representation Theorems for MV_δ Algebras with Additional Operators"*, J. of Algebra **238**, 99-125 (2001).

18. Nielsen M.A., Chuang I.L.: Quantum Computation and Quantum Information, Cambridge University Press, Cambridge (2000).

19. T. Sagawa, "Thermodynamic and Logical Reversibilities Revisited", *Journal of Statistical Mechanics: Theory and Experiment* **3**, P03025 (2014).

20. G. Sergioli, H. Freytes, "Fuzzy approach to quantum Fredkin gate", *Journal of Logic and Computation*, **28,1**, 245-263 (2017).

21. T. Toffoli, "Reversible computing", *Proceedings of the 7th Colloquium on Automata, Languages and Programming*, Springer-Verlag London, 632–644 (1980).

22. R. Venkatrama, G. Sergioli, R. Leporini. H. Freytes : "Fuzzy type representation of the Fredkin gate in quantum computation with mixed states", *International Journal of Theoretical Physics* **56,12**, 3860-3868 (2017).

PHASE SYMMETRIES OF COHERENT STATES IN GALOIS QUANTUM MECHANICS

JULIEN PAGE

Laboratoire SPHERE (UMR 7219), Université Paris Diderot
CNRS, 5 rue Thomas Mann, 75205 Paris Cedex 13, France.
E-mail: ju.page@hotmail.fr

GABRIEL CATREN

Laboratoire SPHERE (UMR 7219), Université Paris Diderot
CNRS, 5 rue Thomas Mann, 75205 Paris Cedex 13, France.
E-mail: gabrielcatren@gmail.com

In this paper, we study the symmetries of (a particular kind of) coherent states defined in the framework of the *Galois quantum theory* introduced in a previous publication. The configuration and the momentum spaces of this theory are given by finite and discrete abelian groups, namely the Galois group $G = Gal(L : K)$ of a Galois field extension $(L : K)$ and its unitary dual $\widehat{G} \doteq Hom_{gr}(G, U(1))$. The main interest of this quantum theory is that it is possible to define coherent states with indeterminacies in the position $q \in G$ and the momentum $\chi \in \widehat{G}$ encoded by subgroups of G and $\widehat{G}$ respectively. First, we show that the group of automorphisms of a coherent state with indeterminacy $H \subseteq G$ in the position is $H \times H^{\perp}$, where $H^{\perp}$ is the annihilator of H in $\widehat{G}$ and encodes the corresponding indeterminacy in the momentum. Second, we show that the quantum numbers that completely define such coherent states fix an irreducible unitary representation of $H \times H^{\perp}$. These results generalize the group-theoretical interpretation of the limit cases of Heisenberg indeterminacy principle proposed in previous publications to states with non-zero indeterminacies in both q and p. According to this interpretation, a quantum coherent state describes a structure-endowed system characterized by a *group of automorphisms* acting in an irreducible unitary representation fixed by the quantum numbers that define the state.

Keywords: Quantum mechanics; phase symmetries; coherent states; Galois theory.

1. Introduction

The construction of a satisfactory conceptual interpretation of Heisenberg indeterminacy principle is one of the central problems that any interpretation of the quantum formalism must treat. In previous publications [2–6], one of the authors has proposed a group-theoretical interpretation of the limit cases of Heisenberg indeterminacy principle, i.e. the cases in which either the position or the momentum is sharply defined. From a conceptual viewpoint, we have interpreted these ideas in the wake of (what Mackey dubbed [14]) *Weyl's program*, being the aim of this program to understand the commutation relations and Heisenberg indeterminacy principle from a group-theoretical perspective [2]. The starting point of this program is the discovery (by Weyl and Wigner) of the fact that "*all quantum numbers* [...] *are indices characterizing representations of groups*" ([23], p.xxi).[a] In what follows we shall combine this fact from the heuristic idea that grounds Klein's Erlangen program, namely the idea according to which a structure-endowed entity (like a Klein geometry) is somehow determined by its group of automorphisms. This idea was informally generalized by Weyl in the following terms: "*Whenever you have to do with a structure-endowed entity Σ try to determine its group of automorphisms, the group of those element-wise transformations which leave all structural relations undisturbed. You can expect to gain a deep insight into the constitution of Σ in this way.*" ([Weyl 1983], p.144). Now, if we understand the group whose unitary irreducible representation (*unirrep* in short) is fixed by the quantum numbers as the group of automorphisms of the corresponding state, then this heuristic idea coming from Klein's Erlangen program acquires a precise formal implementation. Indeed, we can say that a quantum state is completely determined by its concrete (i.e. acting in a particular unirrep) group of automorphisms. It is worth stressing that this implementation of the fundamental idea of Klein's Erlangen program in the framework of quantum mechanics relies on a linearization of the notion of group of automorphisms in the sense that these groups are given by concrete (rather than abstract) groups. In order to summarize these ideas, we shall paraphrase the fundamental idea of Klein's Erlangen program in the form of the following principle:

Klein-Weyl principle: a homogeneous structure-endowed entity is completely determined by its (concrete) group of automorphisms.

---

[a]This interpretation of the quantum numbers in terms of indices of unitary representations was developed by Weyl and Wigner in the treatises *The Theory of Groups and Quantum Mechanics* [23] and *Group Theory and Its Application to the Quantum Mechanics of Atomic Spectra* [24] respectively.

Let us consider the simplest case of the abelian group $G = \mathbb{R}_q$ of translations in the position q acting on the symplectic manifold $M = \{(q,p)\} = \mathbb{R}^2$. In this case, each value p_0 of the momentum p defines a 1-dimensional unirrep of G given by

$$\rho_{p_0} : G \to U(1)$$
$$q \mapsto e^{2\pi i q p_0}.$$

In turn, the quantum state determined by the index p_0 is given by an equivalence class of normalized vectors in the 1-dimensional Hilbert space $\mathbb{C}|p_0\rangle$ defined by this unirrep, where two vectors are equivalent if they differ by a translation in q (by an amount q_0) acting in the unirrep ρ_{p_0}. In other terms, if we multiply one of such vectors $|p_0\rangle$ with a phase factor $e^{2\pi i q_0 p_0}$ we obtain another representative of the same quantum state (that we shall denote $\overline{|p_0\rangle}$). Therefore, translations in the position — far from physically changing the state $\overline{|p_0\rangle}$ — merely interchanges the different representatives of $\overline{|p_0\rangle}$. We can rephrase this by saying that the position q of the state $\overline{|p_0\rangle}$ is completely "undetermined". Now, if we interpret the state $\overline{|p_0\rangle}$ as a description of a structure-endowed entity and the translations in q as automorphisms of this structure, then we can say that this structure is completely determined by its *concrete* group of automorphisms.

The rationale behind the fact that the position q of the state $\overline{|p_0\rangle}$ is completely undetermined can be understood in the light of Weyl's statement:

- "[...] *objectivity means invariance with respect to the group of automorphisms.*" (Weyl [1983], p.132).

In the present context, this means that the observables that are not invariant under the action of the group of automorphisms of a quantum state cannot define "objective" properties of the latter. In particular, q cannot be an "objective" property of the state $\overline{|p_0\rangle}$ since it is not invariant under the group of automorphisms of $\overline{|p_0\rangle}$, namely the group of translations in q. We could say that the position q, far from being an "external" variable necessary to individualize the state (i.e. to separate it from other states), has to be understood as an "internal" variable spanning the intrinsic structure of the state. In this way, the proposed conceptual framework entails a straightforward group-theoretical interpretation of Heisenberg indeterminacy principle for the cases in which one of the conjugate variables is sharply determined.

44

In Ref.[18], we moved a step forward in this research program by analyzing the extension of the proposed interpretation of Heisenberg indeterminacy principle to cases in which both conjugate variables have a non-zero degree of indeterminacy. This was done in the restricted formal framework provided by (what we have called) *Galois quantum theory*. Briefly, the (finite and discrete) configuration space of such a theory is given by a finite abelian group G, which can always be interpreted as the Galois group of a Galois field extension $(L : K)$. In turn, the momentum space is given by the unitary dual $\hat{G} = \text{Hom}_{\{Group\}}(G, U(1))$ parameterizing the 1-dimensional unirreps of G. The interest of this quantum theory is that it contains coherent states supported by subgroups H of G, i.e. states represented by "wave functions" given by normalized indicator functions of the subgroups $H \subseteq G$.[b] The main property of these states which is relevant in the present context is that they minimize the corresponding Heisenberg indeterminacy principle. While the subgroup $H \subseteq G$ measures the indeterminacy in the "position" g in the configuration space G, the annihilator of H in $\hat{G}$ — namely the subgroup $H^{\perp} \doteq \{\chi \in \hat{G}/\chi(h) = 1, \forall h \in H, \} \subseteq \hat{G}$ — measures the indeterminacy in the conjugate "momentum" χ.[c] In other terms, the (finite and discrete version of the) Fourier transform sends a normalized indicator function of a subgroup H of G into a normalized indicator function of $H^{\perp} \subseteq \hat{G}$.

In the present article, we extend the analysis proposed in Ref.[18] by studying the groups of automorphisms of the H-coherent states of the Galois quantum theory. More precisely, we investigate whether these H-coherent states also satisfy the Klein-Weyl principle, i.e. whether a H-coherent state is completely determined by (the indices that fix the unirreps of) its concrete group of automorphisms. To do so, we shall need to

---

[b]It is worth stressing the difference with respect to the continuous case. In the case of coherent states in a continuous configuration space, the indeterminacies in the position and the momentum — rather than being supported by subgroups of the groups of translations in q and p — are supported by fuzzy sets described by Gaussian probability distributions. Coherent states in finite discrete configuration spaces can also be characterized as discrete approximations of Gaussian functions on continuous configurations spaces (this is the strategy followed in Ref.[20], where the authors define the coherent states as the elements in the kernel of the annihilation operator $\hat{a} = \hat{q} + i\hat{p}$.) In Ref.[18] we have characterized the coherent states as the states given (in the position representation) by the indicator functions of subgroups of G (modulo possible translations and modulations).

[c]Strictly speaking, the coherent states that we consider in this paper should be called *squeezed coherent states*, since they can have different indeterminacies in the position and the momentum. For the sake of simplicity, we shall just call them coherent states.

pass from the configuration space description used in Ref.[18] to a phase space description. We shall first show that the group of automorphisms of a H-coherent state supported (in the position representation) by $H \subseteq G$ is $H \times H^{\perp}$, which is a subgroup of $\mathcal{P} \doteq G \times \widehat{G}$.[d] We shall then show that the parameters that permit us to completely individualize such a coherent state are indices that fix a unirrep of this group of automorphisms.

In Section N°2, we recall the basic features of the Galois quantum theory introduced in Ref.[18] and we analyze the relations between the different relevant dualities, namely the Galois-Grothendieck duality, the Pontryagin duality, the duality between subgroups $H \subseteq G$ and theirs annihilators $H^{\perp}$ in $\widehat{G}$, and the Fourier duality. In Section N°4, we pass from the position and the momentum representations to the phase space representation. We then discuss the relations between the "external" Pontryagin duality between G and $\widehat{G}$ and the corresponding "internal" self-duality of the phase-space $\mathcal{P} = G \times \widehat{G}$, which is encoded by the natural symplectic structure of $\mathcal{P}$. We also introduce the Heisenberg group and we argue that the non-commutativity of this group is a consequence of the anti-symmetric character of the symplectic structure. In Section N°5, we show that the group of automorphisms of a H-coherent state is $H \times H^{\perp}$. We then propose an interpretation of the phase invariance of quantum states (i.e. of the fact that quantum states are given by normalized vectors in a Hilbert space modulo phase factors) in terms of these groups of automorphisms. In Section N°6, we continue the analysis of the H-coherent states by considering theirs associated Wigner functions on the phase-space. We show that these functions are indicator functions of the Lagrangian "submanifolds" $H \times H^{\perp}$ (modulo possible translations) in the phase space $\mathcal{P}$. We also show that the H-coherent states satisfy the Klein-Weyl principle, i.e. that the quantum numbers that completely determine these states are nothing but indices that fix a unirrep of theirs groups of automorphisms $H \times H^{\perp}$. In the final Section, we discuss the obtained results.

---

[d]It is worth noting that $H \times H^{\perp}$ can also be viewed as a subgroup of the corresponding Heisenberg group $H(G)$ (that we shall define in Section N°4). In fact, the stabilizer in $H(G)$ of the ray associated to a H-coherent state is $H \times H^{\perp}$ modulo an overall phase factor (which is "invisible" at the level of the rays).

2. The Galois-Grothendieck Duality and the Pontryagin Duality

The Galois-Grothendieck duality is a duality between intermediate fields F of a finite Galois extension $(L : K)$ (or, more generally, finite commutative K-algebras split by L) and the spaces

$$Spec_K(F) \doteq \mathrm{Hom}_{K-alg}(F, L)$$

that parameterize the L-representations of F, i.e. the sets of K-algebra morphisms $F \to L$ (see Ref.[1] for a technical presentation and Ref.[8] for a conceptual discussion). If F is a field, the space $Spec_K(F)$ is a G-homogeneous space, where $G = Gal(L : K)$ is the Galois group of the field extension $(L : K)$. This means that $Spec_K(F) \simeq G/H$ considered as G-sets, where $H = Gal(L : F)$. In this way, we have a correspondence between intermediate fields F of a Galois extension $(L : K)$ and certain G-homogenous spaces $Spec_K(F)$. The so-called *Gelfand transform*

$$F \to \mathrm{Hom}_{Set}(Spec_K(F), L) \tag{1}$$

associates to each element f in the field F an L-valued function $\hat{f}$ on $Spec_K(F)$ defined by the expression

$$\hat{f}([g]) = [g](f),$$

where $[g]$ denotes the H-class of $g \in G$ considered as a point in $Spec_K(F)$. It is worth noting that the Gelfand transform is injective (which results from the fact that L splits F) but not surjective in general. However, we can obtain an isomorphism by extending the scalars of F from K to L. More precisely, it can be shown that

$$L \otimes_K F \simeq \mathrm{Hom}_{Set}(Spec_K(F), L) \tag{2}$$

as L-algebras ([1], pp.23–24).

The Galois-Grothendieck duality can be understood as a particular instantiation of the duality between algebra and geometry, i.e. of the duality by means of which an algebraic structure A is isomorphic to an algebraic structure of functions on a given space (that we can generally call the *spectrum of A*). In turn, this duality between algebraic and geometric structures can be understood in representation-theoretic terms. Indeed, the Gelfand transform associates to each element a in the algebraic structure a function $\hat{a}$ on the space that parameterizes the representations of

A into a "dualizing object" ([13], pp.121–122).[e] In the present case, the Galois-Grothendieck duality assigns to each field F a G-homogenous space $Spec_K(F)$ on which the elements f of F induce (by means of the Gelfand transfrom) L-valued functions $\hat{f}$. The dualizing object is the field L and the function $\hat{f}$ encodes the L-representations of f in all the representations parameterized by $Spec_K(F)$. Indeed, by evaluating $\hat{f}$ on $[g] \in Spec_K(F)$ we obtain the element in L that represents f in the representation $[g]$. The new feature of the Galois-Grothendieck duality with respect to other representation theories is that it encodes the "Galois correspondence" between controlled variations of the K-algebras to be represented on the one hand and the concomitant variations in the corresponding L-representation theories (i.e. in theirs L-spectra) on the other.

Let us consider now the *Pontryagin duality* for locally compact abelian groups (see for instance [11], §24). Given a locally compact abelian group G, we can define its unitary dual as the group of continuous morphisms $\chi : G \to U(1)$:

$$\widehat{G} \doteq Hom_{Top-Group}(G, U(1)).$$

Each element in $\widehat{G}$ is a character $\chi : G \to U(1)$ that defines an unirrep of G in a 1-dimensional Hilbert space. It is easy to see that $\widehat{G}$ is an abelian group with respect to the pointwise product $(\chi_1 \cdot \chi_2)(g) = \chi_1(g)\chi_2(g)$, where the unity and the inverse elements are given by $g \mapsto 1$ and $\chi^{-1}(g) = \overline{\chi(g)}$ respectively. It is also worth noting that $\widehat{G}$ is locally compact for the compact-open topology. Once again, the elements of G induce $U(1)$-valued functions on $\widehat{G}$ given by the Gelfand transform

$$G \xrightarrow{\simeq} \widehat{\widehat{G}} = Hom_{Top-Group}(\widehat{G}, U(1)),$$
$$g \mapsto \hat{\hat{g}}, \tag{3}$$

where the duality operation $(\hat{-})$ is given by

$$(\hat{-}) = Hom_{Top-Group}(-, U(1))$$

---

[e]Important examples of representation theories are the Stone representation theorem for Boolean algebras (representations of Boolean algebras into the dualizing object $\{0, 1\}$) [12], the duality between finitely generated integral domains A over an algebraically closed field K and the corresponding algebraic affine varieties (representations of A into K) ([10] p.20), the Gelfand-Naimark duality between commutative $\mathbb{C}^*$-algebras and locally compact Hausdorff spaces (representations of commutative $\mathbb{C}^*$-algebras into $\mathbb{C}$) [9], and the Pontryagin duality (representations of locally compact abelian groups into $U(1)$) [11,19].

and

$$\hat{g}(\chi) = \chi(g).$$

The Pontryagin duality states that we can recover G from the set of its $U(1)$-valued unirreps, i.e. that the Gelfand transform defines an isomorphism

$$G \simeq \hat{\hat{G}}.$$

Now, the configuration space G of our Galois quantum theory is a locally compact (since it is finite) abelian group.[f] Therefore, G can be considered both in the framework of the Galois-Grothendieck duality and in the framework of the Pontryagin duality. On the one hand, $G \simeq Spec_K(L)$, i.e. G parameterizes the representations of L into L as K-algebras (i.e. the automorphisms of L leaving the elements of K invariant). On the other hand, G can be considered the character group of $\hat{G}$.[g] In this way, the group G can be understood either

- as the space that parameterizes the representations of the field L into L as K-algebras,
- or as the space that parameterizes the 1-dimensional unirreps of $\hat{G}$.

In turn, we have two sets of observables on G. First, we have the set of observables defined by the Gelfand transform $L \otimes_K L \to L^{\mathrm{Hom}_{L-Alg}(L \otimes_K L, L)}$ for L-algebras. It can be shown that this Gelfand transform yields all the L-valued observables on G, i.e. that $L \otimes_K L \simeq L^{\mathrm{Hom}_{\{L-Alg\}}(L \otimes_K L, L)} = L^{\mathrm{Hom}_{K-Alg}(L,L)} = L^G$.[h]

Secondly, we have the observables on G defined by the elements in $\hat{G}$. Condition (ii) of the Galois quantum theory (see appendix (8.1)) implies that these observables are L-valued. Among all the L-valued observables on G the observables defined by the characters $\chi \in \hat{G}$ satisfy $\chi(gg') = \chi(g)\chi(g')$. It is a standard result of Fourier theory that these observables define an orthonormal basis of L^G as a L-vector space (with respect to the inner product that we shall define in the next section). All in all, while

---

[f]Since G is also discrete every morphism $\chi : G \to U(1)$ is necessarily continuous. We can then forget the topology on G and write $\hat{G} = \mathrm{Hom}_{Group}(G, U(1))$.

[g]Let's note that the roles of G and $\hat{G}$ can be interchanged. By doing so, it is also true that $\hat{G} \subseteq L^G$ can be recovered from the set of $U(1)$-valued unirreps of G.

[h]It is worth noting that the Galois-Grothendieck duality applied to the K-algebra L yields the G-set G. In turn, the Gelfand transform (1) applied to the elements of L yields the subset $\mathrm{Hom}_G(G, L)$ of $\mathrm{Hom}_{Set}(G, L) = L^G$ containing the G-morphisms (i.e. the $f \in L^G$ such that $\forall g, g' \in G, f(g.g') = g[f(g')]$).

the Gelfand transform applied to the elements of $L \otimes_K L$ yield the whole algebra L^G, the Pontryagin duality only gives an orthonormal basis of L^G. Hence, we could say that the Pontryagin duality between $\hat{G}$ and G "spans" in a sense the Galois-Grothendieck duality between L and $Spec_K(L) = G$.

3. Galois Quantum Mechanics

In Ref.[18] we used this formal and conceptual framework to address an important problem in the foundations of quantum mechanics, namely the interpretation of Heisenberg indeterminacy principle. To do so, we have studied a quantum theory defined on a discrete and finite configuration space given by a finite abelian group G. This kind of groups can always be interpreted as the Galois groups $Gal(L : K)$ of a Galois abelian field extension $(L : K)$. In order to reobtain the usual features of a quantum theory (e.g. inner product, Fourier transform between the position and the momentum representations, etc.), we have to impose a condition on G, namely that its cardinal is odd. Then, we can extract from G a field extension $(L : K)$, where K and L are subfields of $\mathbb{C}$. In what follows, the field L will play the role of the field $\mathbb{C}$ of complex numbers used in quantum mechanics (see Ref.[18] and the appendix (8.1) for details).

The resulting quantum theory is given by a normed L-vector space $\mathcal{H}$ composed of L-valued "wave functions" $\psi : G \to L$ on the configuration space G (i.e. $\mathcal{H} = L^G$). The inner product and the "norm" are given by the following standard expressions:

$$\langle \psi_1, \psi_2 \rangle_G \doteq \frac{1}{n} \sum_{g \in G} \psi_1(g) \overline{\psi_2(g)}, \qquad \| \psi \|_G = \sqrt{\langle \psi, \psi \rangle_G},$$

where the bar denotes complex conjugation (condition (iii) in the appendix (8.1) guarantees that the inner product is L-valued). One can prove that $\langle ., . \rangle_G : \mathcal{H} \times \mathcal{H} \to L$ is L-sesquilinear definite positive, in the sense that $\langle \psi, \psi \rangle_G \in L \cap \mathbb{R}_+$ and $\| \psi \|_G = 0 \Leftrightarrow \psi = 0$. The normed L-vector space $\mathcal{H}$ is a "restriction" of the Hilbert space $\mathbb{C}^G$ from $\mathbb{C}$ to L. The vector space $\mathcal{H}$ is not a Hilbert space since it is not complete (it is not even a $\mathbb{C}$-vector space). The property of completeness is required to prove the convergence of infinite series of the form $\psi = \sum_{n=0}^{+\infty} a_n e_n$, where $\{e_n\}_n$ is an infinite basis of the Hilbert space. In the present context, completeness is not necessary since $\mathcal{H}$ is of finite dimension. Moreover, every wave function $\psi : G \to L$ is square integrable so that $\mathcal{H} = L^G = L^2(G, L)$. As usual, we define the projective vector space $P\mathcal{H} = (\mathcal{H} - \{0\})/\sim_L$, where $\psi \sim_L \varphi$ if there is an element $\lambda \in L^\times$ such that $\psi = \lambda \varphi$.

A general vector $|\psi\rangle$ can be decomposed in the position representation by means of the expression $|\psi\rangle = \sum_{g\in G}\psi(g)|g\rangle$, where $\psi \in L^G$ and $|g\rangle$ represents a state sharply localized at $g \in G$. The wave functions that describe the vectors of the form $|g\rangle$ in the position representation are given by the indicator wave function of the trivial subgroup id_G of G translated to g. In order to remain within the Galois-Grothendieck framework, we shall consider indeterminacies in both the position g and the momentum χ encoded by subgroups of G and $\widehat{G}$. Let us consider a vector $|H\rangle = 1_H^{nor}$ defined in the position representation by the normalized indicator function of the subgroup $H \subseteq G$. The vector $|H\rangle$ represents a H-coherent state centered at id_G. There is a duality between subgroups of G and subgroups of $\widehat{G}$ given by the following arrow-reversing isomorphism

$$Subgp(G) \simeq Subgp(\widehat{G})$$
$$H \mapsto H^{\perp} \doteq \{\chi \in \widehat{G}/\chi(h) = 1, \forall h \in H,\}$$
$$J^{\perp} \doteq \{h \in G/\chi(h) = 1, \forall \chi \in J\} \leftarrow\!\shortmid J \tag{4}$$

where $Subgp(-)$ is the category of subgroups of $(-)$ with inclusions as morphisms. The duality between H and $H^{\perp}$ can be understood in terms of the inclusion-reversing duality between families of equations and the "varieties" defined by the common "zeros" of these equations. The subgroup $H^{\perp}$ is by definition the subgroup of elementary observables χ in $\widehat{G}$ that vanish on the "variety" $H \subseteq G$, i.e. that satisfy the system of equations $\{\chi(h) = 1\}_{h\in H}$. In other terms, H can be interpreted as the "subvariety" of G defined by the system of equations $\{\chi(g) = 1\}_{\chi\in H^{\perp}}$. The bigger $H^{\perp} \subseteq \widehat{G}$ the smaller the "variety" in G whose points satisfy the corresponding equations. Moreover, we have the following natural group isomorphisms[i]:

$$\widehat{H} \simeq \widehat{G}/H^{\perp}, \tag{5}$$

$$\widehat{J} \simeq G/J^{\perp}. \tag{6}$$

Since the characters in $\widehat{G}$ are by definition $U(1)$-valued functions on G and $H^{\perp}$ is the subgroup of functions that "vanish" on H, expression (5) means that $\widehat{H}$ can be interpreted as the "coordinate group" of the subvariety H of G, i.e. as the group of characters on H (since the roles played by G and $\widehat{G}$ can be interchanged, expression (6) can be interpreted in an analogous manner).

---

[i]The first isomorphism is given by the surjective restriction $r : \widehat{G} \to \widehat{H}$, where $ker(r) = H^{\perp}$ (and analogously for the second).

The duality (4) can also be interpreted in representation-theoretic terms. A representation theorem amounts to reconstruct an algebraic structure (such as for instance the group G) from a well-chosen family of representations of the former into a suitable dualizing object ($U(1)$ in the present case). In particular, the unitary dual $\hat{G}$ parameterizes the representations (i.e. the group homomorphisms) of G into $U(1)$. In order to individualize a single element $g \in G$, we have to consider *all* the 1-dimensional unirreps parameterized by $\hat{G}$.[j] Let us now consider only the unirreps of G in a subgroup $J \subseteq \hat{G}$. Since J can be interpreted as the unitary dual of $G/J^\perp$ — by taking duals on both sides of expression (6) —, the unirreps in J permit to individualize the elements in the quotient group $G/J^\perp$, i.e. the elements of G modulo $J^\perp$. Roughly speaking, if we only consider the unirreps of G contained in $J \subseteq \hat{G}$, then we can only separate the elements in G up to $J^\perp$.

Let us consider now the quantum-mechanical meaning of the duality (4). The H-coherent state represented by $|H\rangle$ can be labeled by a single element of the quotient group G/H, namely the identity $id_{G/H} = [1_G]$. Analogously, the H-coherent states represented by vectors $|gH\rangle$ obtained by translating $|H\rangle$ by g are labeled by the elements $[g] \in G/H$. In turn, expression (2) says that the algebra $L^{G/H}$ of L-valued observables on the space $Spec_K(F) \simeq G/H$ is isomorphic to the L-algebra $F \otimes_K L$, where $F = Fix(H) = \{x \in L/g \cdot x = x, \forall g \in H\} \subseteq L$. Roughly speaking, only the elements in L that are H-invariant induce well-defined L-valued observables on the quotient space G/H. As an L-algebra, $L^{G/H}$ is spanned by the group of characters $\widehat{G/H} \simeq H^\perp$ (by expression (5)). In this way, if we want to separate coherent states characterized by an indeterminacy in the position given by H, we have to use an algebra of observables spanned by a range of momenta given by $H^\perp$. In other terms, $H^\perp$ is the minimal group of characters in $\hat{G}$ that can separate the states in G modulo H (i.e. with "indeterminacy" H in the position). Reciprocally, expression (5) means that H is the minimal subgroup of elements in $G \simeq \hat{\hat{G}}$ that can separate the states in $\hat{G}$ modulo $H^\perp$ (i.e. with "indeterminacy" $H^\perp$ in the momentum).

From a more conceptual perspective, we could say that only the elements in L that are H-invariant induce well-defined "objective" properties of the H-coherent states represented by vectors of the form $|gH\rangle$ (in other terms, only the H-invariant elements in L define observables in $L^{G/H}$). In other

---

[j]In other terms, $\hat{G}$ is the smallest group of characters necessary to separate the elements of G. This means that for all pairs of distinct group elements $g, g' \in G$ there is a character χ such that $\chi(g) \neq \chi(g')$.

terms, *only the observables that are compatible with the automorphisms of the state — encoded by the subgroup H — induce well-defined properties on the state.*[k] The larger the group H (i.e. the larger the indeterminacy in the position), the stronger the constraint on the observables, i.e. the fewer the elements in L^G that satisfy this compatibility condition. In other terms, the larger the group of automorphisms of the state, the fewer the observables that can be evaluated on the states. In this way, the introduction of an indeterminacy H in the position amounts to descend in the Galois tower of fields from L to the intermediate field F (where $K \subseteq F \subseteq L$). We could say that the considerations of subgroup-supported indeterminacies amounts to explore the different levels of the Galois-Grothendieck duality.

We have thus far considered H-coherent states represented by vectors of the form $|gH\rangle$, where the duality (4) between H and $H^\perp$ encodes the inverse relation between the indeterminacies in the position and the momentum. The Fourier transform generalizes these dualities from indicator functions of subgroups H of G to any L-valued function on G. Indeed, the fact that G is finite and abelian (and therefore locally compact) allows us to introduce a Fourier duality between the position representation in $L^2(G)$ and the momentum representation in $L^2(\hat{G})$. The (discrete) Fourier transform is given by the following isometry:

$$(L^G, \langle \cdot, \cdot \rangle_G) \to (L^{\hat{G}}, \langle \cdot, \cdot \rangle_{\hat{G}})$$
$$f \mapsto \hat{f},$$

where

$$\hat{f}(\chi) \doteq \frac{1}{\sqrt{n}} \sum_{g \in G} f(g) \overline{\chi(g)} = \sqrt{n} \langle f, \chi \rangle_G.$$

In turn, we can introduce the usual Fourier decomposition of any $f \in L^G$:[l]

$$f = \frac{1}{\sqrt{n}} \sum_{\chi \in \hat{G}} \hat{f}(\chi) \chi$$

The arrow-reversing isomorphism (4) is a particular case of the Fourier duality. Every subgroup H (of cardinal h) of G (of cardinal n) defines the

---

[k]We shall see in the next section that the group of automorphisms of a H-coherent state represented by a vector $|gH\rangle$ is in fact larger than H, namely $H \times H^\perp$. The subgroup H only encodes the automorphisms of the state given by the translations in the position.
[l]Since $n \in K$, $\sqrt{n} \in K \subseteq L$ by condition (iv) in the appendix (8.1). Hence, $\hat{f}$ takes values in L. It is worth noting that the Fourier transform can be understood as an isomorphism of Hopf algebras between L^G and $L[\hat{G}]$ (see Ref.[16] for details).

normalized wave function $1_H^{nor} \doteq \sqrt{\frac{n}{h}} 1_H \in L^G$. Analogously, $H^\perp$ defines the normalized wave function $1_{H^\perp}^{nor} \doteq \sqrt{h} 1_{H^\perp} \in L^{\hat{G}}$. One can now check that

$$\widehat{1_H^{nor}} = 1_{H^\perp}^{nor}.$$

In other words, the following diagram commutes:

$$
\begin{array}{ccc}
Subgp(G) & \xleftrightarrow{\ \ \perp\ \ }_{\simeq} & Subgp(\hat{G}) \\[2pt]
\Big\uparrow & & \Big\uparrow \\[2pt]
L^G & \xrightarrow[\ \simeq\ ]{\ \hat{}\ } & L^{\hat{G}}
\end{array}
\qquad (7)
$$

Hence, the duality (4) can also be understood as a Fourier duality between the position representation and the momentum representation of a H-coherent state represented by a vector $|H\rangle$. While the indeterminacy in the position of this state is encoded by $H \subseteq G$, the indeterminacy in the momentum is encoded by $H^\perp \subseteq \hat{G}$.

4. From Dual Groups to Self-Dual Symplectic Structures

The group of automorphisms of a state sharply localized in G or $\hat{G}$ is given by the whole dual group, namely $\hat{G}$ or G respectively. We are now interested in studying the groups of automorphisms of the H-coherent states. To do so, we shall first argue that the "external" Pontryagin duality between G and $\hat{G}$ can be "internalized" within a single self-dual structure. This process of self-dualization proceeds as follows (see Ref.[15], pp.398–399, for a conceptual discussion of this point). The fact that G is canonically isomorphic to $\hat{\hat{G}}$ implies that the product group $\mathcal{P} = G \times \hat{G}$ is canonically self-dual in the sense that

$$\hat{\mathcal{P}} = \widehat{(G \times \hat{G})} \simeq \hat{G} \times \hat{\hat{G}} \simeq \hat{G} \times G \simeq \mathcal{P},$$

where all the isomorphisms are canonical. The conjugate variables $g \in G$ and $\chi \in \hat{G}$ — which belong to different dual spaces — become coordinates of a unique self-dual *phase space* $\mathcal{P}$.

The self-duality of $\mathcal{P}$ attests the fact that G and $\hat{G}$, far from simply being two *independent* structures, are unitary dual spaces (where each space parameterizes the unirreps of the other one). Now, the duality between the factors G and $\hat{G}$ that compose $\mathcal{P}$ induce an additional structure on $\mathcal{P} \doteq G \times \hat{G}$. This additional structure is a "symplectic" structure given by the following application

$$\omega : \mathcal{P} \times \mathcal{P} \to K,$$

$$((g,\chi);(g',\chi')) \mapsto \frac{\chi'(g)}{\chi(g')}.$$

The application ω is an antisymmetric non-degenerate bi-character in the sense that it satisfies the following properties

(1) the map

$$\iota_\omega : \mathcal{P} \to \hat{\mathcal{P}},$$

$$X = (g,\chi) \mapsto \omega(\cdot, X) \tag{8}$$

is an isomorphism of abelian groups,

(2) $\omega(X, X') = \omega(X', X)^{-1}$ for every X, X' in $\mathcal{P}$.[m]

In a quantum context, we could expect the group $\mathcal{P}$ to encode the translations in both the position and the momentum. If we work in the position representation $\mathcal{H} = L^G$, we could expect the elements in $G \subseteq \mathcal{P}$ to act on $\mathcal{H}$ by translations in the position and the elements in $\hat{G} \subseteq \mathcal{P}$ to act on $\mathcal{H}$ by phase modulations (which correspond, via the Fourier transform, to translations in the momentum in $L^{\hat{G}}$). However, this is not the case. To see this, let us define an action of $\mathcal{P}$ on $\mathcal{H}$ by means of the map [17]

$$U : \mathcal{P} \to U(\mathcal{H})$$

where

$$U((g,\chi)) \cdot f(g') \doteq \frac{\chi(g)}{\chi(g')^2} f(g^{-1}g')$$

This map does not define a linear representation of $\mathcal{P}$ on $\mathcal{H}$, where the corresponding obstruction is given by ω^{-1}. Indeed, for every X, X' in $\mathcal{P}$, we have

$$U(XX') = \frac{1}{\omega(X, X')} U(X)U(X').$$

Since $XX' = X'X$, we obtain the commutation relations

$$U(X')U(X) = \frac{1}{\omega(X, X')^2} U(X)U(X').$$

---

[m]If $Card(G)$ is odd, this is equivalent to $\omega(X, X) = 1$ for every X in $\mathcal{P}$. It is worth noting that in the case $G = \mathbb{R}$, the symplectic bilinear form is defined by the expression $\omega_0((q,p);(q,p)) = p'q - pq'$. The relation between this $\mathbb{R}$-valued bilinear form ω_0 and the $U(1)$-valued "symplectic" structure ω (adapted to the case $G = \mathbb{R}$) that we have introduced is given by the expression $\omega = e^{i\omega_0}$.

In particular, for $X = (g, 1)$ and $X' = (1, \chi)$ we obtain (up to the exponential $-2)^n$ the well-known Weyl exponential form of the canonical commutations relations

$$U(\chi)U(g) = \chi(g)^{-2}U(g)U(\chi).$$

The associativity and commutativity of $\mathcal{P}$ implies that ω^{-1} is a 2-cocycle in the group cohomology of $\mathcal{P}$ with coefficients in $U(1)$.[o] After the work of Wigner it can be shown that U can be lifted to a true unitary representation $\widetilde{U}$ of a central extension of $\mathcal{P}$ by the abelian group $U(1)$ (or simply $\mu_n(\mathbb{C})$ in our Galois quantum model) (see Ref.[21], theorem 2.1 and corollary 2.2, p.210.). This central extension (known as *Heisenberg group*) is defined by the inverse of the obstruction that forbids U to be a unitary representation, namely the 2-cocycle ω ([21], pp.210–211).[p]

We can now define the Heisenberg group $H(G)$ of G as $U(1) \rtimes_\omega \mathcal{P}$ (or $\mu_n(\mathbb{C}) \rtimes_\omega \mathcal{P}$ if we want to remain in the category of finite groups[q]). The underlying set of the Heisenberg group is

$$U(1) \times \mathcal{P}$$

and the group law

$$(u_1, X_1)(u_2, X_2) = (u_1 u_2 \omega(X_1, X_2), X_1 X_2).$$

It is worth noting that this group is not commutative, which is a direct consequence of the fact that the symplectic structure ω is not symmetric. This means that elements in $\mathcal{P}$ which are *simplectically intertwined* so to

---

[n]It is worth noting that the map $\chi \mapsto \frac{1}{\chi^2}$ is an automorphism of $\widehat{G}$, being this result a consequence of the fact that $Card(G) = Card(\widehat{G}) = n$ is supposed to be odd.

[o]A function $\eta : \mathcal{P}^2 \to U(1)$ defines a 2-cocycle if it satisfies the following cocycle condition

$$\eta(X_1, X_2)\eta(X_1 X_2, X_3) = \eta(X_1, X_2 X_3)\eta(X_2, X_3)$$

for every $X_1, X_2, X_3 \in \mathcal{P}$.

[p]A 2-cocycle η defines a central extension $U(1) \rtimes_\eta \mathcal{P}$ of $\mathcal{P}$ by $U(1)$ given by the underlying set $U(1) \times \mathcal{P}$ and the group law $(u_1, X_1)(u_2, X_2) = (u_1 u_2 \eta(X_1, X_2), X_1 X_2)$. In other terms, we have the following short exact sequence:

$$1 \to U(1) \hookrightarrow U(1) \rtimes_\eta \mathcal{P} \twoheadrightarrow \mathcal{P} \to 1.$$

Any map $U : \mathcal{P} \to U(\mathcal{H})$ such that $U(XX') = \eta(X, X')U(X)U(X')$ can be lifted to an unitary representation $\widetilde{U}$ of $U(1) \rtimes_{1/\eta} \mathcal{P}$ where $\widetilde{U}(u, X) = uU(X)$. More precisely, the non-equivalent central extensions of $\mathcal{P}$ by $U(1)$ are classified by the second cohomology group of $\mathcal{P}$ with coeficients in $U(1)$ (that is $H^2_{Gr}(\mathcal{P}, U(1))$), which is the group of non-equivalent 2-cocycles ([21], pp.213–214).

[q]It is worth noting that this definition uses the fact that — by condition (ii) in the appendix (8.1) — $Im(\omega) \subseteq \mu_n(\mathbb{C}) \subseteq U(1)$.

speak (i.e. elements X_1 and X_2 such that $\omega(X_1, X_2) \neq 1$) define transformations in the Heisenberg group that do not commute. It is a remarkable fact that the quantum non-commutativity — which could be legitimately understood as the hallmark of quantum mechanics — is defined by the "classical" symplectic structure ω.

The action $\tilde{U}$ of $H(G)$ on the Hilbert space $\mathcal{H} = L^2(G, d\mu)$ is given by the expression

$$\tilde{U}((u, X)) \cdot f(g') \doteq u \frac{\chi(g)}{\chi(g')^2} f(g^{-1}g'), \qquad (9)$$

where $f \in \mathcal{H}$, $u \in U(1)$, and $X = (g, \chi) \in \mathcal{P}$.

The "symplectic" application ω defines a notion of *symplectic orthogonality*: two points X and X' in $\mathcal{P}$ are said to be ω-orthogonal if $\omega(X, X') = 1$. It is worth noting that two points X and X' that are orthogonal in this sense define group elements in $H(G)$ that commute. For every subgroup $\mathcal{G}$ of $\mathcal{P}$ we can define the symplectic orthogonal of $\mathcal{G}$ as

$$\mathcal{G}^{\perp} \doteq \{X \in \mathcal{P}/\omega(X, X') = 1, \forall X' \in \mathcal{G}\},$$

which is also a subgroup of $\mathcal{P}$. The "external" duality (4) between subgroups of G and subgroups of $\hat{G}$ now appears as a special case of the "internal" *symplectic self-duality* defined within $\mathcal{P}$. Indeed, one can prove that the functor $Subgp(\mathcal{P}) \to Subgp(\mathcal{P})$ given by $\mathcal{G} \mapsto \mathcal{G}^{\perp}$ is a bijection, and even an involution in the sense that $(\mathcal{G}^{\perp})^{\perp} = \mathcal{G}$. We have thus the following commutative diagram:

$$
\begin{array}{ccc}
Subgp(G) & \xleftrightarrow{\;\perp\;} & Subgp(\hat{G}) \\
\Big\downarrow{\scriptstyle i} & & \Big\downarrow{\scriptstyle \hat{i}} \\
Subgp(\mathcal{P}) & \xleftrightarrow{\;\perp\;} & Subgp(\mathcal{P}),
\end{array}
$$

where $i : H \mapsto H \times 1$ and $\hat{i} : J \mapsto G \times J$.

Analogously to the continuous case in symplectic geometry, a subgroup $\mathcal{G} \subseteq \mathcal{P}$ will be called *Langrangian* if $\mathcal{G}^{\perp} = \mathcal{G}$. In particular, $G \times 1$, $1 \times \hat{G}$, and $H \times H^{\perp}$ (for every subgroup $H \subseteq G$) are all Lagrangian subgroups of $\mathcal{P}$.[r]

---

[r]It is worth noting that we can also define a notion of Lagrangian submanifold as follows. The symplectic group $(\mathcal{P}, \omega)$ can be considered a symplectic manifold by defining at each point X_0 of $\mathcal{P}$ the tangent space $T_{X_0}\mathcal{P} = X_0\mathcal{P}$ endowed with the group law $X_0X \cdot X_0X' = X_0XX'$ for every $X, X' \in \mathcal{P}$ (so that $T_{X_0}\mathcal{P} \simeq \mathcal{P}$ as abelian groups) and the symplectic

Now, the Lagrangian subgroups $1 \times \hat{G}$ and $G \times 1$ are the groups of automorphisms of the quantum states represented by vectors of the form $|g\rangle$ and $|\chi\rangle$, i.e. by states sharply localized in G and $\hat{G}$ respectively. We can then guess that the Lagrangian subgroups of the form $H \times H^{\perp} \subseteq \mathcal{P}$ are the natural generalization of $1 \times \hat{G}$ and $G \times 1$ to H-coherent states. In order to prove this, we shall use the Heisenberg group $H(G)$ of the Galois quantum theory. By using the action of $H(G)$, we can generate all the wave functions $\{\psi_X^H\}_{X \in \mathcal{P}}$ representing H-coherent states from the wave function $1_H^{nor} \doteq \sqrt{\frac{n}{h}} 1_H \in L^G$. Indeed,

$$\psi_{(g,\chi)}^H(g') \doteq (1, g, \chi) \cdot 1_H^{nor}(g') = \frac{\chi(g)}{\chi(g')^2} 1_H^{nor}(g'g^{-1}) = \frac{\chi(g)}{\chi(g')^2} 1_{gH}^{nor}(g'). \quad (11)$$

Hence, each wave function $\{\psi_X^H\}_{X \in \mathcal{P}}$ representing a H-coherent state can be labeled by a point $X = (g, \chi) \in \mathcal{P} = G \times \hat{G}$.[s] The element $(1, g, 1)$ of $H(G)$ translates the wave function 1_H^{nor} to the normalized indicator function of the H-class gH. In turn, the element $(1, 1, \chi)$ modulates the wave function $1_H^{nor}(g')$ by the factor $\chi^{-2}(g')$. This modulation corresponds — via the Fourier duality — to a translation of χ^{-2} in $\hat{G}$. The factor $\chi(g)$ in the formula (11) guarantees that the combination of the actions defined by g and χ respects the multiplication law of the Heisenberg group (indeed, $(1, g, 1) \cdot (1, 1, \chi) = (\chi(g), g, \chi)$). In general, the action of the Heisenberg group can be transferred from the position to the momentum representations by means of the following canonical isomorphism

$$\mathcal{D} : H(G) = \mu_n(\mathbb{C}) \rtimes_{\omega} G \times \hat{G} \to H(\hat{G}) = \mu_n(\mathbb{C}) \rtimes_{\omega} \hat{G} \times \hat{\hat{G}}$$

$$\simeq \mu_n(\mathbb{C}) \rtimes_{\omega} \hat{G} \times G$$

$$(u, g, \chi) \mapsto (u, \chi^{-2}, \sqrt{g}),$$

where $\sqrt{\cdot}$ denotes the inverse of the group morphism $G \to G$ given by $g \mapsto g^2$ (the presupposition according to which the cardinal of G is odd condition guarantees that this morphism is an isomorphism).

"2-form" given by the formula

$$\omega_{X_0}(X_0 X, X_0 X') = \omega(X, X'), \quad (10)$$

for all $X_0 X$ and $X_0 X'$ in $T_{X_0} \mathcal{P} = X_0 \mathcal{P}$. In this way, for every $X_0 \in \mathcal{P}$ and every subgroup $\mathcal{G}$ of $\mathcal{P}$, $X_0 \mathcal{G}$ will be considered a submanifold of $\mathcal{P}$ with tangent space $T_{X_0} X_0 \mathcal{G} = X_0 \mathcal{G}$. We shall say that $X_0 \mathcal{G}$ is a Lagrangian submanifold of $\mathcal{P}$ if and only if $\mathcal{G}$ is a Lagrangian subgroup of $\mathcal{P}$. In particular, $G \times \{\chi\}$, $\{g\} \times \hat{G}$, and $gH \times \chi H^{\perp} = (g, \chi) H \times H^{\perp}$ (for every subgroup $H \subseteq G$, and every $(g, \chi) \in \mathcal{P}$) are all Lagrangian submanifolds of $\mathcal{P}$.
[s]Let us note that the map that assigns to each point in $\mathcal{P}$ a H-coherent state is not injective. As we shall see, the space of H-coherent states is isomorphic to $\mathcal{P}/(H \times H^{\perp})$.

One can prove that the Fourier transform is a $H(G)$-isomorphism in the sense that the Heisenberg group action commutes–modulo the application $\mathcal{D}$–with the Fourier transform. Indeed, for every $\vec{X} \in H(G)$ and every $f \in L^G$, we can derive the following expression:

$$\widehat{\vec{X} \cdot f} = \mathcal{D}(\vec{X}) \cdot \hat{f}. \tag{12}$$

In particular, from expressions (11) and (12) we can derive the following relation

$$\widehat{\psi^H_{(g,\chi)}} = \varphi^{H^\perp}_{(\chi^{-2}, \sqrt{g})},$$

where $\varphi^J_{(\chi,g)} \in L^2(\hat{G})$ represents in the momentum representation the J-coherent state (for a subgroup J of $\hat{G}$) that is centered in $(\chi, g) \in \hat{G} \times G$.

The subgroups $H \subseteq G$ (resp. $J \subseteq \hat{G}$) define coherent states $1^{nor}_H = \psi^H_{(1,1)}$ (resp. $1^{nor}_J = \varphi^J_{(1,1)}$) of indeterminacy H (resp. J) in the position (resp. momentum) centered at id_G (resp. $id_{\hat{G}}$). As we have shown, the other H-coherent states (resp. J-coherent states) can be obtained by acting on $\psi^H_{(1,1)}$ (rep. on $\varphi^J_{(1,1)}$) by elements in $H(G)$. Therefore, we can extend the commutative diagram (7) by introducing general coherent states:

$$
\begin{array}{ccc}
Subgp(G) & \xleftrightarrow[\simeq]{\perp} & Subgp(\hat{G}) \\
\cap \downarrow & & \cap \downarrow \\
Coh_G & \xrightarrow[\simeq]{\wedge} & Coh_{\hat{G}} \\
\cap \downarrow & & \cap \downarrow \\
L^G & \xrightarrow[\simeq]{\wedge} & L[\hat{G}]
\end{array}
$$

We can then claim that in the context of the Galois quantum theory, the Fourier duality is an extension of the orthogonal duality (4) between subgroups $H \subseteq G$ and $H^\perp \subseteq \hat{G}$ to the whole algebras L^G and $L[\hat{G}]$.

It is also worth noting an important property of the H-coherent states, namely that they span the whole "Hilbert spaces" L^G and $L^{\hat{G}}$. Indeed, one can check that — for each H — we can express every $f \in L^G$ by means of the following expression (where we have used that the cardinal of G is odd):

$$f = \frac{1}{n} \sum_{X \in \mathcal{P}} \langle f, \psi^H_X \rangle \psi^H_X.$$

5. Phase groups of H-coherent states

According to the Klein-Weyl principle, a state (describing a non-rigid structure) should be completely determined by its concrete group of automorphisms, i.e. by the indices that fix the corresponding unirrep. We shall now check that H-coherent states fulfil the Klein-Weyl principle, i.e. they can be completely determined by specifying theirs concrete groups of automorphisms. It can be proved that the stabilizer in $H(G)$ of a H-coherent state $\psi^H_{X=(g,\chi)}$ is

$$Stab(\psi^H_X) = \{(\omega(X,X')^2, X')/X' = (g',\chi') \in H \times H^\perp\}. \tag{13}$$

Let us consider for instance the momentum wave function $\varphi^{\{1\}}_{(\chi,1)}$ in $L^2(\widehat{G})$ defined by a sharp value $\chi \in \widehat{G}$ of the momentum. In the position representation, this vector is described by the position wave function $\psi^G_{(1,\sqrt{\chi}^{-1})}$ (where we have used the inverse Fourier transform $L^2(\widehat{G}) \to L^2(G)$). According to expression (13), the stabilizer of such a wave function is

$$Stab(\psi^G_{(1,\sqrt{\chi}^{-1})}) = \{(\chi(g'),g',1)/g' \in G\}.$$

Now, the equation

$$(\chi(g'),g',1) \cdot \psi^G_{(1,\sqrt{\chi}^{-1})} = \psi^G_{(1,\sqrt{\chi}^{-1})}$$

that defines this invariance can be recast as

$$(1,g',1) \cdot \psi^G_{(1,\sqrt{\chi}^{-1})} = \chi(g')^{-1}\psi^G_{(1,\sqrt{\chi}^{-1})}.$$

In other terms, a translation in $g' \in G$ of the wave function $\psi^G_{(1,\sqrt{\chi}^{-1})}$, far from leaving it invariant, multiplies it by a χ-dependent phase factor. This means that the translations in g do not define a subgroup of the stabilizer $Stab(\psi^G_{(1,\sqrt{\chi}^{-1})})$. In this way, whereas the wave function $\psi^G_{(1,\sqrt{\chi}^{-1})}$ is completely determined by a character χ fixing a unirrep of G, the action of G does not leave the wave function invariant. This counter example shows that vectors in the Hilbert space do not fulfill the Klein-Weyl principle. Now, we can force the validity of this principle by defining a quantum state as a normalized vector in the Hilbert space *modulo overall phase factors*, i.e. as an element of the projective "Hilbert space" $P\mathcal{H}$.[t] We can then define the *group of automorphisms* of a state as the subgroup of the Heisenberg

---

[t]The Heisenberg group $H(G)$ naturally acts on the projective "Hilbert space" $P\mathcal{H}$ by means of the expression $(u,g,\chi) \cdot \overline{f}^L = \overline{(u,g,\chi) \cdot f}^L$, where $\overline{f}^L$ denotes the class of $f \in L^G$ with respect to the equivalence relation defined by $f \sim_L f'$ if there is an element $\lambda \in L^\times$ such that $f = \lambda f'$.

group that leaves the corresponding vector in $\mathcal{H}$ invariant modulo an overall phase factor. We can then claim that a state represented by a wave function $\varphi_{(\chi,1)}^{\{1\}}$ defined by a sharp value $\chi \in \hat{G}$ of the momentum is invariant under the action of the group whose unirreps are fixed by the momentum (i.e. the group of translations in the position). In other terms, the state represented by $\varphi_{(\chi,1)}^{\{1\}}$ is completely determined by the index χ that defines its concrete group of automorphisms. This means that the state represented by $\varphi_{(\chi,1)}^{\{1\}}$ fulfills the Klein-Weyl principle.

In the general case, expression (13) means that the wave function ψ_X^H that represents a H-coherent state is invariant under the action of an element $X' \in H \times H^{\perp}$ modulo a phase factor given by $\omega(X; X')^2$. This means that the action of the elements $u \in \mu_n(\mathbb{C})$ in $H(G)$ is "invisible" at the level of the quantum states given by the rays in $P\mathcal{H}$. In other terms, the action of $H(G)$ on the states in the projective Hilbert space coincides with the action of the group $\mathcal{P}$ acting by translations in the indices. Indeed,

$$(u, X') \cdot_{H(G)} \overline{\psi_X^H}^L = (1, X') \cdot_{H(G)} \overline{\psi_X^H}^L = X' \cdot_{\mathcal{P}} \overline{\psi_X^H}^L = \overline{\psi_{X'X}^H}^L,$$

where $\cdot_{H(G)}$ and $\cdot_{\mathcal{P}}$ denote the group actions of $H(G)$ and $\mathcal{P}$ respectively. It can then be proved that $H \times H^{\perp}$ is the group of automorphisms of the H-coherent states in the sense that

$$\overline{\psi_{X'X}^H}^L = \overline{\psi_X^H}^L$$

for any $X' \in H \times H^{\perp}$. At the level of the vectors in the Hilbert space, we obtain the expression[u]

$$\psi_{X'X}^H = \omega(X, X')\psi_X^H.$$

We can schematize the internal symmetries of a H-coherent state represented by a vector $\psi_{X_1}^H$ by using the analogy between Galois theory and Klein's Erlangen program explored in Refs.[8,18]. Such a state can be understood as a sort of Kleinian "figure" in the phase space $\mathcal{P}$, i.e. as

---

[u]Let us consider for instance the wave functions $\psi_{(1_G,\chi)}^G$ and $\psi_{(g,\chi)}^G$. Theirs groups of automorphisms is $G \times 1$. Indeed, there exists an element $(g, 1_{\hat{G}}) \in G \times 1_{\hat{G}}$ such that $(g, \chi) = (g, 1_{\hat{G}}) \cdot (1_G, \chi)$. Therefore, $\psi_{(g,\chi)}^G = \omega((g, 1); (1, \chi))\psi_{(1,\chi)}^G = \chi(g)\psi_{(1,\chi)}^G$. Therefore, the two states differ in the phase factor $\chi(g)$. This formula is similar to the expression $T_x \cdot |p\rangle = e^{iPx} \cdot |p\rangle = e^{ipx} \cdot |p\rangle \sim |p\rangle$ that connects, in the continuous case $G = \mathbb{R}$, the translations in x in the position and the multiplications by the overall phase factors e^{ipx} for the state $|p\rangle$. If we put $\omega((x_1, p_1); (x_2, p_2)) = e^{i(p_2x_1-p_1x_2)}$ for every $((x_1, p_1); (x_2, p_2)) \in \mathbb{R}^2 \times \mathbb{R}^2$, the overall phase factor is exactly $\omega(X, X') = e^{ipx}$ for the particular case $X = (x, 0)$, $X' = (0, p)$.

a subset of points of $\mathcal{P}$ endowed with a transitive action of the subgroup $1 \times H \times H^\perp \subseteq H(G)$. The singular feature of quantum mechanics is that the "figures" in $\mathcal{P}$ described by the quantum states are Lagrangian submanifolds of $\mathcal{P}$. The indeterminacies in the position and the momentum — described by the groups $H \subseteq G$ and $H^\perp \subseteq \hat{G}$ respectively —, far from being interpreted in terms of some kind of epistemic restriction, acquire an intrinsic meaning, namely that of describing the internal structure of the "figure". Since the action is transitive, such a "figure" can be individualized by selecting a point X_1 in the figure to "represent" the latter. Any other point in the "figure" can be reached by acting on $X_1 \in \mathcal{P}$ with an element in the group of automorphisms $H \times H^\perp$ of the "figure". Of course, the election of the point X_1 in the "figure" is completely arbitrary. This means that the vectors $\psi_{X_1}^H$ and $\psi_{X_1 X}^H$ with $X \in H \times H^\perp$ must describe the same "figure" in $\mathcal{P}$. In fact, we have shown that two such vectors differ by an overall phase factor $\lambda = \omega(X, X_1) \in \mu_n(\mathbb{C})$, which means that they are representatives of the same quantum state. In Klein's Erlangen program, the fact that any point in a homogeneous geometry can be reached by acting with an element of the "*principal group*" of the geometry on a base point which can be arbitrarily chosen — the "representative" of the geometry — does not mean that the multiplicity of points is superflous. If the principal group were understood as a mere epistemic artifact by means of which we can get rid of the multiplicity of frames of reference and pass to an intrinsic description (by quotiening the geometry by the group action), then we would arrive to the fallacious conclusion according to which the Klein geometry consiste of a unique point. Analogously, the fact that the different representatives of a H-coherent state can be reached from a given vector by multiplying it with a phase factor does not mean that the phase group and the multiplicity of representatives is deprived of any physical significance. In the wake of Klein's Erlangen program we have interpreted these phase groups as groups of automorphisms of structure-endowed systems.

The fact that quantum states are defined by the rays in the projective Hilbert space — i.e. by normalized vectors in $\mathcal{H}$ modulo phase factors — does not necessarily mean that phase factors are just irrelevant. According to the Kleinian interpretation that we have proposed, the fact that the H-coherent states are defined modulo overall phase factors encodes the fact that these states describe homogeneous (Lagrangian) "figures" of $\mathcal{P}$. It is also worth noting that the existence of these state-dependent phase factors — far from being physically irrelevant — implies that superposed states of the form $\sum_\chi c_\chi |\chi\rangle$ are no longer phase-invariant under the action of G. In other terms, the state-dependence of the phase factors implies that the phase symmetry can be broken by simply superposing phase-invariant states, being the interference effects the physical consequence of this "symmetry breaking".

6. Wigner Functions and Unitary Representations of the Phase Groups

In the Section $N^{\circ}4$, we have argued that the Pontryagin duality between the configuration space G and the momentum space $\hat{G}$ can also be understood as a self-duality of the phase space $\mathcal{P} = G \times \hat{G}$. In turn, we can pass from the position or the momentum representations of quantum states to a phase space description by using the quasiprobability distribution known as Wigner function [25].[v] In analogy with the continuous case, we can associate to a wave function $f \in L^2(G)$ a Wigner function $f^W : \mathcal{P} \to \mathbb{R}$ by means of the following expression:

$$L^G \to L^{\mathcal{P}}$$

$$f \mapsto (f^W : (g, \chi) \mapsto f^W(g, \chi) = \frac{1}{n^2} \sum_{g' \in G} \overline{f(gg')} f(gg'^{-1}) \chi(g'^2)).$$

Let us consider for instance a H-coherent state represented by a vector $\psi^H_{(g,\chi)}$. As we have argued before, the group of automorphisms of this quantum state is $H \times H^{\perp}$, which means that the indeterminacies in the position and the momentum are described by the groups $H \subseteq G$ and $H^{\perp} \subseteq \hat{G}$ respectively. We can verify that the corresponding Wigner function on $\mathcal{P}$ is given by

$$(\psi^H_{(g,\chi)})^W = \frac{1}{n} 1_{gH \times \chi^{-2}H^{\perp}},$$

As expected, the Wigner description yields an indicator function of the cell in phase space defined by the product of the H-class gH and the $H^{\perp}$-class $\chi^{-2}H^{\perp}$ (it is worth reminding that in the model that we are using the map $\hat{G} \to \hat{G}$ given by $\chi \mapsto \chi^{-2}$ is an automorphism of $\hat{G}$). In the simplest case of a Lagrangian "figure" characterized by an indeterminacy H in the position and indeterminacy $H^{\perp}$ in the momentum, the Wigner function is given by $(1_H^{nor})^W = \frac{1}{n} 1_{H \times H^{\perp}}$.[w] The other H-coherent states can

---

[v]Strictly speaking, the Wigner functions are not probability distributions since they can take negative values.
[w]Let us note that in the continuous case the Wigner functions of coherent states — given by Gaussian distributions in the position or the momentum — are Gaussian distributions in both the position and the momentum.

be simply obtained by translating the indicator $1_{H \times H^\perp}$ of the Lagrangian subgroup $H \times H^\perp \subseteq \mathcal{P}$ around $\mathcal{P}$ by means of group elements of the form $(g, \chi) \in \mathcal{P}$. In this way, a H-coherent state describes a Lagrangian submanifold of $\mathcal{P}$ spanned by the group $H \times H^\perp$. While $H \times H^\perp$ acts transitively on this Lagrangian submanifold, the corresponding elements in the Heisenberg group acts on the corresponding H-coherent state ψ_X^H by means of overall state-dependent phase factors according to the expression

$$(1, X') \cdot \psi_X^H = \omega(X; X')^{-2} \psi_X^H, \qquad \forall X' \in H \times H^\perp.$$

In this way the quantum phases just reflect at the level of the quantum states the internal symmetry of the Lagrangian "submanifolds" that support the states.

We also now show that the H-coherent states satisfy the Klein-Weyl principle, i.e. that the quantum numbers that completely determine these states are indices that fix an unirrep of theirs groups of automorphisms $H \times H^\perp$. To do so, we shall identify the moduli space $Pcoh^H$ of H-coherent states with the quotient group $\mathcal{P}/(H \times H^\perp)$ by means of the following bijection:

$$\mathcal{P}/(H \times H^\perp) \to Pcoh^H$$
$$(gH, \chi H^\perp) \mapsto \overline{\psi_{(g,\chi)}^H}^L$$

Now, according to the Klein-Weyl principle, the group of automorphisms $H \times H^\perp$ of a H-coherent state should act on the representatives of the state in the particular unirrep of $H \times H^\perp$ defined by the quantum numbers that determine the state. Let us consider this point in detail.

First, let us note that the isomorphism of abelian groups $\iota_\omega : \mathcal{P} \to \hat{\mathcal{P}}$ (8) identifies $\mathcal{P}$ with its unitary dual $\hat{\mathcal{P}}$. If we restrict the characters of $\mathcal{P}$ to $H \times H^\perp$, we obtain a surjective morphism $\mathcal{P} \to (\widehat{H \times H^\perp})$ given by $X \mapsto \omega(\cdot, X)|_{H \times H^\perp}$. The kernel of this morphism is $H \times H^\perp$. Hence, we have the following isomorphism:

$$\mathcal{P}/(H \times H^\perp) \xrightarrow{\simeq} (\widehat{H \times H^\perp})$$
$$(gH, \chi H^\perp) \mapsto \omega(\cdot, (g, \chi))|_{H \times H^\perp}.$$

Therefore, the moduli space $\mathcal{P}/(H \times H^\perp)$ that parameterizes the H-coherent states is isomorphic to the space that parameterizes the unirreps of the group of automorphisms $H \times H^\perp$ of these states.[x] We can thus

---

[x]This relation can also be obtained by simply multiplying the relations $G/H \simeq \widehat{H^\perp}$

completely determine each coherent state $\overline{\psi_{X_1}^H}^L$ by specifying its concrete group of automorphisms, i.e. the 1-dimensional unirrep of the group $H \times H^\perp$ (which is defined by the character $\omega(\cdot, X_1)\,|_{H \times H^\perp}$).

In this way, the fact that the phase group of automorphisms of a H-coherent state cannot be simply discarded as a sort of surplus structure associated to the arbitrary election of a representative in the Hilbert space acquires a formal realization. Indeed, the H-coherent states can be completely determined — in agreement with what we have called Klein-Weyl principle — by specifying theirs concrete groups of automorphisms.

7. Conclusion

The main objective of this paper was to explore whether the group-theoretical interpretation of the extreme cases of the Heisenberg indeterminacy principle proposed in Refs.[3–6] could be extended to states with non-zero indeterminacies in both the position and the momentum. To do so, we have continued the analyses of the Galois quantum model introduced in Ref.[18]. The quantum states of this model are given by L-valued "wave functions" on a finite and discrete configuration space provided by the abelian group Galois $G = Gal(L : K)$ of a Galois field extension $(L : K)$. The important feature of this model that is relevant in the present context is that we can define coherent states whose indeterminacies in the position and the momentum are encoded by subgroups of G and $\hat{G}$ respectively.

We have shown that the vectors that represent a H-coherent state in the corresponding Hilbert space are invariant modulo overall phase factors under the action of the subgroup $1 \times H \times H^\perp \simeq H \times H^\perp$ of the Heisenberg group $H(G)$. The fact that quantum states are given by the rays of the vector space $L^2(G)$ (rather than by the vectors themselves) naturally results from the interpretation of the transformations in $H \times H^\perp$ as *automorphisms* (i.e. as transformations acting on the "internal" structure of the states) that should not modify the states as such.

The interpretation of the groups $H \times H^\perp$ as group of automorphisms of the corresponding quantum H-coherent states is non straightforward. According to the standard interpretation of phase symmetries, the invariance of quantum states under multiplication (of the vectors that represent the state) by overall phase factors can be understood as a way to get rid of

---

and $\hat{G}/H^\perp \simeq \hat{H}$ (see expressions (3) and (4) in Ref.[18]). Indeed $(G \times \hat{G})/(H \times H^\perp) \simeq G/H \times \hat{G}/H^\perp \simeq \widehat{H^\perp} \times \hat{H} \simeq \widehat{(H \times H^\perp)}$.

the (physically irrelevant) "surplus" structure resulting from the description of quantum states by means of normalized vectors in a Hilbert space.[y] This interpretation faces the important difficulty of explaining the non-trivial physical consequence resulting from the existence of such phase factors, namely the interference effects. On the contrary, the interpretation of the groups $H \times H^{\perp}$ as *group of automorphisms of structure-endowed systems* attributes a fundamental physical importance to such groups. In particular, this interpretation implies that only the observables that are invariant under the action of the group of automorphisms of the state induce well-defined *objective properties* on the latter. At least the extreme cases of Heisenberg indeterminacy principle naturally follow from this compatibility requirement between the automorphisms of a state and its objective properties.

In order to clarify this ontologic (rather than epistemic) interpretation of phase symmetries, we have proposed an analogy with Klein's Erlangen program. In the framework of this program, the fact that the intrinsic properties of a circle in a plane are invariant under rotations around its center does not mean that the group of rotations just eliminates "surplus" structure. In fact, the existence of such a symmetry group reflects the "internal" structure of the circle as such. Analogously, the fact that the intrinsic properties of a H-coherent state are invariant under the action of $H \times H^{\perp}$ (and the concomitant fact that the system as such is described by rays — rather than vectors — in the Hilbert space) does not necessarily mean that the phase transformations merely eliminate surplus structure. We have indeed argued that the phase symmetry group of a quantum state carries physically meaningful information about the intrinsic structure of the state *to the extent that it determines the latter*. We could say that in this sense quantum mechanics realizes Klein's ideas beyond Klein himself. Indeed, we have shown (by generalizing results and ideas presented in Refs.[3–6,18]) that the "quantum numbers" by means of which we can individualize a H-coherent state are "indices" fixing an unirrep of $H \times H^{\perp}$. In this way, the thesis according to which the group of automorphisms of a state carries non-trivial physical information about the state acquires a striking realization: the very properties that allows us to individualize the state are nothing but labels that fix the state's concrete (i.e. acting in a particular unirrep) group of automorphisms. Briefly, the specification of the concrete

---

[y]See Ref.[7] for a critic of this "epistemic" interpretation of gauge symmetries in the framework of Yang-Mills theory.

group of automorphisms of a state — far from being physically irrelevant — amounts to completely determine the state. We can consider this result as a witness of the importance of Weyl's heuristic prescription: "*Whenever you have to do with a structure-endowed entity Σ try to determine its group of automorphisms, the group of those element-wise transformations which leave all structural relations undisturbed. You can expect to gain a deep insight into the constitution of Σ in this way.*" ([22], p. 144).

8. Appendix

8.1. *Galois Quantum Theory*

We recall here the conditions that define the Galois quantum theory introduced in Ref.[18]. If G is a finite abelian group with an odd cardinal n, then one can prove that there are two subfields K and L of $\mathbb{C}$ such that $(L : K)$ is an abelian Galois extension with $G = Gal(L : K)$ and such that the following conditions are satisfied:

(i) $\mathbb{Q}(e^{\frac{2i\pi}{n}}) \subseteq K \subseteq L \subseteq \mathbb{C}$,

(ii) for every $\chi \in \hat{G}$, $Im(\chi) \subseteq \mu_n(\mathbb{C}) \doteq \{e^{\frac{2imn}{n}}\}_{0 \leqslant m \leqslant n-1} = \mu_n(K)$.[z] Thus $\hat{G}^{(\mathbb{C})} = \hat{G}^{(L)} = \hat{G}^{(K)} = \mathrm{Hom}_{\{group\}}(G, \mathbb{U}(1)) = \mathrm{Hom}_{\{group\}}(G, \mu_n(\mathbb{C}))$, where $\hat{G}^{(F)}$ denotes $\mathrm{Hom}_{\{group\}}(G, F^\times)$ for any field F,

(iii) the fields $K \subseteq F \subseteq L$ (for every intermediat field F) are globally invariant under complex conjugation,

(iv) the field K can be defined in such a way that $\sqrt{x} \in K$ for all $x \in K$.

Acknowledgments

The research leading to these results has received funding from the European Research Council under the European Community's Seventh Framework Programme (FP7/2007-2013 Grant Agreement n°263523, *Philosophy of Canonical Quantum Gravity*).

References

1. F. Borceux, G. Janelidze, Galois theories, Cambridge University Press, Cambridge, 2001.

---

[z]Note that $\mu_n(\mathbb{C}) = \{e^{\frac{2imn}{n}}\}_{0 \leqslant m \leqslant n-1}$ is the multiplicative group of nth-roots of unity. It is a finite subgroup of $(\mathbb{C}^\times, \times)$ and even of $(L^\times, \times)$ and $(K^\times, \times)$. The fact that $Im(\chi) \subseteq \mu_n(\mathbb{C})$ simply follows from the fact that $\forall g \in G, g^n = 1$, and thus $\chi(g)^n = \chi(g^n) = \chi(1) = 1$.

2. G. Catren, Klein–Weyl's Program and the Ontology of Gauge and Quantum Systems, submitted, 2016.
3. G. Catren, "On the Relation Between Gauge and Phase Symmetries", *Found. Phys.* **44**(12), 1317–1335, 2014.
4. G. Catren, "Quantum ontology in the light of gauge theories", in: de Ronde, C., Aerts, S., Aerts, D. (eds.) Probing the Meaning of Quantum Mechanics: Physical, Philosophical, and Logical Perspectives, World Scientific Publishing, Singapore, 2014.
5. G. Catren, "Can classical description of physical reality be considered complete?", in: Bitbol, M., Kerszberg, P., Petitot, J. (eds.) Constituting Objectivity: Transcendental Perspectives on Modern Physics, The Western Ontario Series in the Philosophy of Science, vol. 74, pp. 375–386. Springer-Verlag, Berlin, 2009.
6. G. Catren, "On classical and quantum objectivity", *Found. Phys.* **38**, 470–487, 2008.
7. G. Catren, "Geometric foundations of classical Yang-Mills theory", *Studies in History and Philosophy of Modern Physics*, **39**, 511–531, 2008.
8. G. Catren, J. Page, "On the notions of indiscernibility and indeterminacy in the light of the Galois-Grothendieck theory", *Synthese* **191**(18), 4377–4408, 2014.
9. J. Dixmier, C^*-Algebras, North-Holland Publishing Compagny, Amsterdam, 1977.
10. R. Hartshorne, Algebraic Geometry, Springer-Verlag, New-York, 1977.
11. E. Hewitt, K.A. Ross, Abstract Harmonic Analysis I, Second edition, Springer–Verlag, New York Heidelberg Berlin, 1979.
12. P.T. Johnstone, Stone spaces, Cambridge University Press, Cambridge, UK, 1986.
13. F.W. Lawvere, R. Rosebrugh, Sets for mathematics, Cambridge University Press, Cambridge, 2003.
14. G.W. Mackey, Weyl program and Modern Physics, in K. Bleuler, M. Werner (Eds.) "Differential Geometrical Methods in Theoretical Physics", Volume 250 of the series NATO ASI Series, pp. 11–36, Kluwer Academic Publishers, Dordrecht, 1980.
15. S. Majid, "Principle of Representation-Theoretic Self-Duality", *Physics Essays*, volume 4, number 3, 395–405, 1991.
16. S. Majid, Foundations of Quantum Group Theory, Cambridge University Press, Cambridge, 1995.
17. D. Mumford, M. Nori, P. Norman, Tata Lectures on Theta III [1991], Birkhauser, Boston Basel Berlin, 2000.
18. J. Page, G. Catren, "Towards a Galoisian Interpretation of Heisenberg Indeterminacy Principle", *Found. Phys.* **44**(12), 1289–1301, 2014.
19. L.S. Pontryagin, "The theory of topological commutative groups", *Ann. of Math.* **35**(2), 361–388, 1934.
20. J. Tolar, G. Chadzitaskos, "Quantization on Z_M and coherents states over $Z_M \times Z_M$", *J. Phys. A.: Math. Gen.* **30**, 2509–2517, 1997.
21. G.M. Tuynman, W.A.J.J Wiegerinck, "Central extensions and physics", *JGP* **4**(2), 207–258, 1987.

22. H. Weyl, Symmetry, Princeton University Press, Princeton, 1952.
23. H. Weyl, The Theory of Groups and Quantum Mechanics [1931], Dover Publications, Mineola, 1950.
24. E.P. Wigner, Group Theory and Its Application to the Quantum Mechanics of Atomic Spectra, trans. by J.J. Griffin, Academic Press, New-York, 1959.
25. E.P. Wigner, "On the quantum correction for thermodynamic equilibrium", Phys. Rev. **40**, 749–759, 1932.

Metaphors in Science and in Music. A Quantum Semantic Approach

M.L. Dalla Chiara*, R. Giuntini** and E. Negri***

* Dipartimento di Lettere e Filosofia, Università di Firenze, Firenze, Italy
** Dipartimento di Pedagogia, Psicologia, Filosofia, Università di Cagliari, Cagliari, Italy
*** Scuola di Musica di Fiesole, San Domenico di Fiesole, Fiesole

Our current use of metaphors generally involves special forms of allusions that are based on similarity-relations. Such semantic phenomena can be naturally investigated in the framework of quantum computational semantics, where meanings are dealt with as pieces of quantum information. We apply this semantics to a formal analysis of music and discuss the possibility of representing some particular examples of musical themes as *musical metaphors* for extra-musical meanings.

Keywords: Quantum semantics; music.

1. Introduction

Metaphors and allusions play an important role in our current use of natural languages and in the languages of art. At the same time, in the tradition of scientific thought metaphorical arguments have been often regarded as fallacious and dangerous. There is a deep logical reason that justifies such "suspicions". Allusions and metaphors are generally based on *similarity-relations*: when an idea A is used as a metaphor for another idea B, the two ideas A and B are supposed to be similar with respect to something. As is well known, similarity-relations are *weak* relations: they are reflexive and symmetric; but generally they are not transitive and they do not preserve the properties of the objects under investigation. If Alice is similar to Beatrix and Alice is clever, it is not guaranteed that Beatrix also is clever. Wrong extrapolations of properties from some objects to other similar objects are often used in rethoric contexts, in order to obtain a kind of *captatio benevolantiae*. We need only think of the soccer-metaphors that are so frequently used by many politicians!

In spite of their possible "dangers", metaphors have sometimes played an important role even in exact sciences. An interesting example in logic is represented by the current use of the metaphor of *possible world*, based on a general idea that has been deeply investigated by Leibniz. In some situations possible worlds, which correspond to special examples of *semantic models*, can be imagined as a kind of "ideal scenes", where abstract objects behave as if they were playing a theatrical play. And a "theatrical imagination" has sometimes represented an important tool for scientific creativity, also in the search for solutions of logical puzzles and paradoxes. A paradigmatic case can be recognized in the discussions about a celebrated set-theoretic paradox, the Skolem's paradox. Consider an axiomatic version of set theory **T** (say, Zermelo-Fraenkel theory) formalized in first-order logic and assume that **T** is non-contradictory. By purely logical reasons, we know that **T** has at least one "strange" model $\mathcal{M}^*$, where both the domain and all its elements are denumerable sets. In this model $\mathcal{M}^*$ the *continuum* (the set $\mathbb{R}$ of all real numbers) seems to be, at the same time,

- denumerable, because everything is denumerable in $\mathcal{M}^*$;
- non-denumerable, because $\mathcal{M}^*$ must verify Cantor's theorem, according to which the continuum $\mathbb{R}$ is non-denumerable.

In order to "see" a possible way-out from this paradoxical conclusion, we can imagine an ideal scene where all actors are denumerable sets (Fig. 1). Some actors are supposed to wear a mask, playing the role of non-denumerable sets (Fig. 2). As happens in real theatrical plays, characters and actors do not generally share the same properties. The actor who plays the role of Othello is not necessarily jealous himself! In the same way, a denumerable set can play the role of the non-denumerable continuum on the stage represented by the non-standard model $\mathcal{M}^*$.

The Skolem-paradox is one of the possible examples that show us how a recourse to a "metaphorical thinking" may sometimes improve abstract imagination-capacities even in the field of exact sciences.

2. Ambiguities and allusions in the quantum computational semantics

To what extent is "a logic of metaphors" possible? A useful tool for discussing this question is represented by a special form of quantum semantics that has been suggested by the theory of quantum computation.[a] In this

---

[a]See [3,4,6,7]. Some basic intuitive idea of the quantum computational semantics are close to the "quantum cognition approach" that has been extensively developed in recent

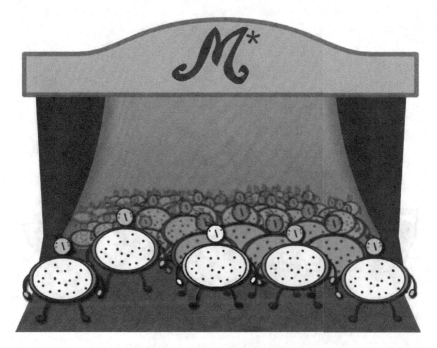

Fig. 1. A set-theoretic theatre

semantics linguistic expressions are supposed to denote *pieces of quantum information*: possible *states* of quantum systems that store the information in question. In the simplest situations one is dealing with a single particle S (say, an electron or a photon), whose "mathematical environment" is the two-dimensional Hilbert space $\mathbb{C}^2$ (based on the set of all ordered pairs of complex numbers). The canonical (orthonormal) basis of $\mathbb{C}^2$ consists of the two following unit-vectors:

$$|0\rangle = (1,0); \quad |1\rangle = (0,1),$$

which represent, in this framework, the two classical bits (0 and 1), or (equivalently) the two classical truth-values (*Falsity* and *Truth*). A *qubit* (or *qubit-state*) is a possible *pure state* of S: a *maximal information* that cannot be consistently extended to a richer knowledge. Such state is represented as a unit-vector $|\psi\rangle$ that can be expressed as a *superposition* of the two elements of the canonical basis of $\mathbb{C}^2$:

$$|\psi\rangle = c_0|0\rangle + c_1|1\rangle,$$

---

times. See, for instance, [1].

Fig. 2. Set-theoretic masks

where c_0 and c_1 (also called *amplitudes*) are complex numbers such that $|c_0|^2 + |c_1|^2 = 1$.

The physical interpretation of $|\psi\rangle$ is the following: the physical system S in state $|\psi\rangle$ might satisfy the physical properties that are *certain* for the bit $|0\rangle$ with probability $|c_0|^2$ and might satisfy the physical properties that are *certain* for the bit $|1\rangle$ with probability $|c_1|^2$. Due to the characteristic indeterminism of quantum theory, the pure state $|\psi\rangle$ is at the same time a *maximal and logically incomplete piece of information* that cannot *decide* some important physical properties of the system S. Accordingly, from an intuitive point of view, one can say that $|\psi\rangle$ describes a kind of *cloud of potential properties* that might become *actual* when a measurement is performed. Measuring a physical quantity (by means of an apparatus associated to the canonical basis) determines a sudden transformation of the qubit $|\psi\rangle$ either into the bit $|0\rangle$ or into the bit $|1\rangle$. Such transformation is usually called *collapse of the wave-function*.

As happens in classical information theory, quantum computation also needs complex pieces of information, which are supposed to be stored by composite quantum systems (generally consisting of n subsystems).

Accordingly, one can naturally adopt the quantum-theoretic formalism for the mathematical representation of composite physical systems, based on the use of *tensor products*. While a single qubit is a unit-vector of the space $\mathbb{C}^2$, a pure state representing a complex piece of information can be identified with a unit-vector of the n-fold tensor product of $\mathbb{C}^2$:

$$\mathcal{H}^{(n)} = \underbrace{\mathbb{C}^2 \otimes \ldots \otimes \mathbb{C}^2}_{n-times} \text{ (with } n \geq 1).$$

Such vectors are called *quregisters*. The canonical basis of the space $\mathcal{H}^{(n)}$ consists af all *registers*, products of bits that have the following form:

$$|x_1\rangle \otimes \ldots \otimes |x_n\rangle \quad \text{(where any } x_i \text{ is either 0 or 1)}.$$

Instead of $|x_1\rangle \otimes \ldots \otimes |x_n\rangle$, it is customary to write $|x_1, \ldots, x_n\rangle$. Any quregister can be represented as a superposition of registers:

$$|\psi\rangle = \sum_i c_i |x_{i_1}, \ldots, x_{i_n}\rangle,$$

where c_i are complex numbers such that $\sum_i |c_i|^2 = 1$.

As is well known, not all states associated to a physical system S are pure. Non-maximal pieces of information can be represented as *mixtures of pure states* (special examples of operators called *density operators*). In the space $\mathcal{H}^{(n)}$ a density operator ρ can be represented as a convenient finite sum of projection-operators:

$$\rho = \sum_i w_i P_{|\psi_i\rangle},$$

where w_i are real numbers such that $\sum_i w_i = 1$, while each $P_{|\psi_i\rangle}$ is a projection-operator that projects along the direction of the vector $|\psi\rangle$ (of $\mathcal{H}^{(n)}$). Notice that such representation is not generally unique. A density operator that cannot be represented as a projection $P_{|\psi\rangle}$ is called a *proper mixture*. Unlike pure states (which always satisfy some well-determined properties), there are mixtures that cannot decide any (non-trivial) property of the associated system. An example of this kind is the state $\rho = \frac{1}{2}\mathrm{I}$, where I is the identity operator of the space $\mathbb{C}^2$.

Quantum computation makes essential use of some characteristic quantum states that are called *entangled*. In order to illustrate the concept of entanglement from an intuitive point of view, let us refer to a simple paradigmatic case. We are concerned with a composite physical system S consisting of two subsystems S_1 and S_2 (say, a two-electron system). By the quantum-theoretic rules that concern the mathematical description of composite systems, all states of S shall live in the tensor product $\mathcal{H} = \mathcal{H}_1 \otimes \mathcal{H}_2$, where $\mathcal{H}_1$ and $\mathcal{H}_2$ are the Hilbert spaces associated to the systems S_1 and S_2, respectively. The observer has a *maximal information* about S: a *pure state* $|\psi\rangle$ of $\mathcal{H}$. What can be said about the states of the two subsystems? Due to the form of $|\psi\rangle$, such states cannot be pure: they are represented by two identical *mixtures*, which codify a "maximal degree of uncertainty". A typical possible form of $|\psi\rangle$ is the following *Bell-state*:

$$|\psi\rangle = \frac{1}{\sqrt{2}}(|0,0\rangle + |1,1\rangle),$$

which lives in the space $\mathbb{C}^2 \otimes \mathbb{C}^2$, whose canonical basis consists of the four vectors $|0,0\rangle, |0,1\rangle, |1,0\rangle, |1,1\rangle$. This gives rise to the following physical interpretation: the global system S might satisfy the properties that are *certain* either for the state $|0,0\rangle$ or for the state $|1,1\rangle$ with probability-value $\frac{1}{2}$. At the same time, $|\psi\rangle$ determines that the *reduced state* of both subsystems (S_1 and S_2) is the mixture $\frac{1}{2}\mathrm{I}$. Although it is not determined whether the state of the global system S is $|0,0\rangle$ or $|1,1\rangle$, the two subsystems S_1 and S_2 can be described as "entangled", because in both possible cases they would satisfy the same properties, turning out to be *indistinguishable*. Apparently, the information about the global system (S) cannot be reconstructed as a function of the pieces of information about its parts (S_1, S_2). For, the composition of two proper mixtures cannot be a pure state. In such cases, information seems to flow from the *whole* to its *parts* (and not the other way around). Phenomena of this kind give rise to the so called *holistic features* of quantum theory that violate a basic assumption of classical semantics, the *compositionality-principle*, according to which the meaning of a compound expression should be always represented as a function of the meanings of its parts.

Let us now briefly recall the basic features of a quantum computational semantics for a first-order language $\mathcal{L}$, whose non-logical alphabet contains individual terms (variables and names), predicates and sentential constants.[b] Interpreting the language $\mathcal{L}$ means associating to any formula α a *meaning*, identified with a piece of quantum information that can be stored by a quantum system. Accordingly, any possible meaning of α is represented by a possible (pure or mixed) state of a quantum system: generally, a density operator ρ_α that lives in a Hilbert space $\mathcal{H}^\alpha$, whose dimension depends on the linguistic complexity of α.

---

[b]See [6].

The logical operators of $\mathcal{L}$ are associated to special examples of Hilbert-space operations that have a characteristic dynamic behavior, representing possible computation-actions. The logical connectives are interpreted as particular (reversible) unitary quantum operations, corresponding to some important quantum-logical gates. At the same time, the logical quantifiers ($\forall$, $\exists$) are interpreted as possibly irreversible quantum operations. Since the universe of discourse (which the language refers to) may be indeterminate, the use of quantum quantifiers may give rise to a reversibility-breaking, which is quite similar to what happens in the case of measurement-phenomena.

Due to the characteristic features of quantum holism, meanings turn out to behave in a *holistic* and *contextual* way: the density operator ρ_α (which represents the *global* meaning of a formula α) determines the *contextual meanings* of all parts of α. As expected, the contextual meaning of any part of α (whose global meaning is ρ_α) can be obtained by applying the *reduced-state function* to ρ_α. As a consequence, it may happen that the meaning of a formula is an entangled pure state, while the meanings of its parts are proper mixtures. In such cases, the meaning of a global expression turns out to be more precise than the meanings of its parts (against the compositionality-principle). It is also admitted that one and the same formula receives different contextual meanings in different contexts.

As an example, consider the atomic sentence "Alice is pretty" (formalized as $\mathbf{Pa}$). In order to store the information expressed by this sentence, we need three quantum objects whose states represent the pieces of information corresponding, respectively, to the predicate $\mathbf{P}$, to the name $\mathbf{a}$ and to the truth-degree according to which the individual denoted by the name $\mathbf{a}$ satisfies the property denoted by the predicate $\mathbf{P}$. Accordingly, the meaning of the sentence $\mathbf{Pa}$ can be identified with a (pure or mixed) state $\rho_{\mathbf{Pa}}$ living in the tensor-product space $\mathcal{H}^{\mathbf{Pa}} = \mathbb{C}^2 \otimes \mathbb{C}^2 \otimes \mathbb{C}^2$. In order to obtain the contextual meanings of the linguistic parts of $\mathbf{Pa}$ it is sufficient to consider the two reduced states $red_1(\rho_{\mathbf{Pa}})$ and $red_2(\rho_{\mathbf{Pa}})$, which describe (respectively) the states of the first and of the second subsystem of the quantum object that stores the information expressed by the sentence $\mathbf{Pa}$. From a logical point of view, $red_1(\rho_{\mathbf{Pa}})$ and $red_2(\rho_{\mathbf{Pa}})$ can be regarded as two *intensional meanings*: a property-concept and an individual concept, respectively; while $\rho_{\mathbf{Pa}}$ represents a *propositional concept* (or *event*).

Notice that, unlike most semantic approaches, we do not assume here any ontological hierarchy between individuals and properties: states of the space $\mathbb{C}^2$ can store either individual names or predicates. For instance, the meaning $\rho_{\mathbf{Pa}}$ might correspond to the following factorized pure state:

$$|\Psi\rangle_{\mathbf{Pa}} = |\psi\rangle \otimes |\varphi\rangle \otimes \frac{1}{\sqrt{2}}(|0\rangle + |1\rangle).$$

Since $\frac{1}{\sqrt{2}}(|0\rangle + |1\rangle)$ corresponds to an intermediate truth-degree between the Falsity and the Truth, one obtains that Alice (the individual described by the qubit $|\varphi\rangle$) satisfies the property described by the qubit $|\psi\rangle$ with probability-value $\frac{1}{2}$. In the context $|\Psi\rangle_{\mathbf{Pa}}$, "prettiness" turns out to be a fully vague property of Alice. In a similar way, one can assign meanings to molecular formulas that may contain either logical connectives or quantifiers.

Like formulas, sequences of formulas also can be interpreted according to the quantum computational rules. As expected, a possible meaning of the sequence $(\alpha_1, \ldots, \alpha_n)$ will be a density operator $\rho_{(\alpha_1, \ldots, \alpha_n)}$ living in a Hilbert space $\mathcal{H}^{(\alpha_1, \ldots, \alpha_n)}$, whose dimension depends on the linguistic complexity of the formulas $\alpha_1, \ldots, \alpha_n$. In this framework one can develop an abstract theory of *vague possible worlds*. Consider a pair

$$W = ((\alpha_1, \ldots, \alpha_n), \, \rho_{(\alpha_1, \ldots, \alpha_n)}),$$

consisting of a sequence of formulas and of a density operator that represents a possible meaning for our sequence. It seems reasonable to assume that W describes a *vague possible world*, a kind of *abstract scene* where most events are characterized by a "cloud of ambiguities", due to quantum uncertainties. In some cases W might be exemplified as a "real" scene of a theatrical play or as a vague situation that is described either in a novel or in a poem. And it is needless to recall how ambiguities play an essential role in literary works.

As an example, consider the following vague possible world:

$$W = ((\mathbf{Pab}), \rho_{(\mathbf{Pab})}),$$

where $\mathbf{Pab}$ is supposed to formalize the sentence "Alice is kissing Bob", while $\rho_{\mathbf{Pab}}$ corresponds to the pure state

$$|\Psi\rangle_{\mathbf{Pab}} = |\varphi\rangle \otimes \frac{1}{\sqrt{2}}(|0, 1\rangle) + |1, 0\rangle) \otimes |1\rangle,$$

where $|\varphi\rangle$ lives in the space $\mathbb{C}^2$, while $|\Psi\rangle_{\mathbf{Pab}}$ lives in the space $\mathbb{C}^2 \otimes \mathbb{C}^2 \otimes \mathbb{C}^2 \otimes \mathbb{C}^2$. Here the reduced state of $|\Psi\rangle_{\mathbf{Pab}}$ that describes the pair (Alice, Bob) is an entangled Bell-state; consequently, the states describing the two individuals Alice and Bob are two identical mixed states. In the context $|\Psi\rangle_{\mathbf{Pab}}$ Alice and Bob turn out to be indistinguishable: it is not determined "who is who" and "who is kissing whom". It is not difficult to imagine some "real" theatrical scenes representing ambiguous situations of this kind.

The quantum semantics can be naturally applied to an abstract analysis of metaphors. Both in the case of natural languages and of literary contexts metaphorical correlations generally involve some allusions that are based on particular similarity-relations. Of course the inverse relation does not generally hold: similarities do not necessarily give rise either to allusions or to metaphors. Ideas that are currently used as possible metaphors are often associated with concrete and visual features. As observed by Aristotle, a characteristic property of metaphors is "putting things under our eyes".[c] Let us think, for instance, of a visual idea that is often used as a metaphor: the image of the sea, correlated to the concepts of immensity, of infinity, of obscurity, of pleasure or fear, of places where we may get lost and die.

The concept of quantum superposition can represent a natural and powerful semantic tool in order to represent the ambiguous allusions that characterize metaphorical correlations. Consider a quregister

$$|\psi\rangle = \sum_i c_i |\psi_i\rangle, \quad \text{where} \quad c_i \neq 0.$$

In such a case any $|\psi_i\rangle$ turns out to be *non-orthogonal to* $|\psi\rangle$. We have:

$$|\psi_i\rangle \not\perp |\psi\rangle$$

(i.e. the inner product of $|\psi_i\rangle$ and $|\psi\rangle$ is different from 0). As is well known, the non-orthogonality relation $\not\perp$ represents a typical similarity-relation (which is reflexive, symmetric and generally non-transitive). Hence, in particular semantic applications, the idea $|\psi_i\rangle$ (which $|\psi\rangle$ alludes to) might represent a *metaphor* for $|\psi\rangle$, or viceversa.

3. Metaphors in music

An abstract version of the quantum computational semantics can be applied to a formal analysis of musical compositions, where both *musical ideas* and *extra-musical meanings* are generally characterized by some essentially vague and ambiguous features.[d]

Any musical composition (say, a sonata, a symphony, a lyric opera,...) is, generally, determined by three elements:

- a *score*;

---

[c]See Aristotle, *Meteorologica*, 357.
[d]See [5].

- a set of *performances*;
- a set of musical *thoughts* (or *ideas*), which represent possible *meanings* for the *musical phrases* written in the score.

While scores represent the syntactical component of musical compositions, performances are physical events that occur in space and time. From a logical point of view, we could say that performances are, in a sense, similar to *extensional meanings*, i.e. well determined systems of objects which the linguistic expressions refer to.

Musical thoughts (or ideas) represent, instead, a more mysterious element. Is it reasonable to assume the existence of such ideal objects that are, in a sense, similar to the *intensional meanings* investigated by logic? Is there any danger to adhere, in this way, to a form of *Platonism*? When discussing semantic questions, one should not be "afraid" of Platonism. In the particular case of music, a composition cannot be simply reduced to a score and to a system of sound-events. Between a score (which is a system of signs) and the sound-events created by a performance there is something intermediate, represented by the musical ideas that underlie the different performances. This is the abstract environment where normally live both composers and conductors, who are accustomed to study scores without any help of a material instrument.

Following the rules of quantum semantics, *musical ideas* can be naturally represented as superpositions that ambiguously describe a variety of co-existent thoughts. Accordingly, we can write:

$$|\mu\rangle = \sum_i c_i |\mu_i\rangle,$$

where:

- $|\mu\rangle$ is an abstract object representing a musical idea that *alludes* to other ideas $|\mu_i\rangle$ (possible *variants* of $|\mu\rangle$ that are, in a sense, all co-existent);
- the number c_i measures the "importance" of the component $|\mu_i\rangle$ in the context $|\mu\rangle$.

As happens in the case of composite quantum systems, musical ideas (which represent possible meanings of *musical phrases* written in a score) have an essential *holistic* behaviour: the meaning of a global musical phrase determines the *contextual meanings* of all its parts (and not the other way around).

As an example, we can refer to the notion of *musical theme*. What exactly are musical themes? The term "theme" has been used for the first time in a musical sense by Gioseffo Zarlino, in his *Le istitutioni harmoniche* (1558), as a melody that is repeated and varied in the course of a musical work. Generally a theme appears in a musical composition with different "masks". In some cases it can be easily recognized even in its transformations; sometimes it is disguised and can be hardly discovered. Of course, a theme cannot be identified with a particular (syntactical) phrase written in the score; for, any theme essentially alludes to a (potentially) infinite set of possible variants. One is dealing with a vague musical idea that cannot be either played or written. At the same time, it is interesting to investigate (by scientific methods) the musical parameters that represent a kind of *invariant*, characteristic of a given theme. In different situations the relevant parameters may concern the melody or the harmony or the rhythm or the timbre.

The ambiguous correlations between a theme and its possible variants turn out to be exalted in the fascinating musical form that is called *Theme and Variations*. By using the superposition-formalism, we can represent the abstract form of a theme as follows:

$$|\mu\rangle = c_0|\mu_0\rangle + c_1|\mu_1\rangle + \ldots + c_n|\mu_n\rangle,$$

where:

- $|\mu_0\rangle$ represents the *basic theme* (a sharp musical idea, precisely written in the score).
- $|\mu_1\rangle, \ldots, |\mu_n\rangle$ represent the variations of $|\mu_0\rangle$.
- $|\mu\rangle$ represents an ambiguous musical idea that is correlated to the basic theme and to all its variations.

Of course the basic theme $|\mu_0\rangle$ has a privileged role, while the global theme $|\mu\rangle$ seems to behave like a kind of "ghost", which is somehow mysteriously present even if it appears hidden.

As is well known, an important feature of music is the capacity of *evoking* extra-musical meanings: subjective feelings, situations that are vaguely imagined by the composer or by the interpreter or by the listener, real or virtual theatrical scenes (which play an essential role in the case of lyric operas and of *Lieder*). The interplay between musical ideas and extra-musical meanings can be naturally represented in the framework of our quantum semantics, where extra-musical meanings can be dealt with as special examples of vague possible worlds.

We can refer to the tensor product of two spaces

$$M\,Space \otimes W\,Space,$$

where:

- $M\,Space$ represents the space of musical ideas $|\mu\rangle$.
- $W\,Space$ represents the space of vague possible worlds, dealt with as special examples of abstract objects $|w\rangle$ that can be evoked by musical ideas.

Following the quantum-theoretic formalism, we can distinguish between *factorized* and *non-factorized* global musical ideas. As expected, a factorized global musical idea will have the form:

$$|M\rangle = |\mu\rangle \otimes |w\rangle.$$

But we might also meet "Bell-like" entangled global musical ideas, having the form:

$$|M\rangle = c_1(|\mu_1\rangle \otimes |w_1\rangle) + c_2(|\mu_2\rangle \otimes |w_2\rangle).$$

In the case of lyric operas and of *Lieder* musical ideas and vague possible worlds are, in fact, always *entangled* (in an intuitive sense). We need only think how some opera-librettos may appear naive and, in some parts, even funny, if they are read as pieces of theatre, separated from music. Also *Lieder*, whose texts have often been written by great authors (Goethe, Schiller, Heine, etc.) give rise to similar entangled situations. Generally a musical intonation of a given poem transforms the text into a new *global semantic object* that somehow absorbs and renews all meanings of the original literary work.

To what extent can some musical ideas be interepreted as *musical metaphors* for *extra-musical meanings*? Is it possible to recognize any natural similarity-relations that connect ideal objects living in two different worlds that seem to be deeply far apart? In order to discuss this question it is expedient to refer to some interesting musical examples. Significant cases can be found in the framework of Schubert's *Lieder*, where some musical figures and themes based on *sextuplets* often evoke images of water and of events that take place in water. Let us refer, for instance, to the celebrated *Lieder*-cycle *Die Schöne Müllerin* (*The Beautiful Miller's Daughter*). The story told in the poems of the German poet Wilhelm Müller is very simple. A young man, a miller, falls in love with *die Schöne Müllerin*, the beautiful daughter of the mill's owner. But the girl refuses him and prefers a wild hunter. The young miller cannot overcome his love's pains and finally dies.

During his *Wandern* (wandering) his only true friend is *der Bach*, the mill's brook that has a constant dialog with him. The flowing of the brook's water represents a clear poetic and musical metaphor for the flowing of time and for the changing feelings of the young lover.

When in the second *Lied* of the cycle, *Wohin* (*Whereto*), the miller meets the brook for the first time, singing "Ich hört' ein Bächlein rauschen wohl aus dem Felsenquell" ("I heard a brooklet rushing right out of the rock's spring"), the piano-accompaniment begins playing a sequence of sextuplets that will be never interrupted until the end of the *Lied*. Even the graphical shape of the sextuplets in the score suggests a natural similarity with a sinusoidal form representing the water's wave-movement (Fig. 3). This creates a complex network of dynamic interactions among different elements:

- the musical thoughts that become "real" musical events during a performance of the *Lied*;
- the graphic representation of the musical phrases written in the score;
- the poetic metaphors, suggested both by the text and by the music, that allude to the flowing of time, to changing subjective feelings and to a mysterious fear for an uncertain future.

In many of his *Lieder* Schubert has often associated sextuplet-figures with images of water and with abstract ideas that refer to the flowing of time. Wonderful (and famous) examples are, for instance, the two *Lieder Auf dem Wasser zu singen* (*Singing on the water*) and *Die Forelle* (*The Trout*).

We will now consider another significant case that concerns Robert Schumann's compositions. We will refer to a very special musical theme that has been called "Clara's theme". Clara is Clara Wieck, the great pianist and composer who has been the wife of Schumann. One is dealing with a somewhat mysterious theme that appears as a kind of "hidden thought" in different works by Schumann, by Clara herself and by Johannes Brahms, three great musicians whose lives have been in a sense "entangled" even outside the sphere of music.

Unlike the basic theme of a "Theme and Variations"- composition, Clara's theme cannot be identified with a precise musical phrase written in a particular score: many different variants of this theme have been recognized in different contexts, associated to different semantic connotations. It is well known that Schumann liked the use of "secret codes": special musical ideas whose aim was an ambiguous allusion to some extra-musical situations. The code of Clara's theme is based on the letters that occur

82

Fig. 3. Sextuplets in the *Lied Wohin*

in the name "CLARA", where "A" and "C" correspond to musical notes, while "L" and "R" do not have any musical correspondence. In spite of this, one can create some interpolation, giving rise to different variants, all inspired by the name "Clara". An interesting example is the following note-sequence, which belongs to the F sharp minor-tonality:

C♮ (B) A (G♯) A

[C   L   A   R   A]

Like in the case of Schubert's sextuplets we can ask: is it reasonable to interpret Clara's theme as a kind of *musical metaphor*? Using a code (in a musical form) clearly suggests a reference to some extra-musical ideas. But what exactly is evoked by means of this special code? Of course, the aim cannot be a realistic description of the person denoted by the name "Clara" (a kind of *extensional reference* in logical sense). Let us consider some significant examples where Clara's theme has played an important role. In 1853 Clara Wieck composed the piano-piece *Variationen op. 20, über ein Thema von Robert Schumann, ihm gewidmet*, dedicated to her husband in occasion of his birthday. One year later Brahms wrote his own *Variations* on the same theme and dedicated his composition to Clara. Schumann's theme, which Clara and Brahms present exactly in the same way, is drawn from *Bunte Blätter*, a composition that Schumann wrote in 1841 (Fig. 4).

One can easily see that this "Schumann's theme" is based on one of the possible *variantas* of Clara's theme (in *F sharp* minor):

C♯ C♯ C♯ (B) A (G♯) A.

The melodic line is developed as a descending sequence of joint grades of the F sharp minor scale (from the fifth to the second grade). Then the melody ascends and remains somehow suspended on the third grade of the scale. Soon after the musical phrase is repeated, concluding on a A major chord. The dynamic and agogic indications are "piano" and "Ziemlich langsam". After a short digression to the C sharp minor-tonality, Clara's theme appears again and finally concludes, *pianissimo*, on the tonic chord of the basic tonality.

Is it possible to recognize, in a natural way, some extra-musical meanings, connected with Clara's personality, that might be correlated as vague allusions to the musical features of Schumann's theme? A resonable conjecture seems to be the following: Clara is here evoked as a kind of "consoling figure", who inspires serene and peaceful feelings. It is not a chance that in

Fig. 4. Schumann's theme

one of the most famous Schumann's *Lieder*, *Widmung* (*Dedication*), composed in the year 1840, when Robert and Clara got married, the voice sings with the words of the poet Rückert "Du bist die Ruh, du bist der Frieden" ("You are the rest, you are the peace"), while in the piano conclusion the consoling theme of Schubert's *Ave Maria*, which is repeated twice, suddenly appears as a somewhat hidden quotation. The hypothesis that a vague consolation-idea represents an important semantic connotation associated to Clara seems to be confirmed by some *Lieder* where Clara's theme can be easily recognized. Of course, metaphorical correlations that emerge in *Lieder* are often somewhat cryptic, also because musical metaphors turn out to be ambiguously interlaced with the poetic metaphors that are expressed in the literary text. An interesting example is represented by the eighth *Lied* (*Und wüssten's die Blumen*) of the famous *Lieder*-cycle *Dichterliebe* Op. 48, based on Heine's poems. Clara's theme appears here at the very beginning of the first phrase sung by the voice. In the version "für mittlere und tiefe Stimme" (baritone and bass) we find the same tonality of Schumann's theme (*F* sharp minor) and the same descending note-sequence that in this case reaches the tonic (Fig. 5).

Fig. 5. *Und wüssten's die Blumen*

The leading idea expressed by Heine's poem is the search for a consolation that might be offered by a friendly Nature:

> *Und wüssten's die Blumen, die kleinen,*
> *Wie tief verwundet mein Herz,*
> *Sie würden mit mir weinen,*
> *Zu heilen meinen Schmerz.*[e]

One first addresses the flowers that could "heilen meinen Schmerz", but then the same request is turned to the nightingales and to the golden stars:

> *Sie kämen aus ihrer Höhe,*
> *Und sprächen Trost mir ein.*[f]

And significantly enough the first three stanzas of Heine's poem are all set to music by means of one and the same musical phrase (based on Clara's theme) that is repeated three times.

We have seen how metaphorical correlations can be described, from an abstract point of view, as very special cases where ideas belonging to different conceptual domains are connected by means of vague allusions. The

---

[e] *If the little flowers knew / How deeply wounded is my heart, / They would weep with me, / to soothe my pain.*
[f] *They would come down from their height, /and speak words of comfort to me.*

occurrence of a metaphor in a given context is generally characterized by a "cloud" of ambiguity and indetermination that can be naturally analyzed by using quantum theoretic concepts. The strength of quantum semantics depends on the fact that meanings are, in this framework, represented as relatively simple and cognitively accessible ideal objects that ambiguously allude to a potentially infinite variety of alternative ideas. In fact, any pure state of a Hilbert space can be represented as a superposition of elements of infinitely many possible bases of the space. And from an intuitive point of view any choice of a particular basis can be regarded as a possible *perspective* from which we are looking at the phenomena under investigation.

As is well known, semantic phenomena of ambiguity and vagueness have been investigated in the literature by a number of different approaches. In classical logical frameworks one has often referred to complex systems of possible worlds, where each particular world is characterized by sharp and deterministic features, according to the excluded-middle principle. This gives rise to a "multiplication of entities" that may represent a shortcoming from a cognitive point of view. More natural theories of vagueness have been developed in the framework of fuzzy logics. But what is generally missing in the standard many-valued semantics is the capacity of representing holistic aspects of meanings, which instead play an important role either in natural languages or in the languages of art. Of course, recognizing the advantages of quantum semantics does not imply an "ideological" conclusion, according to which the quantum theoretic formalism should have a kind of privileged position in the rich variety of semantic theories that have been proposed in the contemporary literature.

References

1. Aerts, D., Sozzo, S., *Quantum Entanglement in Conceptual Combinations*, International Journal of Theoretical Physics **53** (2014), 3587–3603.
2. Black, M., *Metaphors*, Proceedings of the Aristotelian Society, New Series, Vol. 55 (1954-1955).
3. Dalla Chiara, M. L., Giuntini, R., Leporini, R., *Logics from quantum computation*, International Journal of Quantum Information **3** (2005), 293–337.
4. Dalla Chiara, M.L., Freytes, H., Giuntini, R., Ledda, A., Leporini, R., Sergioli, G., *Entanglement as a semantic resource*, Foundations of Physics **40** (2011), 1494–1518.
5. Dalla Chiara, M.L., Giuntini, R., Luciani, A.R., Negri, E., *From Quantum Information to Musical Semantics*, College Publications, London, 2012.

6. Dalla Chiara, M.L., Giuntini, R., Leporini, R, Sergioli, G., *A first-order epistemic quantum computational semantics with relativistic-like epistemic effects*, Fuzzy Sets and Systems, **298** (2016), 69–90.
7. Gudder, S., *Quantum computational logics*, International Journal of Theoretical Physics **42** (2003), 39–47.
8. Nielsen, M., Chuang, I., *Quantum Computation and Quantum Information*, Cambridge University Press, Cambridge, 2000.
9. Spitzer, M., *Metaphors and Musical Thoughts*, The University of Chicago Press, Chicago, 2004.

WHY PROTECTIVE MEASUREMENT IMPLIES THE REALITY OF THE WAVE FUNCTION: FURTHER CONSOLIDATION

SHAN GAO

Research Center for Philosophy of Science and Technology
Shanxi University, Taiyuan 030006, P. R. China.
Centre for Philosophy of Natural and Social Science
London School of Economics and Political Science, UK.
E-mail: gaoshan2017@sxu.edu.cn

The existing ψ-ontology theorems are based on a simplified assumption of the ontological models framework, according to which when a measurement is performed the behaviour of the measuring device is determined by the ontic state of the measured system immediately before the measurement. In this paper, I give an argument for the reality of the wave function in terms of protective measurements under a more reasonable assumption, according to which the behaviour of the measuring device during a measurement is determined by the total evolution of the ontic state of the measured system during the measurement. In addition, I present a new analysis of how a protective measurement obtains the expectation value of the measured observable in the measured wave function. The analysis strengthens my argument by further clarifying the role the protection procedure plays in a protective measurement.

Keywords: Wave function; protective measurement; ψ-ontology theorems.

1. Introduction

In a previous paper [1], I gave a proof of the reality of the wave function in terms of protective measurements [2,3].[a] The proof does not rely on auxiliary assumptions. This improves the Pusey-Barrett-Rudolph theorem [5] and other ψ-ontology theorems [6–8]. This new proof, like these ψ-ontology theorems, is also based on the second assumption of the ontological models framework [5,9], according to which when a measurement is performed, the behaviour of the measuring device is determined by the ontic state of the measured system (along with the physical properties of the measuring

---

[a]Recently the first protective measurement has been realized in experiments [4].

device) immediately before the measurement, whether the ontic state of the measured system changes or not during the measurement. As noted by Gao [1], however, this is a simplified assumption which may be not valid in general. A more reasonable assumption is that the ontic state of the measured system may be disturbed and thus evolve in a certain way during a measurement, and the behaviour of the measuring device is determined by the total evolution of the ontic state of the system during the measurement, not simply by the initial ontic state of the system. In this paper, I will give an argument for the reality of the wave function in terms of protective measurements under this new assumption. Moreover, I will also clarify the role the protection procedure plays in a protective measurement, which may help understand this result.

2. The argument

As pointed out by Gao [1], the proofs of existing ψ-ontology theorems such as the Pusey-Barrett-Rudolph theorem will be invalid under the new assumption. The reason is that under this assumption, even if two nonorthogonal states correspond to the same ontic state initially, they may correspond to different evolution of the ontic state, which may lead to different probabilities of measurement results. Then the proofs of the ψ-ontology theorems by reduction to absurdity cannot go through. According to [1], his direct argument for ψ-ontology in terms of protective measurements can go through under the new assumption. He said:

> First, according to this assumption, the evolution of the ontic state of a physical system during a protective measurement determines the result of the protective measurement, namely the expectation value of the measured observable in the measured quantum state. Next, since the quantum state of the system keeps unchanged, the evolution of the ontic state of the system is still compatible with the quantum state. This means that even when the system being in the quantum state is not measured, its ontic state may also evolve in this way and such evolution is then a realistic property of the system. Therefore, the expectation value of the measured observable is determined by a realistic property of the measured system, and it is also a realistic property of the system. Then similar to the direct argument given in the last section, we can also prove the reality of the quantum state.

However, this argument is seriously flawed. It is true that during a protective measurement the disturbed evolution of the ontic state of the measurd system is compatible with the wave function of the system. But this does not mean that when the system is not measured its ontic state may also evolve in this disturbed way. Thus it still needs to be argued that the disturbed evolution is a realistic property of the system. In the following, I will fix this loophole.

For a protective measurement, there are two sources which may interfere with the spontaneous evolution of the ontic state of the measured system: one is the protection procedure, and the other is the measuring device. However, no matter how they influence the evolution of the ontic state of the measured system, they cannot generate the definite result of the protective measurement, namely the expectation value of the measured observable in the measured wave function, since they contain no information about the measured wave function.[b] The measuring device only contains information about the measured observable, and it does not contain information about the measured wave function. Compared with the measuring device, the protection procedure "knows" less. The protection procedure is either a protective potential or a Zeno measuring device. In each case, the protection procedure contains no information about both the measured observable and the measured wave function.[c] For example, in the case of Zeno protection, the protection procedure only "knows" the information about an observable, of which the measured wave function is a nondegenerate eigenstate. Thus, if the information about the measured wave function is not contained in the measured system, then the result of a protective measurement cannot be the expectation value of the measured observable in the measured wave function.

---

[b]In other words, the properties of the protection setting and the measuring device and their time evolution do not determine the measured wave function.

[c]Certainly, the measurer who does the protective measurement knows more information than that contained in the measuring device and protection procedure. Besides the measured observable, the measurer also knows the measured wave function is one of infinitely many known states (but she needs not know which one the measured wave function is). In the case of protective potential, the measurer knows that the measured wave function is one of infinitely many nondegenerate discrete energy eigenstate of the Hamiltonian of the measured system. In the case of Zeno protection, the measurer knows that the measured wave function is one of infinitely many nondegenerate eigenstates of an observable. Note that this permits the possibility that the measurer can cheat us by first measuring which one amongst these infinitely many states the measured wave function is (e.g. by measuring the eigenvalue of energy for the case of protective potential) and then calculating the expectation value and outputing it through a device. Then the result will have no implications for the reality of the wave function. But obviously this is not a protective measurement.

On the other hand, if the result of a protective measurement is also determined by the ontic state of the measuring device or the protection procedure through their influences on the spontaneous evolution of the ontic state of the measured system, then the result may be different for the same measured observable and the same measured wave function. This contradicts the predictions of quantum mechanics, according to which the result of a protective measurement is always the expectation value of the measured observable in the measured wave function.

Therefore, the definite result of a protective measurement, namely the expectation value of the measured observable in the measured wave function, is determined by the spontaneous evolution of the ontic state of the measured system during the measurement. Since the spontaneous evolution of the ontic state of the measured system is an intrinsic property of the system independent of the protective measurement, the expectation value of the measured observable in the measured wave function is also a property of the system. This then proves the reality of the wave function, which can be constructed from the expectation values of a sufficient number of observables.

3. How does a protective measurement obtain the expectation value?

In the following, I will present a new analysis of how a protective measurement obtains the expectation value of the measured observable in the measured wave function. The analysis may help understand the above result by further clarifying the role the protection procedure plays in a protective measurement.

By a projective measurement on a single quantum system, one obtains one of the eigenvalues of the measured observable, and the expectation value of the observable can only be obtained as the statistical average of eigenvalues for an ensemble of identically prepared systems. Thus it seems surprising that a protective measurement can obtain the expectation value of the measured observable directly from a single quantum system. In fact, however, this result is not as surprising as it seems to be. The key point is to notice that according to the linear Schrödinger evolution the pointer shift rate at any time during a projective measurement is proportional to the expectation value of the measured observable in the measured wave function at the time. Concretely speaking, for a projective measurement of an observable A, whose interaction Hamiltonian is given by the usual form $H_I = g(t)PA$, where $g(t)$ is the time-dependent coupling strength of the interaction, and P is the conjugate momentum of the pointer variable, the pointer shift rate at each instant t during the measurement is:

$$\frac{d\langle X\rangle}{dt} = g(t)\langle A\rangle, \qquad (1)$$

where X is the pointer variable, $\langle X\rangle$ is the center of the pointer wavepacket at instant t, and $\langle A\rangle$ is the expectation value of the measured observable A in the measured wave function at instant t. This pointer shift rate formula indicates that at any time during a projective measurement, the pointer shift after an infinitesimal time interval is proportional to the expectation value of the measured observable in the measured wave function at the time. This result may be more surprising for some people. As is well known, however, since the projective measurement changes the wave function of the measured system greatly, and especially it also results in the pointer wavepacket spreading greatly, the point shift after the measurement does not represent the actual measurement result, and it cannot be measured either. Moreover, even if the point shift after the measurement represents the actual measurement result (e.g. for collapse theories), the result is not definite but random, and it is not the expectation value of the measured observable in the initial measured wave function either.

Then, how to make the expectation value of the measured observable in the measured wave function, which is hidden in the process of a projective measurement, visible in the final measurement result? This requires that the pointer wavepacket should not spread considerably during the measurement so that the final pointer shift is qualified to represent the measurement result, and moreover, the final pointer shift should be also definite. A direct way to satisfy the requirement is to protect the measured wave function from changing as a protective measurement does. Take the Zeno protection scheme as an example. We make frequent projective measurements of an observable O, of which the measured state $|\psi\rangle$ is an nondegenerate eigenstate, in a very short measurement interval $[0, \tau]$. For instance, O is measured in $[0, \tau]$ at times $t_n = (n/N)\tau, n = 1, 2, ..., N$, where N is an arbitrarily large number. At the same time, we make the same projective measurement of an observable A in the interval $[0, \tau]$ as above. Different from the usual derivation [3,10,11],[d] here I will calculate

---

[d]Note that in the usual derivation, the measurement result of a protective measurement, namely the expectation value of the measured observable in the measured wave function, is already contained in the measurement operator which describes the measurement

the post-measurement state in accordance of the order of time evolution. This will let us see the process of protective measurement more clearly.

The state of the combined system immediately before $t_1 = \tau/N$ is given by

$$e^{-\frac{i}{\hbar}\frac{\tau}{N}g(t_1)PA}|\psi\rangle|\phi(x_0)\rangle = \sum_i c_i|a_i\rangle\left|\phi(x_0 + \frac{\tau}{N}g(t_1)a_i)\right\rangle$$

$$= |\psi\rangle\left|\phi(x_0 + \frac{\tau}{N}g(t_1)\langle A\rangle)\right\rangle$$

$$+ \frac{\tau}{N}g(t_1)(A - \langle A\rangle)|\psi\rangle\left|\phi'(x_0 + \frac{\tau}{N}g(t_1)\langle A\rangle)\right\rangle$$

$$+ O(\frac{1}{N^2}), \tag{2}$$

where $|\phi(x_0)\rangle$ is the pointer wavepacket centered in initial position x_0, $|a_i\rangle$ are the eigenstates of A, and c_i are the expansion coefficients. Note that the second term in the r.h.s of the formula is orthogonal to the measured state $|\psi\rangle$. Then the branch of the state of the combined system after $t_1 = \tau/N$, in which the projective measurement of O results in the state of the measured system being in $|\psi\rangle$, is given by

$$|\psi\rangle\langle\psi|e^{-\frac{i}{\hbar}\frac{\tau}{N}g(t_1)PA}|\psi\rangle|\phi(x_0)\rangle = |\psi\rangle\left|\phi(x_0 + \frac{\tau}{N}g(t_1)\langle A\rangle)\right\rangle + O(\frac{1}{N^2}). \tag{3}$$

Thus after N such measurements and in the limit of $N \to \infty$, the branch of the state of the combined system, in which each projective measurement of O results in the state of the measured system being in $|\psi\rangle$, is

$$|t = \tau\rangle = |\psi\rangle\left|\phi(x_0 + \int_0^\tau g(t)dt\langle A\rangle)\right\rangle = |\psi\rangle|\phi(x_0 + \langle A\rangle)\rangle. \tag{4}$$

Since the modulus squared of the amplitude of this branch approaches one when $N \to \infty$, this state will be the state of the combined system after the protective measurement.

By this derivation, it can be clearly seen that the role of the protection procedure is not only to protect the measured wave function from the change caused by the projective measurement, but also to prevent the

---

procedure. But this does not imply that what the measurement measures is not the property of the measured system, but the property of the measurement procedure such as the protection procedure, cf. [11]. Otherwise, for example, diseases will exist not in patients, but in doctors or expert systems for disease diagnosis.

pointer wavepacket from the spreading caused by the projective measurement. As a result, the pointer shift after the measurement can represent a valid measurement result, and moreover, it is also definite, being natually the expectation value of the measured observable in the *initial* measured wave function.

4. Further consolidation

The above analysis of how a protective measurement obtains its result will strengthen my previous argument for ψ-ontology in terms of protective measurements.

Since the width of the pointer wavepacket keeps unchanged during a protective measurement, and the pointer shift rate at any time during the measurement is proportional to the expectation value of the measured observable in the measured wave function at the time,[e] which is the same as the initial measured wave function, we can actually obtain the final measurement result at any time during the protective measurement (when the time-dependent coupling strength is known). This indicates that the result of a protective measurement is determined by the initial ontic state of the measured system, not by the evolution of the ontic state of the system during the measurement, whether spontaneous or disturbed. Thus the second, simplified assumption of the ontological models framework is still valid for protective measurements, so does my previous argument for the reality of the wave function based on this assumption [1].

It has been conjectured that the result of a protective measurement is determined not by the ontic state of the measured system but by the protection procedure, which may lead to a certain evolution of the ontic state of the system that may generate the measurement result [11]. If this is true, then protective measurements will have no implications for the reality of the wave function. However, as I have argued in the beginning of

---

[e]Since the pointer shift is always continuous and smooth during a protective measurement, it is arguable that the evolution of the ontic state of the measured system (which determines the pointer shift) is also continuous. Then for an ideal situation where the protective measurement is instantaneous, the ontic state of the measured system will be unchanged after the measurement and my previous argument for ψ-ontology in terms of protective measurements will be still valid [1]. Note that the evolution of the position of the pointer as its ontic state may be discontinuous in an ψ-epistemic model. However, the range of the position variation is limited by the width of the pointer wavepacket, which can be arbitrarily small in principle. Thus such discontinuous evolution cannot be caused by the evolution of the ontic state of the measured system, whether continuous or discontinuous.

this section, this conjecture cannot be correct. The essential reason is that the protection procedure does not "know" the measured wave function, and thus it cannot generate the measurement result, the expectation value of the measured observable in the measured wave function.[f] In addition, the above analysis also clarifies the role of the protection procedure during a protective measurement. The expectation value of the measured observable in the measured wave function is already hidden in the process of the projective measurement, and what the protection procedure does is to make it visible in the final measurement result by keeping the measured wave function unchanged.

5. Conclusion

In this paper, I strengthen my previous argument for the reality of the wave function in terms of protective measurements. The previous argument, like other ψ-ontology theorems, is based on a simplified assumption of the ontological models framework, according to which when a measurement is performed the behaviour of the measuring device is determined by the ontic state of the measured system immediately before the measurement, whether the ontic state of the measured system changes or not during the measurement. This simplified assumption may be not valid in general. A more reasonable assumption is that the ontic state of the measured system may be disturbed and thus evolve in a certain way during a measurement, and the behaviour of the measuring device is determined by the total evolution of the ontic state of the system during the measurement. Although the proofs of the existing ψ-ontology theorems by reduction to absurdity cannot go through under the new assumption, I argue that my previous proof of ψ-ontology in terms of protective measurements can still go through under the assumption. In addition, I present a new analysis of how a protective measurement obtains the expectation value of the measured observable in the measured wave function, and clarify the role the protection procedure plays in a protective measurement. The analysis strengthens my argument for the reality of the wave function in terms of protective measurements.

---

[f]Note that in the ψ-epistemic models given by Combes et al [11], it is implicitly assumed that the protection procedure knows the measured wave function. Thus it is not surprising that the models can reproduce the predictions of quantum mechanics for protetcive measurements.

6. Acknowledgments

I am grateful to Matthew Leifer and Matthew Pusey for helpful discussions. Most of this paper was written when I visited the Centre for Philosophy of Natural and Social Science (CPNSS) at the London School of Economics and Political Science from 14 July 2016 to 23 July 2016. I thank the Center for providing research facilities and Roman Frigg and Laura O'Keefe for their help during my visit. This work is partly supported by the National Social Science Foundation of China (Grant No. 16BZX021).

References

1. S. Gao, "An argument for ψ-ontology in terms of protective measurements", *Studies in History and Philosophy of Modern Physics*, **52**, 198-202, 2015.
2. Y. Aharonov and L.Vaidman, "Measurement of the Schrödinger wave of a single particle", *Physics Letters A*, **178**, 38, 1993.
3. Y. Aharonov, J. Anandan and L. Vaidman, "Meaning of the wave function", *Physical Review A*, **47**, 4616, 1993.
4. Piacentini, F. et al, "Determining the quantum expectation value by measuring a single photon", *Nature Physics*, doi:10.1038/nphys4223, 2017. (arXiv:1706.08918).
5. M. Pusey, J. Barrett, and T. Rudolp, "On the reality of the quantum state", *Nature Physics*, **8**, 475-478, 2012.
6. R. Colbeck and R. Renner, "Is a system's wave function in one-to-one correspondence with its elements of reality?", *Physical Review Letters*, **108**, 150402, 2012.
7. L. Hardy, "Are quantum states real?", *International Journal of Modern Physics B*, **27**, 1345012, 2013.
8. M. S. Leifer, "Is the quantum state real? An extended review of ψ-ontology theorems", *Quanta*, **3**, 67-155, 2014.
9. N. Harrigan and R. Spekkens, "Einstein, incompleteness, and the epistemic view of quantum states", *Foundatinos of Physics*, **40**, 125-157, 2010.
10. S. Gao, "Protective measurement: An introduction", In *Protective Measurements and Quantum Reality: Toward a New Understanding of Quantum Mechanics*, pp.1-12, S. Gao (ed.), Cambridge: Cambridge University Press, 2014.
11. J. Combes, C. Ferrie, M. S. Leifer, and M. Pusey, "Why protective measurement does not establish the reality of the quantum state", *Quantum Studies: Mathematical Foundations*, doi:10.1007/s40509-017-0111-4, 2017. (arXiv:1509.08893)

DOES IDENTITY HOLD *A PRIORI* IN
STANDARD QUANTUM MECHANICS?

JONAS R. BECKER ARENHART

Department of Philosophy, Federal University of Santa Catarina,
Florianópolis, Santa Catarina, Brazil
E-mail: jonas.becker2@gmail.com

DÉCIO KRAUSE

Department of Philosophy, Federal University of Santa Catarina,
Florianópolis, Santa Catarina, Brazil
E-mail: deciokrause@gmail.com

We discuss an argument by Francesco Berto to the effect that quantum particles have well-defined identity conditions: Berto holds they are objects individuated by metaphysical ingredients such as haecceities. With this argument Berto intends to attack a general tenet that characterizes a version of the so-called *Received View* on quantum non-individuality; according to this view, roughly speaking, quantum particles are not individuals, they don't have well-defined identity conditions. Now, the version of the Received View attacked by Berto relates the metaphysical thesis about the lack of individuality with a linguistic thesis that the relation of identity makes no sense for quantum entities; it is this linguistic facet of the view that provides for its own rather direct representation in systems of non-reflexive logics such as quasi-set theory. Recall that non-reflexive logics are logics where, in some cases, the relation of identity may fail to hold between a given kind of terms. We discuss the argument, which Berto intends to be wholly *a priori*, in terms of pure conceptual analysis, and find it at fault. As we see it, the argument begs relevant questions against the friend of the Received View, given that it conceives some of the core concepts involved in the analysis as already involving the notion of identity that the Received View refuses to attribute to quantum entities. We also advance a general framework in which it is presented a more general way of addressing such issues, one which we believe is able to escape from such purely *a priori* analysis and provide for a more cooperative work between metaphysics and science.

Keywords: Identity; quantum mechanics; non-reflexive logics; individuality; conceptual analysis.

1. Introduction

As it is fairly well known, quantum entities give rise to a vast number of delicate issues concerning identity and individuality. Aside from the question itself, one of the pressing issues concerns the very methodology with which to address this problem: it is widely argued that quantum theory cannot decide those issues by itself; in fact, as French and Krause [11, chap.4] put it, the metaphysics of identity and individuality of quantum mechanics is underdetermined by the theory. In other words, quantum theory is compatible with at least two broad metaphysical packages on what concerns identity and individuality: according to one option, its entities are regarded as individuals, while according to another (rival) option, known as the *Received View* on quantum non-individuality, its entities are regarded as non-individuals. Both notions, individuality and non-individuality, are typically framed in terms of identity, so that this relation between identity and (non-)individuality is at the center of the stage too.

Breaking such underdetermination requires, it seems, going much beyond the use of purely quantum mechanical resources. For instance, Dorato and Morganti [10] have argued in favor of a minimal form of identity and individuality by appealing to a mix of quantum theory and use of theoretical virtues in metaphysics. Bueno [6] has argued that identity, even though it is metaphysically innocent, is fundamental in a much stronger — almost transcendental — sense, so that any theory, even quantum mechanics, requires identity in order to be intelligible for us. Those attempts illustrate how the debate has to shift from a rather crude form of naturalism to a more sophisticated combination of metaphysics, linguistic issues, and philosophy of science. However, while those papers are interesting and deserve careful consideration on their own,[a] in this paper we shall focus on the most recent attempt at granting identity and individuality to quantum entities, advanced by Berto in [4].

Following the path of Dorato and Morganti, Berto claims that quantum particles must have identity due to the fact that a well-defined number of them — as is always the case in standard quantum mechanics — always implies attribution of identity, which is then taken as a primitive concept. Now, due to metaphysical underdetermination, Berto radicalizes the approach to attack the problem, and what will concern us as a most pressing issue is Berto's strategy: he proposes to argue for such a view through

---

[a]We have considered Dorato and Morganti's paper in [1]; we addressed Bueno's claim in [12].

a much more powerful move than previous authors; that is, he proposes to argue for that conclusion by "a priori conceptual analysis" [4, p.8]. In a nutshell, the idea is that given the concepts of object, identity, and unity, there is no way to avoid attributing identity to quantum particles. Berto further argues that identity is a primitive concept, and that this primitive commits him with an innocent version of primitive thisness or haecceitism.

Now, if Berto is right and identity holds by the pure force of conceptual analysis, the Received View, at least as it is framed by French and Krause [11], viz. in terms of the idea that identity makes no sense for certain entities, is certainly a non-starter. We shall present and discuss Berto's conceptual analysis and find it at fault in some important points. In particular, some of the crucial steps in the analysis introduce unwarranted content in the concept of unity and object, so that it is not a surprise that identity should hold for everything in the end. In fact, French and Krause [11, chap.4] had already made a case against a similar step while discussing weak discernibility. In the end, the Received View is still respectable, even though we shall not argue for its truth here.

The structure of this paper is as follows. In section 2, we present Berto's argument for the fact that, by pure conceptual analysis, identity holds in the quantum context. We try to present as faithfully as possible all the ingredients in Berto's analysis and point to some aspects we shall challenge in section 3, where we address the question of whether that analysis holds good for the concepts of unity, object, and cardinality. We argue that the analysis provided by Berto is committed with a particular choice of logical foundation for the concepts, and identity is naturally already there. So, some questions are clearly begged against the friend of the Received View. In section 4 we advance an alternative account of how to proceed on a conceptual analysis, which preserves the possibility that identity and the notion of object and unity may be developed independently from identity. On the account sketched here, there is a place for productive interplay between science and metaphysics, so that the very idea of object may not be framed completely in a priori terms. We conclude in section 5.

2. Identity for quantum objects

Berto sets the stage for his analysis by introducing his terminology and the main concepts involved in the argument. According to him, those very concepts are also employed by the defenders of the Received View, so, as a simple matter of rationality, they should also endorse the conclusion that quantum entities have identity. If that is correct, then, the Received View fails even by its own lights: the very concepts employed require identity to hold for everything.

First of all, as we mentioned in the introduction, the Received View is characterized as a view according to which identity somehow fails to make sense for quantum entities in certain situations. The main spokesman of the view is Schrödinger, in a now classic passage, also quoted by French and Krause and in general discussions of the Received View[b]:

> I beg to emphasize this and I beg you to believe it: it is not a question of our being able to ascertain the identity in some instances and not being able to do so in others. It is beyond doubt that the question of 'sameness', of identity, really and truly has no meaning. [17, pp.121-122].

The general interpretation that is typically made of this passage is rather straightforward: for quantum entities, it makes no sense to say that they are equal or different in some circumstances. That is not merely a reflection of epistemic deficiencies, but it is rather an ontological problem. Berto takes that one to be the standard characterization of the Received View ([4, p.4]; for further discussion on that characterization, see also [2]). Due to quantum permutation symmetry, things have no identity in the quantum realm, as Schrödinger remarked. So, despite the fact that there are situations where there may be a plurality of things, a collection with n elements (for some natural number n), there is no fact of the matter that allows one to attribute identity to them. Now, while this is not the only way to characterize the Received View, this is the way it is encompassed in formal theories underpinning the Received View, such as non-reflexive logics and quasi-set theory. Also, in Berto's view this is the thesis to be attacked.

We now go on to deal with the concepts of "entity", "thing", and "object". These notions are all synonymous for Berto, and are understood according to the so-called 'thin' notion of object, due mostly to Quine (see Lowe [14] for a full discussion on the concept of object and for a relevant alternative to the thin notion; Lowe calls that a *syntactic* notion of objects). According to this view, to be an object is to be the value of a variable. An object is anything we quantify over, attribute properties to, speak about, refer to. As Berto [4, p.5] puts it: "[t]hen quantum particles are things, too: physicists talk about particles in a system, quantify over them, and

---

[b]For further discussion on how to understand this passage and its relation to non-reflexive logics, see Arenhart [3].

ascribe properties to them, like having a certain momentum, position, or spin". Notice that this is also the notion of object which is employed by Saunders and Muller in their defense of weak discernibility in quantum mechanics (see Muller and Saunders [15]; for a general discussion of weak discernibility, see Bigaj [5]).

The important point to be kept in mind here is that this is *not* a metaphysically innocent characterization of an object (perhaps there is no such thing as an innocent characterization of an object). As soon as one recognizes that most of the dispute between the Received View (as characterized following Schrödinger's quote, anyway) and its opponents (friends of identity) concern the very idea that a change of logic is required, there is no common ground on which to discuss the nature of objects if those are assumed to be characterized relatively to a logic; indeed, to characterize a concept such as object in terms of quantification will usually depend on the theory of quantification being employed. Intuitionistic quantifiers behave very differently from those of classical logic, to mention a textbook example. The same may be said about quantifiers in quasi-set theory and non-reflexive logics in general. So, given that a non-standard logic is being advocated by the Received View, a non-standard notion of object may be in use there too. The fundamental point is: it is difficult, when dealing with these issues, not to beg any question. Assume a logic that is rejected by the opponent and you have opened the door for unauthorized concepts.

One such concept, that may imperceptibly slip in with the concept of object, is identity. Identity is also mentioned in the definition of the Received View, so it is crucial to present it as clearly as possible. French and Krause [11, chap.6] have characterized identity in various different approaches, depending on the framework in which it appears: first-order logic, higher-order logic, classical set theory. Now, Berto goes on and attributes three features to identity, which he deems as fundamental for this relation:

i) Identity is a relation between objects, not between names of objects.

ii) Identity is not sortal relative; that is, it cannot happen that a is the same F as b, while a is not the same G as b, where F and G are sortal concepts.

iii) Identity is not vague; it does not come in degrees.

Notice that those features of identity do not mention the framework where it is embedded. Also, item i) leaves it open whether the relation of identity needs to hold between every object, or maybe only between some objects. This issue will be relevant very soon, when we point to the weaknesses of Berto's analysis.

For the argument from numerical plurality to identity, it is worth following Berto himself. Suppose that, as French and Krause [11] would certainly agree, there is a well-determined natural number that is the cardinal number for any plurality of particles (in standard quantum theory[c]). Following Berto [4, p.11], suppose the number of particles is two:

> [i]t seems to make sense, then, to claim that one of the particles is not the other, that is, they are different .... Once *identity* has been characterized as per the three features above (objectual, not vague, not sortal relative), this is what claims of difference and identity mean. That a sentence of the form "$a = b$" is true, under this reading of "$=$", means that we need to count one thing: the thing named "a", which happens to be the thing named "b"... That we, instead, count two things, means that that sentence is false. But then, it's negation, "$\neg(a = b)$" is true. So, a and b are different. And if the concept of difference meaningfully applies to a and b, the one of identity does as well.

> When the number of things (in a system) is given by positive integer n, these things cannot lack self-identity.

So, that is the argument from conceptual analysis. From the fact that there are, say, 2 objects, one may with sense say that they are different. Now, difference is the negation of identity, so, given that difference makes sense, identity makes sense as well.

Berto makes a few further remarks about what is achieved. First, identity amounts to unity, to count as one entity [4, p.11]. Now, given that unity is taken as primitive by Berto, and identity is the same as unity, identity is primitive too [4, pp.12-13]. The idea is that nothing can be said without presupposing identity; it is a fundamental concept.[d] Also, by the fact that identity is primitive, Berto [4, p.14] thinks that one is committed with haecceity or primitive thisness. Now, differently from Bueno [6], who holds that identity may be primitive and metaphysically deflated, Berto things

---

[c]For quantum field theory, that assumption may reasonably be challenged, see Domenech and Holik [8]; it would be instructive to query what would be of the argument from cardinality to identity works in this context.

[d]Here, issues of fundamentality and of primitivity entangle. Given that Berto does not advance arguments to the claim that identity is fundamental, we shall not focus on that issue; see Bueno [6] and Krause and Arenhart [12] for a discussion.

that primitive identity implies that haecceity is to be granted, even though it is an innocent, set theoretic version of haecceitism [4, p.14]:

> Although I take unity as a primitive notion, I allow the addition of a set theoretic gloss to it: for a to be one, I claim, is for there to be $\{a\}$: to be one thing is to have one's singleton. Each thing, a, has the property of being identical with a. This property, extensionally, is nothing but $\{a\}$, which manifests the thing's unity. And the existence of its singleton — its haecceity, in this sense — is guaranteed to each thing in standard set theory. Primitive thisness or haecceity, so understood, is kosher to the extent that standard mathematics, that is, standard set theory, is.

By being represented inside standard set theory, then, one has the means available to provide for haecceities for each entity through the singleton of each of those entities. This is on a par with the idea that, as French and Krause themselves adopt, a haecceity is represented by self-identity (see, for instance, [11, p.5;pp.13-14;p.140]). That is, Monica Bellucci's haecceity is represented by her identity with herself: Monica Bellucci is identical with Monica Bellucci. In this sense, a haecceity is a non-qualitative property, and every item instantiates its own haecceity, which is shared with no other item. In standard set theory, for any item a, that property is equivalent to the fact that a is the only member in $\{a\}$. So, as Berto claimed, there is a straightforward way to grant haecceities for everything in standard set theory.

Also, it is possible to treat quantum indiscernibility as merely relative to a structure. Inside a given structure $e = \langle D, R_i \rangle_{i \in I}$, where D is a non-empty domain and R_i is a family of n-ary relations ($0 \leq n, n \in \omega$), items a and b inside D are indiscernible by the relations in R_i when there is an automorphism f of e such that $f(a) = b$. Obviously, as is well known, in standard set theories every structure may be extended to a rigid structure, so that indiscernibility collapses with identity. In this sense, indiscernibility is an epistemic issue: there would always be a possibility, at least in principle, to distinguish any thing from any other thing. The only problem is that some structures are not rich enough; they lack the relations required for the distinctions. However, by set theoretical force such relations may be introduced, and even quantum mechanics could be rigidified in this context (for further discussions on structures and axiomatization of quantum mechanics, see [13]).

3. Resisting the analysis

Let us now discuss Berto's argument and the main concepts involved. Recall that the idea was rather simple: given that there are (say), two quantum objects, there seems to make complete sense in saying that they are different, given that they are objects. Identity applies to objects, and given that identity is not vague, we can plausibly say that those objects are different (the claim of their identity is false). By being different, identity itself makes sense for them, given that difference makes sense. So, there is identity as a meaningful relation holding for everything. In particular, self-identity holds of everything, and is a representation of a haecceity; everything is endowed with its own haecceity. There is a rigorous formal picture that can be developed by employing standard set theory, which accounts for the haecceities and for indiscernibility in quantum mechanics through permutation invariance in a structure. This is clearly a representation of one of the horns of the metaphysical underdetermination described by French and Krause, the one comprising individuals.

Now, as we mentioned, there is a problem with this analysis. It begs the question against the friend of the Received View, at least in the particular version as it is developed by French and Krause in [11]. Our claim in this section is that it allows identity to infiltrate in the concepts employed, so that the friend of the Received View needs not accept those concepts as Berto advances them.

In order to make our point clearer and make most sense of the Received View, let us recall the very definition of the Received View. Recall that in generic terms, the main tenets of the Received View were originally coined by some of the founding fathers of quantum mechanics; it originated by their analyzes of quantum permutation symmetry, and its novelties were expressed by claims to the fact that quantum particles 'have lost their identities', or that they raised a 'problem for identity'. Anyway, at the informal level, there was no clearly articulated thesis about what it amounts to, but it certainly involved identity and individuality (see [11, chap.3] for a historical overview dealing with most of the founding fathers of quantum theory). There is no clear-cut definition, so, one of the possibilities is the one embraced by French and Krause themselves, following Schrödinger: identity does not make sense for quantum entities, at least sometimes.

Due to the vagueness of claims that 'identity is lost', and similar remarks, it should be clear that following Schrödinger here is *not* the only option for characterizing the Received View (again, see [2] for further discussion). Berto himself, while characterizing the Received View, quotes

Herman Weyl [18, p.241]. Here we find Weyl discussing quantum permutation invariance and comparing two indiscernible particles to two identical twins Mike and Ike; unlike the twins, which can always provide for an 'alibi' attesting to each own identity — by saying their own names, for instance —, Weyl holds that quantum particles cannot do such kind of thing. There is nothing in quantum mechanics to individuate such particles, not even their labels can be employed for that purpose. So, while identity is also involved in this situation, the idea of failure of identity here is typically framed in terms of the failure of a version of the *Principle of the Identity of Indiscernibles* (PII), which states, roughly: no two entities differ *solo numero*. That is, there is always a qualitative feature to distinguish *two* entities; if there is no such property, there are no two entities, but rather only one. It is this principle that holds for the twins, but fails for quantum entities.

So, the problem of identity for quantum particles may be framed as i) a failure of the PII,[e] which arguably means trouble for those willing to endorse a reduction of the relation of identity to qualitative properties, or else as ii) a more radical failure of identity even to make sense, as Schrödinger was willing to do. Notice that while the failure of PII does not by itself involve identity losing its sense, the second alternative is defined in terms of this very idea. What is the relation between these two alternatives? French and Krause establish a relation between them that will bring us back to the problem for Berto's analysis.

While failure of PII is tied to failure of the so-called *bundle theory of individuality*, which, roughly speaking, says that an entity is merely a 'bundle' of its properties,[f] failure of identity in Schrödinger's sense, in particular failure of self-identity, is tied by French and Krause to failure of an alternative kind of principle of individuation, the so-called *Transcendental Individuality Principles*. Among such transcendental principles, we find haecceity, primitive thisness, and bare particulars. The idea, very roughly put again, is that it is precisely such Transcendental Principles that ground the individuality of an item, rather than the qualitative features the item may possess; individuality grounded by such principles transcends the item's properties. So, under this picture, while it is still possible that two entities be indiscernible,

---

[e]There is a dispute on whether Weyl really was calling the PII into question for Fermions; see Muller and Saunders [15]; however, even if that particular interpretation of Weyl does not work, other examples may be adduced to the failure of at least some versions of the PII in quantum mechanics.

[f]Obviously, this is a very rough characterization; it will do for our purposes here, however. See French and Krause [11, chap.1] for further information.

they may still be *two* individuals due to the fact that they have distinct Transcendental Principles, distinct haecceities, for instance.

But how is such a Transcendental Individuality to be represented? Precisely as self-identity. To quote from French and Krause [11, pp.13-15]:

> ... the idea is apparently simple: regarded in haecceistic terms, "Transcendental Individuality" can be understood as the identity of an object with itself; that is, '$a = a$'. We shall then defend the claim that the notion of non-individuality can be captured in the quantum context by formal systems in which self-identity is not always well-defined, so that the reflexive law of identity, namely, $\forall x(x = x)$, is not valid in general.
>
> [...] conceiving of individuality in terms of self-identity will allow us to appropriately represent its denial.

So, the point is that as soon as self-identity is selected as the proper form of representation for a haecceity, there is a metaphysical thesis about individuality that gets associated with the relation of identity. Also, the Schrödingerian thesis that identity does not make sense for quantum entities shifts from a linguistic thesis to a more metaphysically loaded one; identity as a relation gets substantiated by a metaphysical counterpart, and the Received View amounts then to the claim that quantum entities lack such a metaphysical counterpart. So, as a consequence, attributing identity to particles, self-identity in particular, amounts to attributing haecceities to everything. That is precisely what this version of the Received View wants to deny. It seems to us that this is where the locus of the controversy should be, before any discussion on formal apparatuses and set theoretical representation begins. Berto's analysis simply makes the claim that it is impossible that some entities may exist without having haecceities right from the start.

Having settled the scenario like that, let us now explore how one may resist Berto's argument. Berto manages to arrive at the conclusion that everything has self-identity by claiming that by being two, the particles involved are different. Now, in the metaphysical reading of identity advanced by French and Krause, and by Berto also, that means that each particle would have its own haecceity. However, as it is usually said, one's modus ponens is the other's modus tollens. Instead of accepting that each entity has it own haecceity, French and Krause go on to deny that one can meaningfully claim that those particles are different. That, as Berto correctly points out, would imply that there is a haecceity for each; that last claim, however, is the one that is at the center of the stage, and cannot be taken for granted. So, once the association of identity and haecceities is accepted, identity cannot be taken for granted!

Berto tries to grant his argument through the concepts of object and identity that are in play. They are supposed by Berto to be accepted by the friends of the Received View. But are they? Do the very notions of object and identity imply that Berto is right, while French and Krause are wrong? Not really; it seems that those concepts are already loaded with a metaphysics of individuality that is the very metaphysics that the Received View, as defined by French and Krause, attempt to resist (at least on what concerns quantum particles). To begin with, there is a rather straightforward way to resist Berto's argument: one may be deflationist about the relation of identity, as Bueno [6] proposes. As Bueno sees identity, there is no metaphysical consequence on the fact that identity, as a relation, holds between a and b, say. However, that move ends up being a threat to the Received View anyway, because it accepts the notion of identity; here we shall accept, for the sake of argument, that identity may be seen as a metaphysically loaded relation, and ask whether we can resist Berto's claims.

So, let us begin with the notion of object, the metaphysically thin notion that is defined in terms of quantification and reference. Could it be that the notion of object, by itself, commits us with identity? It seems fair to say that that notion of object, taken by itself, with no further requirements, is neutral as to what happens with identity. In fact, the thin notion of object by itself, as defined by Berto, does not require anything besides the fact that objects are those things we quantify over. What is relevant is that this definition is typically presented in the context of a Quinean setting, with quantifiers understood as involving a classical set as a domain of quantification. In this setting, it is added the requirement that entities must have identity, and then the notion of objects goes one step further in being metaphysically loaded. However, accepting that classical theory of quantification already goes one step further than the notion of object itself, and seems clearly not warranted when it is identity that is under discussion. So, if Berto wants the claim that identity holds for everything to be a kind of conclusion of an analysis, then it should follow from the notion of object, not to be added by hand through the addition of a classical theory of quantification.

Maybe the notion of object is indeed neutral as to identity. Perhaps the claim that identity holds overall obtains if we add to that quantification theory some set theoretic axioms. That is, once we attempt to go from logic to mathematics, by adding set theoretical axioms, there is no way to avoid identity for everything, and Berto could be right after all. But what kind of set theoretic axioms should we choose? There are many incompatible alternatives. Berto chooses standard Zermelo-Fraenkel axioms, and claims that those axioms are metaphysically innocent. Recall that according to standard set theory, given any item a, one may form the set $\{a\}$, which is equivalent to the haecceity of a. Furthermore, in standard set theory the reflexive law of identity holds, that is, for any a, $a = a$, and that is haecceity again. So, set theory implies haecceitism, or at least has the resources to express that the thesis holds for everything. But given that we are discussing precisely *whether quantum entities should have haecceities*, there is no justification in assuming such a set of axioms as a default position. That clearly is not innocent under the stage that we have set, in which it is haecceity (encapsulated as self-identity) that is at stake! So, if identity is granted through the use of standard set theoretic axioms, that is also a way to beg the question.

In other words, by complementing the concept of object with Quinean strictures or with further set theoretic axioms, one clearly assumes those very notions that are being called into question by the Received View. This shows that one cannot grant those concepts without begging some relevant questions against the Received View.

What Berto does is to try to present a formal system and then read identity for everything. That move fails because identity is already there, present in the formal system he chooses. Notice the difference in French and Krause's move: once one assumes that quantum entities may be understood as non-individuals (they don't claim to have established that once and for all), one may reasonably seek the more adequate formal system to represent those things. The choice falls on the quasi-set theory and non-reflexive systems of logic already developed by Krause and others (see [11, chaps.7-8]). So, the formalism must only be judged as to its adequacy and its consequences may be explored *after* a decision has been made as to the metaphysics being represented. That is: one does not choose a formalism and explores its metaphysical consequences; rather, one sets the metaphysics, in its main outlines, and seeks to provide for a more appropriate formal system which represents it.

But we are not done yet. It seems clear that once the concept of object is complemented and understood as holding in the context of the standard axioms for set theory, Berto's analysis is correct. So, it is not a set of axioms for standard set theory that must be added, if one is not to beg any relevant question. What could it be, then, in order for Berto to grant his point? It seems that it is the concept of identity that Berto needs to add to the concept of object. Once identity is added, it seems, one cannot escape the claim that a plurality of items implies the overall validity of identity. Maybe there is a notion of identity accepted even by the friends of the Received View, a notion that will lead us to Berto's conclusion. Recall that identity, according to Berto, has three main features: it holds between objects, it is not vague and not sortal relative.

Of course, applying to objects, whenever there are objects, it applies, truly or falsely. Not being sortal relative, identity applies to everything without restrictions. Not being vague, there is always a definite truth value to be attributed to an identity statement. So, given objects a and b, identity applies, and it is true or false that $a = b$. Anyway, identity is granted.

How could the friend of the Received View, as defined by French and Krause, resist such a characterization of identity? Well, recall the setting of the discussion: once identity is seen as representing haecceities, it seems perfectly plausible to claim that it should fail for some things if those things do not have haecceities. So, unless one is reading in the very notion of object that they should have identity (the Quinean move we identified above), there is no reason to suppose that identity applies to everything. On this kind of move, it is the notion of object that gets the blame, and, as we have seen, it may be resisted. Under the Received View, identity applies to objects, but not every object needs to be related by identity. In other words: if there is identity, there is object, but if there is an object, there needs not to be identity (it depends on whether the object has haecceity or not).

So, in order to discuss this notion of identity, let us suppose that there is nothing in the notion of object that implies that identity always holds, in order to keep without begging any question against the Received View. The friend of the Received View must grant that identity may hold between objects, not be vague, and still not apply to everything. One must also present a case as to whether that does or does not violate the requirement that identity is sortal relative. In order to do that, let us briefly explain how it is analyzed in non-reflexive logics.

There are at least two distinct approaches to identity respecting the first two features. In typical non-reflexive logics, such as Schrödinger logics, identity is a primitive concept of the language. Its failure for some entities is represented by a syntactical restriction on the formation of formulas; the language is two-sorted, so that one sort of terms represents quantum entities, and for those terms, formulas involving identity are not well-formed. In quasi-set theory, on the other hand, identity is defined. The theory distinguishes two kinds of atoms, classical and quantum atoms. Identity holds between classical atoms (atoms belonging to the same collections are identical), or between quasi-sets, the collections of the theory (collections having the same members are identical). However, it is simply not defined for atoms representing quantum entities (for details, see French and Krause [11, chaps.7-8]). In this case, it is also not expressible in the language of the theory any claim of identity or difference between quantum entities.

Now, these restrictions in the application of identity do not seem to violate the three requirements on identity. It applies to objects, but not to every object. It is not vague: whenever identity applies, it provides for true or false formulas, without indeterminate formulas. What about the sortal relativity? It seems that in quasi-set theory, for instance, the concept of identity is relative to a class of things: the so-called classical things. However, notice that it is hard to claim that this means that identity is sortal relative. The situation is rather different. It is not the case that identity obtains for some sortal concept F while it fails for some other sortal concept G. The fact is that sortal concepts are defined in terms of identity conditions, and identity conditions simply fails to apply to quantum entities under this view. On French and Krause's view, concepts in quantum mechanics, applying to quantum entities, do not constitute sortal concepts in the traditional way. Indeed, quantum entities constitute what they call *quantum-sortal predicates*: items having a cardinality in every circumstance,[g] but no identity conditions (see French and Krause [11, chap.8]). So, identity under this analysis is not really sortal relative. For sortal concepts, identity works just fine, in an absolute sense. So, this requirement is also satisfied.

In this sense, one may see that once identity is taken as provided with the three requirements laid by Berto, the first requirement, that identity holds between objects, contained a hidden assumption that it holds between every object. Given any two objects, identity is true or false of them. It is this assumption that the Received View challenges with non-reflexive systems. To assume that claim is, again, to beg a relevant question against the Received View.

But how are we to understand the restrictions of identity then? Recall that they are representing the fact that quantum entities do not have haecceity. In this case, one suggestion may be as follows: applying identity to

---

[g]Recall that we are not dealing with quantum field theory.

quantum entities is a kind of category mistake; identity is not the kind of relation that applies to quantum entities. It makes no sense to make assertions of identity and of difference for those entities. Of course, because one cannot deny that a has haecceity by claiming that $a \neq a$. So, the reasonable alternative is to shift to the idea that such claims also make no sense.

We hope that this shows that identity may still benefit from the features attributed to it by Berto while still failing to apply to some entities. The notion of category mistake may be a helping hand in illustrating its failure, although this reading is not mandatory.

4. An alternative framework for conceptual analysis

After having provided some means to resist Berto's analysis, perhaps we should provide also for some more positive ways about how to better understand the Received View and how to understand the relevant concepts that Berto employs in his analysis. We shall do precisely that in the present section. The first thing to be recalled, as it was already mentioned, is that the version of the Received View that was attacked by Berto is not the only way to substantiate the idea that some entities may not be individuals; things may be non-individuals in much less revisionary ways than the Schrödingerian suggestion of abolishing identity (see again the discussions in Arenhart [2] for the distinct approaches to the Received View). But here we shall focus on the Received View as it was discussed in this paper, as a thesis about some entities lacking haecceities, where haecceities are framed in terms of self-identity.

Now, how should we understand the general notions of object, quantification, identity, cardinality, given that we do not take them to imply identity, as Berto has proposed? We shall follow the steps of Berto himself, and propose that conceptual analysis helps providing for the proper treatment of such terms. Notice that Berto took for granted the classical apparatuses of quantification, set theory, and the accompanying definition of cardinality, which may be found in standard logic and set theory textbooks. This is one of the possible ways such concepts may be analyzed, but here we shall propose that the strategy of employing formal tools should be more flexible, and we may benefit from being so: in times of logical pluralism and alternative systems of logic, distinct logics may attribute distinct meanings for those concepts. Of course, that does not mean that each such logic is just as legitimate as any other for quantum mechanics, but we shall come back to this issue soon.

In fact, when used in ordinary language, such concepts as quantifiers, identity, cardinal number of a collection (plurality), have very fuzzy borders (just as the typical logical connectives, such as negation and the conditional, have a less than completely clear meaning in natural language). They are endowed with a kind of plasticity in meaning, and by being so, do not determine a fixed univocal meaning. Informal investigation may provide us precious hints as to their semantics and pragmatics, but still it is not possible to completely determine the meaning of those terms once and for all without controversy. Philosophical debates on existence and quantification illustrate that pretty well, just as disputes on negation and conditionals do. Philosophical debates on the identity of quantum entities also enlarge that list, we believe, as the present case illustrates. How to provide for a firmer ground on which to discuss such issues? By furnishing a formal analysis of those concepts. As Berto indicated, the classical analysis, proceeding in a Quinean fashion, leads us directly to the first horn of the metaphysical underdetermination on the metaphysics of quantum mechanics, which is committed with individuality (with haecceities being an extra to the Quinean analysis). However, as French and Krause [11, p.244] have remarked, each of the horns of the underdetermination have their own logic. So, it seems natural to think that, once an alternative logic is assumed to govern the concepts of object, identity, and cardinality, we should have a distinct result in the conceptual analysis.

Now, concepts such as object, identity and quantification are no exception to such a natural less than complete determination or vagueness. They may be endowed with more precise meanings in distinct, incompatible ways; there are incompatible ways to fix the details, and one should notice that complete determination requires some idealization, which already makes concepts lose full contact with reality. Here we shall follow a suggestion by da Costa [7], according to whom those concepts acquire a fixed meaning once they are properly regimented in a specified logical system (for details and further discussion, see [3]). In fact, da Costa calls 'categories' those most general concepts such as objects and properties. In natural language those concepts are very useful, but due to their plasticity, they do not have a well determined meaning. Once a context is selected, which means for da Costa that a scientific theory is selected, such as quantum mechanics or Newtonian mechanics, one is ready to begin a more precise analysis of the workings of those concepts in that context.

How that is supposed to be done? According to da Costa [7, chap.2, sec.V and VI], the meaning of those terms is not given *a priori*, independently of a scientific theory, *pace* Berto. Rather, to acquire a precise meaning they depend on the scientific theory in which they are embedded. It is the scientific theory which fixes the meaning of those terms in details that natural use would leave open. In particular, as an example, da Costa mentions the fact that quantum mechanics seems to require a radical change in some of the features of the traditional Newtonian notion of object. That is precisely one of our terms! But let us keep the general discussion for the moment. In order not to be completely trapped in *a priori* metaphysics, da Costa suggests that a logical analysis of scientific theory helps us determining the laws that entities of that domain do obey. That is done by the linguistic analysis of the theory, through axiomatization and study of the logical foundations of the theory.

Once a scientific theory is axiomatized and has a specific logical theory as the underlying logic determined, we are able to understand more precisely the meaning of quantifiers, objects, and to understand the nature of the attribution of properties. Formal systems provide for a rigorous rational reconstruction of scientific theories; as a by-product, they also provide for a precise meaning for logical constants. The metaphysics gets specified due to a close relation between *syntactical categories* and the *general categories* of the theory. Without going into full details on da Costa's suggestion, we may reasonably claim that the syntactical categories of individual terms (individual variables and individual constants), predicate constants and variables, and atomic sentence, more or less match ontological categories of object, relation and property, and fact, respectively (for details, see da Costa [7, p.53]). As a result of this suggestion, through the regimentation and logical investigation of the language of a scientific theory, when the underlying logic is properly selected and explicitly mentioned, the precise meaning of the ontological categories such as objects and properties gets determined. As da Costa puts it in [7, p.39], "[o]ne may say that the laws of reason are susceptible of being obtained, in large measure, by the critical analysis of the contexts of scientific exposition".

As we have mentioned, one of the possible ways to develop such an analysis in the context of quantum mechanics is through the use of non-reflexive logics, instead of using classical logic. Of course, that does not mean that we adhere to a naive form of naturalism, according to which the idea that quantum entities are non-individuals can be read-off from quantum theory directly. No, the issue is much more complicated. Put rather briefly, once one has conducted an investigation on quantum mechanics and has taken it as plausible that there are no haecceities in quantum theory, and also, that haecceities are represented by self-identity (which really involves a purely metaphysical discussion), one may try to find a underlying logic that is strong enough to ground quantum mechanics and to capture such metaphysical features (see the developments of quantum mechanics in non-reflexive logics in [9]). Of course, that is how non-reflexive logics are actually motivated and developed.

So, overall, this involves a joint work between metaphysics and science. It may not be a definitive answer to the problem of metaphysical underdetermination, as Berto was seeking with his analysis, but at least it provides for fruitful interaction between metaphysics and science that begs no question against relevant alternatives. Furthermore, it takes much of the weight of *a priori* metaphysics and puts some of the responsibility of the proper workings of metaphysical notions to rely on contributions from science itself. So, we think that one of the advantages of this proposal is that it relies less on *a priori* conceptual analysis and brings us closer to scientific theories, which are the fountain of our best contact with reality. Of course, there is no direct route to reality, but even classical metaphysics seems to fare no better in this topic, unless one embraces a dogmatic position.

Perhaps an analogy with a more familiar subject matter may be enlightening. Consider intuitionist mathematics, in the Brouwer-Heyting tradition. Once one adopts the Brouwerian thesis that the behavior of mathematical entities is determined by a particular kind of mental constructions, one has the resources to make an analysis and present the underlying logic of such entities.[h] Concepts such as existence and negation acquire very distinct meaning than they have in classical logic. Also, the very notion of object is different, as the intuitionist theory of real numbers clearly exemplifies. As a remarkable similarity with the non-reflexive case, in the intuitionist real number theory, one of the possible ways to define real numbers is through the so-called free choice sequences; roughly speaking, the creative subject freely chooses the elements of a sequence that will determine a real number. However, due to their constructive character, such sequences are ever being created, they are not given as ready. Some of those choice sequences may be defined in such a way that their $n - th$ members ends up depending on the choice of the $(n - 1) - th$ member. What is relevant for us is that it is possible to have such choice sequences α and β so that in any moment of time it is not determined whether the real number r_α determined by α, is identical with the real number r_β, determined by β. Real numbers need not have identity or difference always well-determined and settled; however, they are

---

[h]Of course, Brouwer thought such a logical analysis a sterile exercise.

objects anyway. Notice that this is also not a mere matter of epistemological limitation: the real numbers must not be conceived as being already there, fully ready for us to make such a claim! For further information on intuitionistic mathematics and philosophy, see Posy [16].

So, just as in the context of intuitionistic mathematics one may need an alternative logic, one which is more fine-tuned with the nature of the entities being presented in this context, in quantum mechanics it could be plausible to ask for a non-reflexive logic, once one has been convinced that a metaphysics of non-individuals is more appropriate to deal with the quantum oddities. Of course, the resulting non-reflexive logics are rather revisionary in that they require that some expressions involving identity should be forbidden, or substituted by cardinality claims. However, quantum mechanics itself is revisionary on many of our concepts. So, it is not strange that a new logic should be developed in order to account for the quantum mysteries.

5. Concluding remarks

The problem of identical particles in quantum mechanics is one of the central philosophical puzzles of quantum theory. Nothing in the theory allows us to choose between two incompatible options: i) quantum entities are individuals, and ii) quantum entities are not individuals. In order to do so, one must transcend the resources of the theory. As we have seen, Berto attempts to pull the discussion one step back; instead of keep expecting for quantum features that would allow us to decide the issue (and that may never come), he proposes that the basic notions of object, cardinality and identity are such that the issue is already settled: quantum entities are individuals, they have identity, which is implied by the very meaning of such concepts.

We hope to have convinced the reader that once the concepts are chosen in a specific way, a conclusion as to whether identity holds or does not hold is going to follow. Berto chose to analyze the concepts of object and identity following the guidance of classical logic. However, as we have argued, it is precisely the possibility of legitimately applying such an apparatus that is in question. So, to begin with those concepts in order to reach the conclusion that identity holds overall begs the question against the defender of the Received View. In this sense, Berto's analysis is not conclusive on the nature of quantum particles. In fact, it would be surprising if such a strong conclusion as the one established by Berto, that quantum entities are individuated by haecceities, could be reached almost with no help of the theory itself.

Our suggestion to address such issues is rather similar to Berto's, but it involves a shift in methodology. Quantum mechanics has something important to say about quantum objects, so the analysis of quantum objects, what they are and how they behave, should explicitly mention quantum mechanics. Now, given that such general notions as 'object' are involved in the use of the logical apparatus of a theory, our suggestion is that the underlying logic is relevant for this discussion, and the investigation should proceed with complement from quantum mechanics itself. Logic is, even if indirectly, in the business of describing reality. That would account for a more collaborative work between metaphysics and scientific theories.

Acknowledgments

Décio Krause is partially supported by CNPq.

Bibliography

1. J. R. B. Arenhart, D. Krause, "From primitive identity to the non-individuality of quantum objects", *Studies in history and philosophy of modern physics* **46**, 273–282, 2014.
2. J. R. B. Arenhart, "The received view on quantum non-individuality: formal and metaphysical analysis", *Synthese* **194**, 1323–1347, 2017.
3. J. R. B. Arenhart, "Newton da Costa on non-reflexive logics and identity", to appear in *Metatheoria*, 2017.
4. F. Berto, "Counting the particles: entity and identity in the philosophy of physics", to appear in *Metaphysica*, 2016.
5. T. Bigaj, "Dissecting weak discernibility of quanta", *Studies in history and philosophy of modern physics* **50**, 43–53, 2015.
6. O. Bueno, "Why identity is fundamental", *American Philosophical Quarterly* **51**(4), 325–332, 2014.
7. N. C. A. da Costa, *Logiques Classiques et Non Classiques. Essai sur les fondements de la logique.* (Masson, Paris, 1997)
8. G. Domenech, F. Holik, "A discussion on particle number and quantum indistinguishability", *Foundations of Physics* **37**(6), 855–878, 2007.
9. G. Domenech, F. Holik, D. Krause, "Q-spaces and the foundations of quantum mechanics", *Foundations of Physics* **38**(11), 969–994, 2008.
10. M. Dorato, M. Morganti, "Grades of Individuality. A pluralistic view of identity in quantum mechanics and in the sciences", *Philosophical Studies* **163**(3), 591–610, 2013.
11. F. French, D. Krause, *Identity in Physics: A historical, philosophical and formal analysis.* (Oxford University Press, Oxford, 2006).
12. D. Krause, J. R. B. Arenhart, Is identity really so fundamental? *Found. Sci.*, 2018, https://doi.org/10.1007/s10699-018-9553-3.
13. D. Krause, J. R. B. Arenhart, *The Logical Foundations of Scientific Theories: Languages, Structures, and Models.* (Routledge, London, 2016).

14. E. J. Lowe, "Objects and Criteria of Identity". In: B. Hale, C. Wright (eds.) *A Companion to the Philosophy of Language*, pp.613-633. Oxford: Blackwell, 1997.

15. F. Muller, S. Saunders, "Discerning Fermions", *British Journal for the Philosophy of Science* **59**, 499–548, 2008.

16. C. Posy, "Intuitionism and philosophy". In: S. Shapiro (ed.) *The Oxford Handbook of Philosophy of Mathematics and Logic*, pp. 318-355. Oxford: Oxford Un. Press, 2005.

17. E. Schrödinger, *Nature and the Greeks and Science and Humanism, with a foreword by Roger Penrose.* (Cambridge Un. Press, Cambridge, 1996).

18. H. Weyl, *The theory of groups and quantum mechanics.* (Methuen & Co., London, 1931)

IMMANENT POWERS VERSUS CAUSAL POWERS (PROPENSITIES, LATENCIES AND DISPOSITIONS) IN QUANTUM MECHANICS

CHRISTIAN DE RONDE*

CONICET, Buenos Aires University - Argentina
Center Leo Apostel and Foundations of the Exact Sciences
Brussels Free University - Belgium
E-mail: cderonde@vub.ac.be

In this paper we compare two different notions of 'power', both of which attempt to provide a realist understanding of quantum mechanics grounded on the potential mode of existence. For this propose we will begin by introducing two different notions of potentiality present already within Aristotelian metaphysics, namely, *irrational potentiality* and *rational potentiality*. After discussing the role played by potentiality within classical and quantum mechanics, we will address the notion of *causal power* which is directly related to irrational potentiality and has been adopted by many interpretations of QM. We will then present the notion of *immanent power* which relates to rational potentiality and argue that this new concept presents important advantages regarding the possibilities it provides for understanding in a novel manner the theory of quanta. We end our paper with a comparison between both notions of 'power', stressing some radical differences between them.

Keywords: Quantum probability; objectivity; immanent powers; potentia.

Introduction

The notion of potentiality has played a major role within the history of quantum physics. Its explicit introduction within the theory goes back to the late 1950' when several authors discussed —independently— the possibility to understand Quantum Mechanics (QM), in close analogy to the Aristotelian hylomorphic metaphysical scheme, through the consideration of a potential realm —different to that of actuality. Werner Heisenberg [46], Gilbert Simondon [68], Henry Margenau [55], and Karl Popper [63], presented different interpretations of QM in terms of potentialities, propensities and latencies. Since then, these interpretations and ideas have been developed in different directions enriching the debate about the meaning

122

and possibilities of the theory [1,14,15,38,42,44,60,61,74]. However, as we shall argue, most of these attempts ground themselves on a one sided view of potentiality, one which understands potentiality as defined exclusively in terms of actual effectuations —i.e., in terms of what Aristotle called *irrational potentiality*. In particular, dispositional and propensity type interpretations of QM have been linked through this understanding of potentiality to the concept of *causal power*. In this paper —continuing with ongoing work [22,25,30]— we investigate a different standpoint, one which attempts to develop potentiality on the lines of *rational potentiality*.[a] In turn, this new potential mode of existence can be related to the physical concept of *immanent power*. We believe that this new understanding of the meaning of both the 'potential realm' and the notion of 'power' can provide us with new insights that might help us to understand what QM is really talking about.

The paper is organized as follows. In the first section we outline the main aspects of the orthodox empiricist project within philosophy of QM. In section 2 we present the representational realist project which attempts to produce a different account of the problem of interpreting quantum theory. Section 3 recalls the metaphysical hylomorphic Aristotelian scheme and the distinction between, on the one hand, actuality and potential modes of existence, and on the other, irrational and rational potentiality. In section 4, we analyze the Newtonian atomistic representation which eliminated the potential realm from the description of physical reality. In section 5, we discuss Heisenberg's return to hylomorphic metaphysics —through the reintroduction of the potential realm— in order to interpret QM. Section 6 presents the continuation of Heisenberg's proposal following the teleological scheme of causal powers, dispositions, propensities, etc. —all of which are based on the notion of irrational potentiality. In section 7 we present a different non-reductionistic scheme based on the notion of rational potentiality which attempts to discuss the conceptual representation of the potential realm in terms of the notion of immanent power. Section 8 analyses and discusses the pros and cons of the notions of causal power and immanent power in order to provide an answer to the question: what is QM really talking about?

---

[a]We might also consider some interpretations of QM (e.g., [2,49,50]) as closer to the development of rational potentiality presented here.

1. The Orthodox (Empiricist) Project in Philosophy of QM

Philosophy of science in general, and philosophy of physics in particular, were developed after the second world war following the logical positivist project proposed by the so called "Vienna Circle". Following the physicist and philosopher Ernst Mach, logical positivists fought strongly against dogmatic metaphysical thought and *a priori* concepts. As they argued in their famous *Manifesto* [12]: "Everything is accessible to man; and man is the measure of all things. Here is an affinity with the Sophists, not with the Platonists; with the Epicureans, not with the Pythagoreans; with all those who stand for earthly being and the here and now." Their main attack against metaphysics was developed taking an empiricist based standpoint, the idea that one should focus in "statements as they are made by empirical science; their meaning can be determined by logical analysis or, more precisely, through reduction to the simplest statements about the empirically given." The positivist architectonic stood on the distinction between *empirical terms*, the empirically "given" through observation,[b] and *theoretical terms*, their translation into simple statements. This separation and correspondence between theoretical statements and empirical observation would have deep consequences, not only regarding the problems addressed within the new born "philosophy of science" but also with respect to the limits in the development of many different lines of research within the theory of quanta itself.

The main enemy of empiricism has been, since its origin, metaphysical thought —understood mainly as a discourse about non-observable entities. Trying to avoid any metaphysical reference beyond observational phenomena, the empiricist perspective attempted to produce a direct link between observation on the one hand, and linguistic statements on the other. But how to do this without entering the field of metaphysical speculation, specially when language and metaphysics are intrinsically related? Indeed, (metaphysical) concepts, as a necessary prerequisite to account for experience, are defined —since Plato and Aristotle— in a systematic and categorical manner; i.e. in a metaphysical fashion. In this respect, we remark that metaphysics has nothing to do with the distinction between observable and non-observable; it deals instead with the very possibility of defining concepts systematically through general principles.

The main problem of empiricism has been clearly exposed by Jorge Luis Borges in a beautiful short story called *Funes the Memorious* [9]. Borges

---

[b]Later on considered as *observational terms*.

recalls his encounter with Ireneo Funes, a young man from Fray Bentos who after having an accident become paralyzed. Since then Funes' perception and memory became infallible. According to Borges, the least important of his recollections was more minutely precise and more lively than our perception of a physical pleasure or a physical torment. However, as Borges also remarked: "He was, let us not forget, almost incapable of general, platonic ideas. It was not only difficult for him to understand that the generic term dog embraced so many unlike specimens of differing sizes and different forms; he was disturbed by the fact that a dog at three-fourteen (seen in profile) should have the same name as the dog at three fifteen (seen from the front). [...] Without effort, he had learned English, French, Portuguese, Latin. I suspect, however, that he was not very capable of thought. To think is to forget differences, generalize, make abstractions. In the teeming world of Funes there were only details, almost immediate in their presence." The problem exposed by Borges is in fact, the same problem which Carnap [13], Nagel [58], Popper [62] and many others —following the positivist agenda— tried to resolve: the difficult relation between, on the one hand, phenomenological experience or observations, and on the other, language and concepts.

The failure of the orthodox empiricist project to define *empirical terms* independently of theoretical and metaphysical considerations was soon acknowledged within philosophy of science itself. In the sixties and seventies important authors within the field —such as Hanson, Kuhn, Lakatos and Feyerabend, between many others— addressed the problem of understanding observation in "naive" terms. This debate, which unmasked the limits of the empiricist project, was known to the community by the name of: "the theory ladenness of physical observation". But regardless of the impossibility to consider observation as a "self evident given" —as the young born positivist epistemology had attempted to do—, philosophy of science in general, and philosophy of physics in particular, have continued anyhow to ground their analysis in "common sense" observability. Indeed, as remarked by Curd and Cover [18, p. 1228]: "Logical positivism is dead and logical empiricism is no longer an avowed school of philosophical thought. But despite our historical and philosophical distance from logical positivism and empiricism, their influence can be felt. An important part of their legacy is observational-theoretical distinction itself, which continues to play a central role in debates about scientific realism." It is important to notice that the orthodox problems of QM have been also constrained by the empiricist viewpoint which understands that observation in the lab, even in the case of considering quantum phenomena, remains completely unproblematic. Within the huge literature regarding the meaning of QM, the empiricist standpoint has constrained philosophical analysis within the walls of a conservative project which attempts to "bridge the gap" between the "weird" mathematical formalism of QM and our "manifest image of the world" [41]. Indeed, going back to Bohr's reductionistic desiderata,[c] the main goal of the orthodox project in philosophy of QM is to understand the theory in strict relation to our classical "common sense" representation of the world. This reductionistic account of QM is the reason why the quantum to classical limit has been considered within the literature as one of the most important problems. The failure to solve this problem through "the new orthodoxy of decoherence" [10, p. 212] remains still today partly unnoticed by the community and even camouflaged through a "FAPP (For All Practical Purposes) justification" which confuses the epistemological and ontological levels of analysis [27]. In this same context, the orthodox perspective has created a set of "no-problems" (non-separability, non-individuality, non-locality, non-distributivity, non-identity, etc.) which discuss the quantum formalism presupposing "right from the start" the classical notions with which we have been able to build our classical (metaphysical) representation of the world (see for a detailed discussion [26]). This "conservative project" within philosophy of QM has silenced a radically different line of research which would investigate the possibilities of a truly non-classical metaphysical representation of QM.

2. The Representational Realist Program

Representational realism understands physics as a discipline which attempts to represent *physis* (reality or nature) in theoretical —both formal and conceptual— terms [17]. Physical theories are capable, through the tight inter-relation of mathematical formalisms and networks of physical concepts, to represent experience and reality. The possibility to imagine and picture reality beyond observation can be only achieved through a conceptual scheme, not through mathematical formalisms. Mathematics is an abstract discipline which contains no physical concept whatsoever. And this is the reason why mathematicians can work perfectly well without learning

---

[c]We refer to the Bhorian presuppositions according to which, first, QM must be related through a limit to classical physics, and second, experience and phenomena will be represented always in terms of classical physics. See for a detailed discussion and analysis [26].

about physical theories or about the relation between certain mathematical formalisms and physical reality. Physical concepts cannot be "found" or "discovered" within any mathematical formalism. For instance, the mathematical theory of calculus does not contain the Newtonian notions of absolute 'space' and 'time', nor does it talk about 'force', 'particles', 'mass' or 'gravity'. Another example is the physical notion of 'field' which cannot be derived through a theorem from Maxwell's equations. Physical concepts are not mathematical entities, but metaphysical elements defined in systematic, categorical terms, through general principles —such as, for example, those proposed by Aristotle in his metaphysics and logic.

According to the representational realist, reality is not something "self-evidently" exposed through observations —as logico-positivists, empricists and even Bohr might have claimed. The representation and understanding of reality can be only achieved through the analysis of metaphysical conceptual schemes which are provided by physical theories themselves. Let us stress that our realist viewpoint is not consistent with scientific realism, phenomenological realism or realism about observables —which we consider to be in fact variants of empiricism grounded on "common sense" observability. As Musgrave [18, p. 1221] makes the point: "In traditional discussions of scientific realism, common sense realism regarding tables and chairs (or the moon) is accepted as unproblematic by both sides. Attention is focused on the difficulties of scientific realism regarding 'unobservables' like electrons." Contrary to the positivist viewpoint which assumes as a standpoint the "common sense" observability of tables and chairs, the goal of physics has been always —since its Greek origin— the theoretical representation of *physis* [17]. This scientific project has nothing to do with "common sense".

According to our viewpoint, there is no physical observation without the aid of a network of adequate concepts. It is important to stress at this point that we use the term 'metaphysics' to refer to the systematic definition of conceptual schemes. A metaphysical scheme is a conceptual net of interrelated concepts. It provides the very preconditions of observability itself. Without concepts —i.e., generalizations that escape differences and particular experiences— it is not possible to provide meaning to experience, nor is it possible to think. From our perspective, metaphysics has nothing to do with the the empiricist based distinction between observable entities (e.g., tables and chairs) and un-observable entities (e.g., atoms).

As Einstein [35, p. 175] made the point: "[...] it is the purpose of theoretical physics to achieve understanding of physical reality which exists independently of the observer, and for which the distinction between 'direct observable' and 'not directly observable' has no ontological significance". Observability, then, is secondary even though "the only decisive factor for the question whether or not to accept a particular physical theory is its empirical success." For the representational realist, empirical adequacy is only part of a verification procedure, not that which "needs to be saved" — as van Fraassen might argue [76]. Observability needs to be developed within each physical theory, it is a theoretical and conceptual development, not "a given" of experience. As obvious as it might sound, one cannot observe a field without the notion of 'field', one cannot observe a dog (or, in general, an entity) without presupposing the notion of 'dog' (or entity). As Einstein [47, p. 63] explained to Heisenberg many years ago: "It is only the theory which decides what we can observe."[d]

In physical theories, it is only through metaphysical conceptual schemes —produced through the tight interrelation of many different physical notions— that we are capable of producing a qualitative representation and understanding of physical reality and experience. *Gedankenexperiments* have many times escaped the observability of their time and adventured into metaphysical debates about possible but still unperformed experiences. Such thought experiments can be only considered and imagined through the creation of adequate conceptual schemes. As remarked by Heisenberg [48, p. 264]: "The history of physics is not only a sequence of experimental discoveries and observations, followed by their mathematical description; it is also a history of concepts. For an understanding of the phenomena the first condition is the introduction of adequate concepts. Only with the help of correct concepts can we really know what has been observed."

To summarize, there are three main points which comprise representational realism:

I. *Physical Theory*: A physical theory is a mathematical formalism related to a set of physical concepts which only together provide a qualitative and quantitative understanding of a specific field of phenomena.

II. *Formal-Conceptual Representation of Reality*: Physics attempts to provide theoretical —both formal and conceptual— representations of physical reality.

---

[d]A very good example of how observability is developed within physical theories is Einstein's analysis of the notion of *simultaneity* in the context of the theory of special relativity.

III. *The Conditions of Observability are Defined within the Theory*: The conditions of observability are dependent of, and constrained by each specific theory. Observation is only possible through the development of adequate physical concepts, and these, as point I mentions, are part of the theory.

By contrast with empiricism, which considers that the world is accessible through observations —which are the key to develop scientific knowledge—, our realist perspective takes the opposite standpoint and argues that it is only through the creation of theories that we can achieve understanding of our experience in the Cosmos. According to representational realism, the physical explanation of our experience goes very much against "common sense" observability. There is no theory of "common sense" and thus, a intrinsic impossibility to discuss its foundation. In fact, we could say that the history of physics is also the history of how "common sense" changes through the development of new theories: it was not evident for the ancient communities that the Earth is a sphere rather than a plane; it was not obvious for the contemporaries of Newton that the force that commands the movement of the moon and the planets is also responsible for the fall of an apple; it was not obvious in the 18th Century that the strange phenomena of magnetism and electricity could be unified through the strange notion of electromagnetic field; and it was far from evident —before Einstein— that space and time are entangled, that objects shrink and time dilates with increasing speed. From our viewpoint, we will only get to understand QM when we develop a new "quantum common sense" which relates the mathematical equations of the theory of quanta to adequate physical notions.

3. Aristotelian Hylomorphic Metaphysics

The debate in Pre-Socratic philosophy is traditionally understood — through the interpretation of both Plato and Aristotle— as the contraposition of the Heraclitean and the Eleatic schools of thought [66]. Heraclitus defended a theory of flux, a doctrine of permanent motion and unstability in the world. He stated that the ever ongoing change or motion characterizes this world and its phenomena. This doctrine precluded, as both Plato and Aristotle, the possibility to develop certain knowledge about the world. As remarked by Verelst and Coecke [77, p.2]: "This is so because Being, over a lapse of time, has no stability. Everything that it is at this moment changes at the same time, therefore it is not. This coming together of Being and non-Being at one instant is known as the principle of coincidence of opposites." Parmenides, as interpreted by Plato and Aristotle, taught the non-existence of motion and change. In his famous poem Parmenides stated maybe the earliest intuitive exposition of the *principle of non-contradiction*; i.e. that which *is* can only *be*, that which *is not, cannot be*. In his own turn, Aristotle developed a metaphysical scheme in which, through the notions of *actuality* and *potentiality*, he was able to articulate both the Heraclitean and the Parmenidean theories. The well known phrase of Aristotle: "Being is said in different ways" refers to the modes of being in which Being itself can be thought to exist in the realm of actuality and in the realm of potentiality.

The conceptual representation of the world produced by Plato —in the *Sophist*— and by Aristotle —in *Metaphysics*— through a specifically designed set of categories is considered by many to be the origin itself of metaphysical thought. In the case of Aristotle, *actuality* provided a metaphysical representation of a mode of existence which —contrary to the empiricist use of the same term— was independent of a *hic et nunc* experience. In this way —through conceptual representation— metaphysical thought was able to go beyond the appearances of particular observations. A represented conceptual world beyond our "common sense" observed world.

In order to solve "the problem of movement" Aristotle crafted a logical scheme in which the principles of *existence, non-contradiction* and *identity* would constitute a realm of actuality in which the notion entity could be considered.[e] Through these principles, the notion of entity was capable of unifying a multiplicity of phenomena in terms of a "sameness", creating the necessary stability for knowledge to be possible. Funes the memorious would of course regard this metaphysical architectonic as inacceptable. By allowing us to use the same name for the dog at three-fourteen (seen in profile) and the dog at three fifteen (seen from the front) we avoided the appreciation of all the differences which Funes so vividly experienced. The (metaphysical) notion of 'dog' provides a *moment of unity* for the multiple

---

[e]There are three main principles which determine classical (Aristotelian) logic, namely, the existence of objects of knowledge, the principle of non-contradiction and the principle of identity. As noticed by Verelst and Coecke, these principles are "exemplified in the three possible usages of the verb 'to be': existential, predicative, and identical. The Aristotelian syllogism always starts with the affirmation of existence: something is. The principle of contradiction then concerns the way one can speak (predicate) validly about this existing object, i.e. about the true and falsehood of its having properties, not about its being in existence. The principle of identity states that the entity is identical to itself at any moment (a=a), thus granting the stability necessary to name (identify) it." [77, p 167].

different phenomena, a unity which erases the subtle differences of our experience. Hence, it becomes evident that even the most common experience with a table or a chair, when considered linguistically, is always metaphysically grounded.

Aristotle also characterized 'potentiality' as a different realm from actuality. In the book Θ of *Metaphysics*, Aristotle [1046b5-1046b24] remarks there are two types of potentiality: "[...] some potentialities will be non-rational and some will be accompanied by reason." In the following we shall expose in some detail these two very different notions.

3.1. *Irrational Potentiality*

In his book, *Potentialities*, Giorgio Agamben discusses the meaning of irrational potentiality in Aristotle's metaphysics. "There is a generic potentiality, and this is the one that is meant when we say, for example, that a child has the potential to know, or that he or she can potentially become the head of the State." The child has the potentiality to become something else than what he is in actuality. Irrational potentiality implies a realm of 'indefiniteness', a realm of 'incompleteness' and 'lack'. It is then, only when turning into actuality, that the potential is fulfilled or completed. The child becomes then a man, the seed can transform into a tree.

> The word 'actuality', which we connect with fulfillment, has, strictly speaking, been extended from movements to other things; for actuality in the strict sense is identified with movement. And so people do not assign movement to non-existent things, though they do assign some other predicates. E.g. they say that non-existent things are objects of thought and desire, but not that they are moved; and this because, while they do not actually exist, they would have to exist actually if they were moved. For of non-existent things some exist potentially; but they do not exist, because they do not exist in fulfillment. Aristotle [1047b3-1047b14]

The path from irrational potentiality into actualization is associated with the process through which *matter* is *formed*. The matter of a substance being the stuff it is composed of; while the form is the way that stuff is put together so that the whole it constitutes can perform its characteristic functions. Through this passage substance gains perfection and, in this way, becomes closer to God, *pure actus* [1051a4-1051a17].[f] But due to this

---

[f]As noticed by Verelst and Coecke [77, p. 168]: "change and motion are intrinsically not

dependence, it makes no sense to consider the realm of irrational potentiality independently of the actual realm. The final cause plays an essential role connecting the potential and the actual realms. As noticed by Smets in [69], the idea of irrational potentiality is directly linked to Aristotle's theory of teleological causality: "the transition from being [irrational] potential to actual has to be placed within the context of [Aristotle's] theory of movement and change, which is embedded in his teleological conception of causality [1050a7]." This teleological aspect shows the delimitation of irrational potentiality with respect to actual realm. That is, irrational potentiality can be only thought in terms of its actualization, i.e., in terms of its passage into the actual realm.

Although Aristotle first argues that both actuality and potentiality must be considered as independent ontological modes of existence, very soon he chose the actual realm as superior to the potential one (see [16, section 12]). However, and independently of this choice, according to Agamben [3, p. 179], it is not this potentiality which seems to interests Aristotle, rather, it is "the one that belongs to someone who, for example, has knowledge or ability. In this sense, we say of the architect that he or she has the *potential* to build, of the poet that he or she has the *potential* to write poems. It is clear that this *existing* potentiality differs from the *generic* potentiality of the child." We shall now turn our attention to this second kind of potentiality which, we believe, can allow us to develop a notion truly independent of the actual realm and actualization —evading at the same time teleological considerations.

3.2. Rational Potentiality

Rational potentiality is characterized by Aristotle as related to the possession of a capability, a faculty [1046b5-1046b24], to the problem of what I mean when I say: "I can", "I cannot". As explicitly noticed by Aristotle, potentiality implies a mode of existence that must be considered as real as that of actual existence. In Chapter 3 of book Θ of *Metaphysics* Aristotle introduces the notion of rational potentiality. In doing so Aristotle goes against the Megarians who, by contrast with him, considered actuality as the *only* mode of existence:

---

provided for in this [Aristotelian logical] framework; therefore the ontology underlying the logical system of knowledge is essentially static, and requires the introduction of a First Mover with a proper ontological status beyond the phenomena for whose change and motion he must account for." This first mover is God, *pure actus*, pure definiteness and form without the contradiction and evil present in the potential matter.

There are some who say, as the Megaric school does, that a thing can act only when it is acting, and when it is not acting it cannot act, e.g. he who is not building cannot build, but only he who is building, when he is building; and so in all other cases. It is not hard to see the absurdities that attend this view. For it is clear that on this view a man will not be a builder unless he is building (for to be a builder is to be able to build), and so with the other arts. If, then, it is impossible to have such arts if one has not at some time learnt and acquired them, and it is then impossible not to have them if one has not sometime lost them (either by forgetfulness or by some accident or by time; for it cannot be by the destruction of the object itself, for that lasts for ever), a man will not have the art when he has ceased to use it, and yet he may immediately build again; how then will he have got the art? [...] evidently potentiality and actuality are different; but these views make potentiality and actuality the same, so that it is no small thing they are seeking to annihilate. [...] Therefore it is possible that a thing may be capable of being and not be, and capable of not being and yet be, and similarly with the other kinds of predicate; it may be capable of walking and yet not walk, or capable of not walking and yet walk. [1046b29 - 1047a10]

While non-rational potentialities which "are all productive of one effect each" rational potentialities "produce contrary effects" [1048a1-1048a24]. This also means that potentiality is capable of 'being' and 'not being' at one and the same time: "Every potentiality is at one and the same time a potentiality for the opposite; for, while that which is not capable of being present in a subject cannot be present, everything that is capable of being may possibly not be actual. That, then, which is capable of being may either be or not be; the same thing, then, is capable both of being and of not being." [1050b7-1050b28] For our purposes, it is important to notice that rational potentiality can become actual only when the state of affairs allows it. Thus, "everything which has a rational potentiality, when it desires that for which it has a potentiality and in the circumstances in which it has it, must do this. And it has the potentiality in question when the passive object is present and is in a certain state; if not it will not be able to act." [1048a1-1048a24] This opens the question of the contextual existence of such potentiality which might be regarded as independent (or not) of the actual state of affairs. While irrational potentialities become more real by passing to the actual realm, a rational agent can choose to withhold the actualization of a power or capability.[g] This, we believe, allows us to think of a realm of potentiality completely independent of actuality. It seems Aristotle chose once again to limit the expressivity of rational potentiality within the limits of the actual realm.

> To add the qualification 'if nothing external prevents it' is not further necessary; for it has the potentiality in so far as this is a potentiality of acting, and it is this not in all circumstances but on certain conditions, among which will be the exclusion of external hindrances; for these are barred by some of the positive qualifications. And so even if one has a rational wish, or an appetite, to do two things or contrary things at the same time, one cannot do them; for it is not on these terms that one has the potentiality for them, nor is it a potentiality for doing both at the same time, since one will do just the things which it is a potentiality for doing. [1048a25-1048b9]

In Chapter 6 of Book Θ, Aristotle articulates on the relation between potentiality and actuality, placing actuality as the cornerstone of his architectonic and relegating potentiality to a mere supplementary role: "We have distinguished the various senses of 'prior', and it is clear that actuality is prior to potentiality. [...] For the action is the end, and the actuality is the action. Therefore even the word 'actuality' is derived from 'action', and points to the fulfillment." [1050a17-1050a23] Aristotle then provides more arguments towards showing "that the good actuality is better and more valuable than the good potentiality is evident" [1051a4-1051a17] (see [16, Section 12]).

The restrictions imposed even to the notion of rational potentiality —as related to the actual realm— constrained the possibilities of a truly independent development of the potential realm beyond actuality. As recognized by Wolfgang Pauli:

> Aristotle [...] created the important concept of *potential being* and applied it to *hyle*. [...] This is where an important differentiation in scientific thinking came in. Aristotle's further statements on matter cannot really be applied in physics, and it seems to me that much of the confusion in Aristotle stems from the fact that being by far the less able thinker, he was completely overwhelmed by Plato. He

---

[g]This is the subject of the famous short story by Herman Melville, *Bartleby, the Scrivener: A Story of Wall Street.*

134

was not able to fully carry out his intention to grasp the *potential*, and his endeavors became bogged down in early stages. [59, p. 93]

4. The Newtonian Atomist Metaphysics of Classical Physics

As remarked by Giorgio Agamben [3]: "The concept of potentiality has a long history in Western philosophy, in which it has occupied a central position at least since Aristotle. In both his metaphysics and physics, Aristotle opposed potentiality to actuality, *dynamis* to *energeia*, and bequeathed this opposition to Western philosophy and science." However, the importance of potentiality, which was first placed by Aristotle on equal footing to actuality, was soon diminished in the development of Western thought. As we have seen above, it could be argued that the seed of this move was already present in the Aristotelian architectonic itself, the focus of which was placed on the actual realm. The realm of potentiality, as a different (ontological) mode of the being was neglected, becoming merely a (logical) *possibility* or process of fulfillment. In relation to development of physics, the focus and preeminence was also given to actuality. The distinction between *res cogitans* and *res extensa* established in the 17th Century played in this respect an important role separating also the realms of actuality and potentiality. As Heisenberg makes the point:

> Descartes knew the undisputable necessity of the connection, but philosophy and natural science in the following period developed on the basis of the polarity between the 'res cogitans' and the 'res extensa', and natural science concentrated its interest on the 'res extensa'. The influence of the Cartesian division on human thought in the following centuries can hardly be overestimated, but it is just this division which we have to criticize later from the development of physics in our time. [46]

The philosophy developed after Descartes kept 'res cogitans' (thought) and 'res extensa' (entities occupying space-time) as separated realms.[h] This materialistic conception of science is based on the idea that extended things exist as being absolutely definite; that is, as existents within the actual realm. The division produced in the XVII century between *res cogitans*

---

[h]While 'res cogitans', the soul, was related to the *indefinite* realm of potentiality and is discussed by Aristotle in *De Anima*, 'res extensa', the entities as characterized by the principles of logic gave place to the actual considered in terms of *definiteness*.

and *res extensa* together with the subsequent preeminence of "extended things" could be understood as the triumph of the actualist Megarian path over Aristotelian Hylomorphic metaphysics. In this respect, it is also true that the transformation from medieval science to modern science coincides with the abolition of Aristotelian metaphysics as the foundation of knowledge. However, the basic structure of Aristotle's metaphysical scheme and logic still remained the basis for correct reasoning, the principle of non-contradiction —as Kant, Leibniz and many others proclaimed— the most certain of all principles.[i]

Isaac Newton was able to translate both the ontological presuppositions present in Aristotelian logic and the materialistic ideal of *res extensa* together with actuality as its mode of existence into a closed mathematical formalism. He did so with the aid of atomistic metaphysics. In the VI Century B.C., Leucipo and Democritus had imagined existence as consisting of small simple bodies with mass. According to their metaphysical theory, atoms were conceived as small individual substances, indivisible and separated by void. Atoms —which means "not divisible"— were, for both Leucipo and Democritus, the building blocks of our material world. Many centuries later, Newton was able not only to mathematize atoms as points in phase space, he had also constructed an equation of motion for the trajectory —within absolute space-time— of such "elementary particles". The obvious conclusion implied by the conjunction of atomism and Newtonian's use of the *effective cause* was derived by Pierre Simon Laplace:

> We may regard the present state of the universe as the effect of its past and the cause of its future. An intellect which at a certain moment would know all forces that set nature in motion, and all positions of all items of which nature is composed, if this intellect were also vast enough to submit these data to analysis, it would embrace in a single formula the movements of the greatest bodies of the universe and those of the tiniest atom; for such an intellect

---

[i]As noticed by Verlest and Coecke [77, p. 7]: "Dropping Aristotelian metaphysics, while at the same time continuing to use Aristotelian logic as an empty 'reasoning apparatus' implies therefore loosing the possibility to account for change and motion in whatever description of the world that is based on it. The fact that Aristotelian logic transformed during the twentieth century into different formal, axiomatic logical systems used in today's philosophy and science doesn't really matter, because the fundamental principle, and therefore the fundamental ontology, remained the same ([40], p. xix). This 'emptied' logic actually contains an Eleatic ontology, that allows only for static descriptions of the world."

nothing would be uncertain and the future just like the past would be present before its eyes." [53, p. 4]

The abolition of free will in the materialistic realm was the highest peak of the division between res cogitans and res extensa. In the XVII Century, in the newly proposed mechanical description of the world, the very possibility of indetermination present before in the potential realm had been erased from (physical) existence.

In classical mechanics, every physical system may be described exclusively by means of its actual properties. A point in phase space is related to the set of values of properties that characterize the system. In fact, an actual property can be made to correspond to the set of states (points in phase space) for which this property is actual. Thus, the change of the system may be described by the change of its actual properties. Potential or possible properties are then considered as the points to which the system might (or might not) arrive in a future instant of time. Such properties are thought in terms of irrational potentiality; as properties which might possibly become actual in the future. As also noted by Dieks [37, p. 124]: "In classical physics the most fundamental description of a physical system (a point in phase space) reflects only the actual, and nothing that is merely possible. It is true that sometimes states involving probabilities occur in classical physics: think of the probability distributions ρ in statistical mechanics. But the occurrence of possibilities in such cases merely reflects our ignorance about what is actual. The statistical states do not correspond to features of the actual system (unlike the case of the quantum mechanical superpositions), but quantify our lack of knowledge of those actual features." Classical mechanics tells us via the equation of motion how the state of the system moves along the curve determined by initial conditions in the phase space and thus, any mechanical property may be expressed in terms of phase space variables. Needless to say, in the classical realm the measurement process plays no distinctive role and actual properties fit the definition of *elements of physical reality* in the sense of the EPR paper [43]. Moreover, the structure in which actual properties may be organized is the (Boolean) algebra of classical logic.

5. Heisenberg's Return to Aristotelian Hylomorphism in QM

The mechanical description of the world provided by Newton can be sketched in terms of static pictures which provide at each instant of time the set of definite actual properties which constitute an actual state of affairs (see [51, p. 609]). Even though the potential has been erased completely, there is in this description an obvious debt to part of the Aristotelian metaphysical scheme. The description of motion is then given, not *via* the path from the irrational potential to the actual, not from *matter* into *formed matter*, but rather *via* the successions of completely defined and determined actual states of affairs (i.e., "pictures" constituted by sets of actual properties with definite values). As we discussed above, potentiality becomes then completely superfluous.

With the advent of modern science and the introduction of mathematical formalisms, physics seemed capable of reproducing the evolution of the universe in a mechanical manner; just like the complicated composition of a clock allows to account for the passage of time. As Heisenberg explains, this materialistic conception of science chose actuality as the main notion to conceive existence and reality:

> In the philosophy of Aristotle, matter was thought of in the relation between form and matter. All that we perceive in the world of phenomena around us is formed matter. Matter is in itself not a reality but only a possibility, a 'potentia'; it exists only by means of form. In the natural process the 'essence,' as Aristotle calls it, passes over from mere possibility through form into actuality. [...] Then, much later, starting from the philosophy of Descartes, matter was primarily thought of as opposed to mind. There were the two complementary aspects of the world, 'matter' and 'mind,' or, as Descartes put it, the 'res extensa' and the 'res cogitans.' Since the new methodical principles of natural science, especially of mechanics, excluded all tracing of corporeal phenomena back to spiritual forces, matter could be considered as a reality of its own independent of the mind and of any supernatural powers. The 'matter' of this period is 'formed matter,' the process of formation being interpreted as a causal chain of mechanical interactions; it has lost its connection with the vegetative soul of Aristotelian philosophy, and therefore the dualism between matter and form [potential and actual] is no longer relevant. It is this concept of matter which constitutes by far the strongest component in our present use of the word 'matter'. [46, p. 129]

As mentioned above, in classical mechanics the mathematical description of the behavior of a system may be formulated in terms of the set of actual properties. The same treatment can be applied to QM. However, the different structure of the physical properties of a quantum system imposes a deep change of nature regarding the meaning of possibility and potentiality.

QM was related to modality since Born's interpretation of the quantum wave function Ψ as a density of probability. As Heisenberg made the point: "[The] concept of the probability wave [in quantum mechanics] was something entirely new in theoretical physics since Newton. Probability in mathematics or in statistical mechanics means a statement about our degree of knowledge of the actual situation. In throwing dice we do not know the fine details of the motion of our hands which determine the fall of the dice and therefore we say that the probability for throwing a special number is just one in six. The probability wave function, however, meant more than that; it meant a tendency for something." [46, p. 42] It was Heisenberg himself who tried to interpret for the first time the wave function in terms of the Aristotelian notion of potentia. Heisenberg [*Op. cit.*, p. 156] argued that the concept of probability wave "was a quantitative version of the old concept of 'potentia' in Aristotelian philosophy. It introduced something standing in the middle between the idea of an event and the actual event, a strange kind of physical reality just in the middle between possibility and reality." According to him, the concept of potentiality as a mode of existence had been used implicitly or explicitly in the development of quantum mechanics: "I believe that the language actually used by physicists when they speak about atomic events produces in their minds similar notions as the concept of 'potentia'. So physicists have gradually become accustomed to considering the electronic orbits, etc., not as reality but rather as a kind of 'potentia'."

But even though Heisenberg criticized the abolition of the potential realm in science and attempted to reintroduce it in order to overcome the interpretational problems of QM, when doing so he restricted potentiality "right from the start" to the sole consideration of *irrational potentiality*. As we have seen above, irrational potentiality is only subsidiary, through the teleological relation of actualization, to the actual realm; it cannot be thought to exist beyond its future actuality. Thus, by restricting his analysis of the potential realm, Heisenberg was trapped "right from the start" within an actualist account of reality.[j] We might remark that in this

---

[j] In this sense it is interesting to take into account the question posed by Heisenberg to Henry Stapp regarding the ontological meaning of ideas: "When you speak about the ideas (especially in [Section 3.4]), you always speak about human ideas, and the question arises, do these ideas 'exist' outside of the human mind or only in the human mind? In

respect atomistic metaphysics has also played a major role constraining the possibilities of analysis. Even though the idea of QM as a theory that described atoms was severly criticized by Heisenberg and many others since the time of its construction, the language used by quantum physicists to refer to the formalism of the theory remained —up to the present day— an inadequate "language of elementary particles".

6. Causal Powers as Future Possible Existents

Closely related to the development of Heisenberg in terms of (irrational) potentialities stands the development of Henri Margenau and Karl Popper in terms of latencies, propensities or dispositions. As recalled by Mauricio Suárez [74], Margenau was the first to introduce in 1954 a dispositional idea in terms of what he called *latencies*. In Margenau's interpretation the probabilities are given an objective reading and understood as describing tendencies of latent observables to take on different values in different contexts [55]. Later, Popper [64], followed by Nicholas Maxwell [56], proposed a propensity interpretation of probability. Quantum reality was then characterized by irreducibly probabilistic real propensity (propensity waves or propensitons).[k] More recently, Suárez has put forward a new propensity interpretation in which the quantum propensity is intrinsic to the quantum system and it is only the manifestation of the property that depends on the context [72–74]. Mauro Dorato has also advanced a dispositional approach towards the GRW theory [38–40]. The GRW theory after their creators: Ghirardi, Grimmini and Weber [45]; is a dynamical reduction model of non-relativistic QM which modifies the linearity of Schrödinger's equation. As remarked by Dorato [38, p. 11]: "According to this reduction model, the fundamentally stochastic nature of the localization mechanism is not grounded in any categorical property of the quantum system: the theory at present stage is purely 'phenomenological', in the sense that no 'deeper mechanism' is provided to account for the causes of the localization. 'Spontaneous', as referred to the localization process, therefore simply means

---

other words: Have these ideas existed at a time when no human mind existed in the world. (Heisenberg, 1972)" [70].

[k]The realist position of Popper attempted to evade the subjective aspect of Heisenberg's interpretation [46, p. 67-69] according to which: "[The quantum] probability function combines objective and subjective elements. It contains statements on possibilities, or better tendencies ('potentiae' in Aristotelian philosophy), and such statements are completely objective, they don't depend on any observer the passage from the 'possible' to the real takes place during the act of observation."

'uncaused'." Dorato continues then to discuss the meaning of dispositions
and reviews the need of different interpretations of QM to account for such
intrinsic tendencies within the theory:

> [...] whether and in what sense QM, in its various interpretations,
> forces us to accept the existence of ungrounded, irreducible, prob-
> abilistic dispositions, i.e. dispositions, that, unlike fragility or per-
> meability, lack any categorical basis to which they can be reduced
> to. My claim is that the presence of irreducible quantum disposi-
> tions in many (but not all) interpretations involves the difficulty of
> giving a spatiotemporal description to quantum phenomena, and is
> therefore linked to our lack of understanding of the theory, i.e., of
> our lack of a clear ontology underpinning the formalism. [*Op. cit.*,
> p. 3]

Dorato also explains very clearly the meaning of dispositional properties as
well as their relation to categorical properties:[1]

> Intuitively, a disposition like permeability is not directly observ-
> able all the times, as is the property given by the form of an object
> ('being spherical'), but becomes observable only when the entity
> possessing it interacts with water or other fluids. [...] From these
> ordinary language examples, it would seem that the function of
> dispositional terms in natural languages is to encode useful infor-
> mation about the way objects around us would behave were they
> subject to causal interactions with other entities (often ourselves).
> This remark shows that the function of dispositional predicates in
> ordinary language is essentially predictive. [...] In a word, disposi-
> tions express, directly or indirectly, those regularities of the world
> around us that enable us to predict the future. Such a predictive
> function of dispositions should be attentively kept in mind when
> we will discuss the 'dispositional nature' of microsystems before
> measurement, in particular when their states is not an eigenstate
> of the relevant observable. In a word, the use of the language of
> 'dispositions' does not by itself point to a clear ontology underlying

---

[1]A very similar criticism to dispositions by Dieks can be found in [37, p. 133]: "we do
not really know what kind of things dispositions are, and it is obscure exactly how a
disposition could take care of the task of arranging for the right relative frequencies to
occur in long series of experiments. Indeed, the very content of the notion of disposition
does not seem to go beyond 'something responsible for the actual relative frequencies
found in experiments'."

the observable phenomena, but, especially when the disposition is irreducible, refers to the predictive regularity that phenomena manifest. Consequently, attributing physical systems irreducible dispositions, even if one were realist about them, may just result in more or less covert instrumentalism. [*Op. cit.*, pp. 2-4]

In favor of dispositions, he argues [*Op. cit.*, p. 5] that contextuality seems to call for dispositional properties: "Within QM, it seems natural to replace 'dispositional properties' with 'intrinsically indefinite properties', i.e. with properties that before measurement are objectively and actually 'indefinite' (that is, without a precise, possessed value). So the passage from dispositional to non-dispositional is the passage from the indefiniteness to the definiteness of the relevant properties, due to measurements interactions." We can see here the direct relation between Aristotle's metaphysics, his potentiality-actuality scheme conceived in terms of causality, and the dispositional account developed in order to understand QM in terms of *causal capacities* [44]. The joint proposal of Dorato and Esfeld regarding the interpretation of QM relies on the a-causal stochastic GRW theory. Going back to dispositions and the remark of Dorato, it is interesting to notice that his idea of 'observability' determines very explicitly the distinction between dispositional and categorical properties. This idea goes against our representational realist stance. But independently of our critical considerations regarding "common sense" observability, it is not at all clear that such dispositions are not simply a "black box" where we can hide the mystery surrounding QM. "It must be granted that introducing irreducible physical dispositions is implicitly admitting that there is something we don't understand. Admitting an in-principle lack of any categorical basis to which dispositions could be reduced, in both the non-collapse views and Bohr's seems a way to surrender to mystery." [*Op. cit.*, p. 9]. As clearly exposed by Dorato:

That the distinction between dispositions and categorical properties cannot be so sharp is further confirmed by Mumford's analysis of the problem of the reducibility of dispositions to their so-called 'categorical basis'. According to Mumford (1998), the difference between a dispositional property like fragility and the microscopic property of glass constituting its categorical basis is merely linguistic, and not ontological. Referring to a property by using a dispositional term, or by choosing its categorical-basis terms, depends on whether we want to focus on, respectively, the functional role of the property (the causal network with which it is connected), or the particular way in which that role is implemented or realized. But notice that if we agree with Mumford's analysis, it follows that it makes little sense to introduce irreducible quantum dispositions as ontological hypotheses. If, by hypothesis, no categorical basis were available, we should admit that we don't not know what we are talking about when we talk the dispositional language in QM, quite unlike the cases in which we refer to 'fragility' or 'transparency', in which the categorical bases are available and well-known. Introducing irreducible quantum dispositions would simply be a black-box way of referring to the functional role of the corresponding property, i.e., to its predictive function in the causal network of events. In a word, the use of the language of 'dispositions' by itself does not point to a clear ontology underlying the observable phenomena. On the contrary, when the dispositions in question are irreducible and their categorical bases are unknown, such a use should be regarded as a shorthand to refer to the regularity that phenomena manifest and that allow for a probabilistic prediction. Consequently, attributing physical systems irreducible dispositions may just result in a more or less covert instrumentalism, unless the process that transforms a dispositional property into a categorically possessed one is explained in sufficient detail. [*Op. cit.*, pp. 8-9] (emphasis added)

Dispositional proposals need thus to provide descriptions of the selecting physical process which takes place during the path from the indefinite level of dispositional properties to the definite level of actual properties. Without such explanation, the measurement problem remains unsolved.

Causal powers have been developed in the context of QM in order to provide an understanding of the multiple terms within a superposition and provide in this way an answer to the infamous measurement problem (see [30] for a detailed analysis of the measurement process in QM). On the one hand, the problem in question assumes an empiricist perspective according to which the observation of 'clicks' in detectors is unproblematic and treated as a "self evident given". On the other hand, even though dispositionalist and propensity type interpretations go as far as claiming that propensitons and dispositions are real, they remain at the same time captive of atomistic metaphysics —they keep holding on to the claim that "QM talks about elementary particles". The propensity and dispositional interpretations of QM rest, thus, on paradoxical tension: on the one hand, causal powers still attempt to describe the metaphysical existence of unobservable elementary particles, while, on the other hand, they define such existence only in relation to the process of actualization and the observability of 'clicks' in detectors. The entanglement between the un-observable metaphysical existence of elementary particles and the observability of 'clicks' in detectors as the foundation for understanding physical theories seems to create a Moebius reasoning strip which, rather than providing rational constraints of debate, it is a source of paradoxes and pseudoproblems.

7. Immanent Powers as Intensive Existents

As we discussed above, representational realism assumes a very different standpoint with respect to the empiricist based characterization of physical theories. While the orthodox viewpoint in philosophy of physics continues to consider theories from an empiricist perspective according to which observability is the basis for the development of science, our neo-spinozist realist perspective returns to the original Greek understanding of physics as a discipline which through theories provide the foundation for the expression of *physis* in representational terms [23]. In this context, our approach stresses the need to provide a conceptual representation of the mathematical formalism, one which need not be constrained or reduced to our "common sense" observability of tables and chairs. Since the conceptual representation of QM seems to escape the limits imposed by classical notions —including that of "atom" or "elementary particle"—, instead of insisting dogmatically in applying the metaphysical worldview inherited from Newton and Maxwell, we believe it might seem wise to start searching — against Bohr's viewpoint[m]— for new, non-classical concepts, in order to make sense of both the quantum formalism and quantum phenomena. It is this non-empiricist viewpoint regarding the problem of QM which allows us to "invert" the measurement problem and replace it by what we call *the superposition problem* [29]. Let us first recall the problem we are dealing with.

---

[m] According to Bohr [79, p. 7]: "[...] the unambiguous interpretation of any measurement must be essentially framed in terms of classical physical theories, and we may say that in this sense the language of Newton and Maxwell will remain the language of physicists for all time." Closing the possibility of creating new physical concepts, Bohr [*Op. cit.*] argued that "it would be a misconception to believe that the difficulties of the atomic theory may be evaded by eventually replacing the concepts of classical physics by new conceptual forms."

Measurement Problem: *Given a specific basis (or context),*[n] *QM describes, mathematically, a quantum state in terms of a superposition of, in general, multiple states. Since the evolution described by QM allows us to predict that the quantum system will get entangled with the apparatus and thus its pointer positions will also become a superposition,*[o] *the question is why do we observe a single outcome instead of a superposition of them?*

The measurement problem attempts to justify the observation of actual measurement outcomes, focusing on the actual realm of experience. This allows us to characterize the measurement problem as an empiricist problem which presupposes "right from the start" the controversial idea that actual observations are perfectly well defined for quantum phenomena. However, as we noticed above, from a representational realist stance things must be analyzed from a radically different perspective for —as Einstein remarked— it is only the theory which can tell us what can be observed. If we are willing to investigate the physical representation of quantum superpositions beyond classical concepts —such as 'elementary particle', 'wave' or 'field'— we then need to take a completely different standpoint. Instead of trying to justify what we observe in classical terms in order to "save the phenomena", we need to "invert" the measurement problem and concentrate on the formal-conceptual level. We need to think differently, we need to ask a different question. According to our novel viewpoint, attention should be focused on the conceptual representation of the mathematical expression, not in the measurement outcomes, not in the actualization process and not in the attempt to justify experience in terms of elementary particles, tables and chairs. In short, we need to create a new physical language with concepts that are adequate to account for the structural relationships implied by the quantum formalism.[p] The solution to this problem must be provided defining new concepts in a systematic manner, beyond the reference

---

[n]It is important to remark that, according to this definition, both superpositions and the measurement problem are basis dependent, they can be only defined in relation to a particular basis. For a detailed analysis of this subtle but most important point see [20].
[o]Given a quantum system represented by a superposition of more than one term, $\sum c_i |\alpha_i\rangle$, when in contact with an apparatus ready to measure, $|R_0\rangle$, QM predicts that system and apparatus will become "entangled" in such a way that the final 'system + apparatus' will be described by $\sum c_i |\alpha_i\rangle |R_i\rangle$. Thus, as a consequence of the quantum evolution, the pointers have also become —like the original quantum system— a superposition of pointers $\sum c_i |R_i\rangle$. This is why the MP can be stated as a problem only in the case the original quantum state is described by a superposition of more than one term.
[p]This is in no way different from what Einstein did for the Lorentz transformations in his theory of special relativity.

to linguistic metaphors that make an inadequate use of concepts (e.g., the notion of atom).

Superposition Problem: Given a situation in which there is a quantum superposition of more than one term, $\sum c_i |\alpha_i\rangle$, and given the fact that each one of the terms relates through the Born rule to a meaningful physical statement, the problem is: how do we conceptually represent this mathematical expression? Which is the physical concept that relates to each one of the terms in a quantum superposition?

The new technological era we are witnessing today in quantum information processing requires that we, philosophers of QM, pay attention to the developments that are taking place. We believe that an important help could be provided by philosophers of physics who should be in charge of trying to develop a conceptual representation of quantum superpositions that would allow us to think in a truly quantum mechanical manner. The first step of this project must be to recognize the inadequacy of the notion of elementary particle to account for what is going on in the quantum realm. In this respect, the superposition problem opens the possibility to discuss a physical representation of reality which goes beyond the classical atomist representation of physics. Instead of keep trying —as we have done for almost a century— to impose dogmatically our "manifest image of the world" to QM, this new realist problem will allow us to reflect about possible truly non-classical solutions to the question of interpretation and understanding of quantum theory.

As the reader might already suspect, all these considerations place us in a radically different standpoint with respect to the previous developments of potentiality in terms of causal powers —which implicitly or explicitly constrain the potential real to actuality. The notion of potentiality we have developed in the course of our investigations [21,22,24,25,30,33] is called *ontological potentiality*. Contrary to the empiricist project that attempts to describe the formalism in terms of actualities, we have developed this new realm of existence in order to match the features and characteristics of the quantum formalism. This implies a path, from the orthodox mathematical formalism of QM to an adequate metaphysical scheme which is capable of representing in qualitative terms what the theory is really talking about.[q]

---

[q]Instead of going from a presupposed metaphysical system —such as atomism— to the development of a new mathematical formalism.

Let us present our scheme of ontological potentiality. The proposal begins with the definition of the mode of existence of ontological potentiality, which is completely independent of the actual realm. It continues by defining two key notions, namely, immanent power and potentia. According to representational realism being is said in many different ways,[r] and just like particles, fields and waves are existents within the actual realm and represented by our classical theories, *immanent powers* with definite *potentia* are existents within the potential realm which require a quantum mechanical description. Our physical representation of QM can be condensed in the following seven postulates which contain the relation between our proposed new physical concepts and the orthodox formalism of the theory.

I. **Hilbert Space:** QM is mathematically represented in a vector Hilbert space.

II. **Potential State of Affairs (PSA):** A specific vector Ψ with no given mathematical representation (basis) in Hilbert space represents a PSA; i.e., the definite potential existence of a multiplicity of *immanent powers*, each one of them with a specific *potentia*.

III. **Quantum Situations, Immanent Powers and Potentia:** Given a PSA, Ψ, and the context or basis, we call a quantum situation to any superposition (of one or more than one terms). In general given a basis $B = \{|\alpha_i\rangle\}$ the quantum situation $QS_{\Psi,B}$ is represented by the following superposition of immanent powers:

$$c_1|\alpha_1\rangle + c_2|\alpha_2\rangle + ... + c_n|\alpha_n\rangle \qquad (1)$$

We write the quantum situation of the PSA, Ψ, in the context B in terms of the order pair given by the elements of the basis and the coordinates in square modulus of the PSA in that basis:

$$QS_{\Psi,B} = (|\alpha_i\rangle, |c_i|^2) \qquad (2)$$

The elements of the basis, $|\alpha_i\rangle$, are interpreted in terms of *powers*. The coordinates of the elements of the basis in square modulus, $|c_i|^2$, are interpreted as the *potentia* of the power $|\alpha_i\rangle$, respectively. Given

---

[r]There is in our neo-spinozist account an implicit ontological pluralism of *multiple representations* which can be related to *one reality* through a *univocity principle*. This is done in analogous manner to how Spinoza considers in his immanent metaphysics the *multiple attributes* as being expressions of the same *one single substance*, namely, nature (see [23,26]). Our non-reductionistic answer to the problem of inter-theory relation escapes in this way the requirement present in almost all interpretations of QM which implicitly or explicitly attempt to explain the formalism in substantialist atomistic terms (see [34]).

the PSA and the context, the quantum situation, $QS_{\Psi,B}$, is univocally determined in terms of a set of powers and their respective potentia. (Notice that in contradistinction with the notion of *quantum state* the definition of a *quantum situation* is basis dependent and thus intrinsically contextual.)

IV. **Elementary Process:** In QM we only observe discrete shifts of energy (this is the quantum postulate) in the actual realm. These discrete shifts are interpreted in terms of *elementary processes* which produce actual effectuations. An elementary process is the path which undertakes a power from the potential realm to its actual effectuation. This path is governed by the *immanent cause* which allows the power to remain potentially preexistent within the potential realm independently of its actual effectuation. Each power $|\alpha_i\rangle$ is univocally related to an elementary process represented by the projection operator $P_{\alpha_i} = |\alpha_i\rangle\langle\alpha_i|$.

V. **Actual Effectuation of an Immanent Power (Measurement):** Immanent powers exist in the mode of being of ontological potentiality. An *actual effectuation* is the expression of a specific power within actuality. Distinct actual effectuations expose the distinct powers of a given QS. In order to learn about a specific PSA (constituted by a set of powers and their potentia) we must measure repeatedly the actual effectuations of each power exposed in the laboratory. (Notice that we consider a laboratory as constituted by the set of all possible experimental arrangements that can be related to the same Ψ.) An actual effectuation does not change in any way the PSA.

VI. **Potentia (Born Rule):** A *potentia* is the intensity of an immanent power to exist (in ontological terms) in the potential realm and the possibility to express itself (in epistemic terms) in the actual realm. Given a PSA, the potentia is represented via the Born rule. The potentia p_i of the immanent power $|\alpha_i\rangle$ in the specific PSA, Ψ, is given by:

$$Potentia\ (|\alpha_i\rangle, \Psi) = \langle\Psi|P_{\alpha_i}|\Psi\rangle = Tr[P_\Psi P_{\alpha_i}] \qquad (3)$$

In order to learn about a QS we must observe not only its powers (which are expressed in actuality through actual effectuations) but we must also measure the potentia of each respective power. In order to measure the potentia of each power we need to invetsigate the QS statistically through repeated series of observations. The potentia, given by the Born rule, coincides with the probability frequency of repeated measurements when the number of observations goes to infinity.

148

VII. Potential Effectuations of Immanent Powers (Schrödinger Evolution): Given a PSA, Ψ, powers and potentia evolve deterministically, independently of actual effectuations, producing *potential effectuations* according to the following unitary transformation:

$$i\hbar\frac{d}{dt}|\Psi(t)\rangle = H|\Psi(t)\rangle \tag{4}$$

While *potential effectuations* evolve according to the Schrödinger equation, *actual effectuations* are particular expressions of each power (that constitutes the PSA, Ψ) in the actual realm. The ratio of such expressions in actuality is determined by the potentia of each power.

Let us now continue to analyze in more detail some important aspects of our interpretation:

The potential state of affairs as a sets of immanent powers with definite potentia. Our choice to develop an ontological realm of potentiality absolutely independent of the actual realm of existence implies, obviously, the need to characterize this realm in an independent manner to classical physical concepts such as 'particles', 'waves' and 'fields' —notions which are defined in strict relation to the actual mode of existence. According to our viewpoint, while classical physics talks about systems with definite properties ('particles', 'waves' and 'fields'), QM talks about the existence of powers with definite potentia. While the classical representation of sets of systems with definite properties can be subsumed under the notion of an *actual state of affairs*, QM provides a representation in terms of a *potential state of affairs*. This representation seeks to define concepts in a systematic categorical manner avoiding metaphorical discourse and, in this way, to provide an *anschaulich* content of the theory. Several examples have been already discussed in [25].

The existence and interaction of quantum possibilities. Considering quantum possibilities as part of physical reality is suggested, in the first place, by the fact that quantum probability resists an "ignorance interpretation". The fact that the quantum formalism implies a non-Kolmogorovian probability model which is not interpretable in epistemic terms is a well known fact within the foundational literature since Born's interpretation of the quantum wave function [65].[s] But more importantly, the quantum

---

[s]It is true that QBism does provide a subjectivist interpretation of probability following the Bayesian viewpoint, however, this is done so at the price of denying the very need of an interpretation for QM. See for a detailed analysis: [25,27].

mechanical formalism implies that projection operators can be understood as *interacting* and *evolving* [33]. In classical mechanics the mathematical and a conceptual levels are interrelated in such a consistent manner that it makes perfect sense to relate mathematical equations with physical concepts. For example, the mathematical account of a point in phase space which evolves according to Newton's equation of motion can be consistently related to the trajectory of a particle in absolute space-time. Let us remark against a common naive misunderstanding between mathematics and physics: a point is not a particle, phase space is not Newtonian space-time. But in QM, while the interaction and evolution of projection operators is represented quantitively through the mathematical formalism we still lack a conceptual qualitative representation of what projection operators really mean. In this respect, the *interaction* in terms of entanglement, the *evolution* in terms of the Schrödinger equation of motion and the *prediction* of quantum possibilities in statistical terms through the Born rule are maybe the most important features pointing in the direction of developing an ontological idea of possibility which is truly independent of actuality. This development is not a mathematical one; rather, it is a metaphysical or conceptual one.

The intensity of quantum possibilities. Another important consequence of the ontological perspective towards quantum possibilities relates to the need of reconsidering the binary existencial characterization of properties in terms of an homomorphic relation to the binary Boolean elements {0, 1} (or truth tables). In [25] we proposed to extend the notion of *element of physical reality* escaping the characterization of existence in terms of certitude (probability = 1) and considering "right from the start" the quantum probabilistic measure in objective terms. This move implies the development of existence beyond the gates of certitude and the complementary need of characterizing the basic elements of our ontology —namely, immanent powers— in intensive terms; i.e. as relating to a value which pertains to the interval $[0, 1]$. In this way, each *immanent power* has an intensive characterization which we call *potentia*. We could say that, unlike properties that pertain to systems either exist or do not exist (i.e., they are related either to 1 or 0), immanent powers have a more complex characterization which requires, apart form its binary relation to existence, a number pertaining to the closed interval $[0, 1]$ which specifies its (potential) existence in an intensive manner. It is through the introduction of an intensive mode of existence that we can understand quantum probability as describing an objective feature of QM and at the same time restore its epistemic role as a way to gain knowledge about a still unknown but yet existent (potential) state of affairs.

Immanent powers and contextuality. It is important to notice that the intensive characterization of immanent powers allows us to escape Kochen-Specker contextuality [31] and restore a global valuation to all projection operators of a quantum state, Ψ. By removing the actualist binary reference of classical properties to $\{0, 1\}$, and implementing instead an intensive valuation of projection operators to $[0, 1]$ we are able, not only to bypass Kochen-Specker theorem [52], but also to restore —through a *global intensive valuation*— an objective representation of the elements the theory talks about. Powers are non-contextual existents which can be defined univocally and globally for any given quantum state Ψ. In this way, just like in the case of classical physics, quantum contextuality can be understood as exposing the epistemic incompatibility of measurement situations and outcomes (see for a detailed discussion and analysis: [25,28,31]).

The contradiction of quantum possibilities. Some quantum superpositions of the "Schrödinger cat type" [67] constituted by two contradictory terms, e.g. '$|+\rangle$' and '$|-\rangle$', present a difficult problem for those who attempt to describe the theory in terms of particles with definite non-contradictory properties. Indeed, as discussed in [19], while the first term might relate to the statement 'the atom possesses the property of being decayed' the second term might relate to the statement 'the atom possesses the property of not being decayed'. Obviously, an atom cannot be 'decayed' and 'not decayed' at the same time —just like a cat cannot be 'dead' and 'alive' simultaneously. Any physical object —an atom, a cat, a table or a chair—, by definition, cannot posses contradictory properties. Physical objects have been always —implicitly or explicitly— defined since Aristotle's metaphysics and logic in terms of the principles of existence, non-contradiction and identity. However, regardless of the manner in which objects are defined in classical physics, QM allows us to predict through the mathematical formalism how these terms will interact and evolve in different situations. The realist attitude is of course to consider that the formalism, and in particular quantum superpositions, are telling us something very specific about physical reality. It is in fact this belief which has allowed us to enter the new technological era of quantum information processing. This is also why one might consider Schrödinger's analysis as an *ad absurdum* proof of the impossibility to describe quantum superpositions in terms of classical notions (i.e., particles, waves, tables or cats). The escape road proposed by some modal interpretations [36,37] and some readings of the many words interpretation [75,78], which attempt to consider the terms of a superposition as *possible future actualizations* misses the point, since the question is not the epistemic prediction or justification of future outcomes, but the understanding of what is really going on even before the measurement process has taken place. Finally, recalling that the interpretation of probability in epistemic terms is untenable within the orthodox formalism —which is only consistent with a non-Kolmogorovian probability measure— there seems to be no escape — at least for a realist which attempts to be consistent with the orthodox formalism— but to confront the fact that classical notions such as 'atom', 'wave', 'table' or 'chair' (i.e., notions categorically constrained by the principles of existence, non-contradiction and identity) are not adequate concepts to account for quantum superpositions 29. In this respect, using a term created by Gaston de Bachelard we might say that the notion of classical entity rather than helping us to understand the quantum formalism has always played the role of an *epistemic obstacle* [32].

The relation and independence of immanent powers with respect to the actual realm. Immanent powers have an independent potential existence with respect to the actual realm. Measurement outcomes are not what potential powers attempt to describe. It is exactly the other way around — at least for a representational realist. For the realist, measurement outcomes are only expressions of a deeper moment of unity which requires a categorical definition. This is completely analogous to the classical case in which the view of a dog at three-fourteen (seen in profile) or the dog at three fifteen (seen from the front) are both particular expressions which find their moment of unity in the notion of 'dog'. However, we still need to provide an answer to the measurement problem and explain in which manner quantum superpositions (in formal terms), and powers with definite potentia (in conceptual terms), relate to actual effectuations. In [30], we have provided a detailed analysis of our understanding of how the measurement process should be understood in QM. Within our approach, the quantum measurement process is modeled in terms of the spinozist notion of *immanent causality*. The immanent cause allows for the expression of effects remaining both in the effects and its cause. It does not only remain in itself in order to produce, but also, that which it produces stays within. Thus, in its production of actual effects the potential does not deteriorate by becoming actual —as in the case of the hylomorphic scheme of causal powers (see section 1, p. 4 of [30]).[t] Immanent powers produce, apart from

---

[t]For a more detailed discussion of the notion of immanent cause we refer to [57, Chapter 2].

actual effectuations, also *potential effectuations* which take place within potentiality and remain independent of what happens in the actual realm. Within our model of measurement, while potential effectuations describe the ontological interactions between immanent powers and their potentia — something known today as *entanglement*—, actual effectuations are only epistemic expressions of the potentia of powers. Actualities are only partial expressions of powers. Just like when observing a dog, a table or a chair we only see a partial perspective of the object —the perceptual adumbration of an object in the phenomenological sense— but never the object itself, measurement outcomes expose only a partial account of the potentia of powers.

Relational definition of powers and their potentia. Against the (classical) substantialist atomist representation through which most present interpretations —implicitly or explicitly— attempt to understand QM, our proposal attempts to consider an ontological relational scheme which understands that QM talks, rather than about "elementary particles" (independent substances), about relational existents, namely, immanent intensive powers (see [34]).

8. Final Remarks: Causal vs Immanent Powers

As we have seen, there are many differences between causal powers and immanent powers. While causal powers are understood —following the empiricist conservative agenda which attempts to "save the phenomena"— as properties that attempt to justify the appearances of observed actualities, immanent powers attempt —following the representational realist more ambitious program— to provide a conceptual representation and intuitive understanding of what is going on beyond measurement outcomes. While causal powers are understood as potential properties of elementary particles which at some point acquire a definite value through their actualization and interaction with the environment, immanent powers are understood as characterized in terms of a specific potentia which allows —through the generalization of reality in intensive terms— to restore a *global intensive valuation.* In turn, it is the possibility of such global valuation which — escaping KS contextuality— allows us to define an objective account of physical reality. While the measurement process still remains a problem within the hylomorphic metaphysics proposed by causal powers, immanent causality implements a novel manner of understanding the process of measurement in QM. It is in this way that immanent powers are able to provide an explanation of actual effectuations without invoking the "collapse" of the quantum wave function or turning possibilities into actualities —as it is the case of Everett original "Megarian interpretation" (see [19]) and some versions of the modal interpretation [36]. Furthermore, while in our approach the reference to the actual realm becomes merely epistemic, it is potential effectuations which become the basis for considering the experience QM is really talking about. In this respect, our intensive approach has clear empirical differences with respect to the causal hylomorphic approaches which make use of causal powers. Immanent powers imply the existence of potential effectuations, a type of experience that we must try to understand beyond the classical actual realm. Actual effectuations are just a way to grasp potential powers when related to actual effectuations. We believe that the differences we have discussed between the orthodox notion of causal power and our proposed immanent power might allow us to better understand and even develop the theory of quanta.

Our theory of intensive immanent powers discusses a new realm of existence which goes beyond visual observability and reconsiders the main problem of contemporary physics, that is, the need to account for a new representation of reality. As remarked by Wolfgang Pauli:

> When the layman says 'reality' he usually thinks that he is speaking about something which is self-evidently known; while to me it appears to be specifically the most important and extremely difficult task of our time to work on the elaboration of a new idea of reality. [54, p. 193]

Acknowledgements

I want to thank Nahuel Sznajderhaus for a careful reading of a previous version of the text. I want to thank Ruth Kastner and Matias Graffigna for discussions on related subjects presented in this manuscript. This work was partially supported by the following grants: FWO project G.0405.08 and FWO-research community W0.030.06. CONICET RES. 4541-12 (2013-2014) and the Project PIO-CONICET-UNAJ (15520150100008CO) "Quantum Superpositions in Quantum Information Processing".

References

1. Aerts, D., 1981, *The one and the many: towards a unification of the quantum a classical description of one and many physical entities*, Doctoral dissertation, Brussels Free University, Brussels.
2. Aerts D., 2010, "A potentiality and Conceptuality Interpretation of Quantum Physics", *Philosophica*, **83**, 15-52.

3. Agamben, G., 1999, *Potencialities*, Stanford, Stanford University Press.
4. Aristotle, 1995, *The Complete Works of Aristotle*, The Revised Oxford Translation, J. Barnes (Ed.), Princeton University Press, New Jersey.
5. Bohr, N., 1935, "Can Quantum Mechanical Description of Physical Reality be Considered Complete?", *Physical Review*, **48**, 696-702.
6. Bohr, N., 1960, *The Unity of Human Knowledge*, In *Philosophical writings of Neils Bohr*, vol. 3., Ox Bow Press, Woodbridge.
7. Bokulich, A., 2014, "Bohr's Correspondence Principle", *The Stanford Encyclopedia of Philosophy (Spring 2014 Edition)*, E. N. Zalta (Ed.), URL: http://plato.stanford.edu/archives/spr2014/entries/bohr-correspondence/.
8. Bokulich, P., and Bokulich, A., 2005, "Niels Bohr's Generalization of Classical Mechanics", *Foundations of Physics*, **35**, 347-371.
9. Borges, J.L., 1989, *Obras completas: Tomo I*, María Kodama y Emecé (Eds.), Barcelona. Translated by James Irby from *Labyrinths*, 1962.
10. Bub, J., 1997, *Interpreting the Quantum World*, Cambridge University Press, Cambridge.
11. Cao, T., "Can We Dissolve Physical Entities into Mathematical Structures?", *Synthese*, **136**, 57-71.
12. Carnap, R., Hahn, H. and Neurath, O., 1929, "The Scientific Conception of the World: The Vienna Circle", *Wissendchaftliche Weltausffassung*.
13. Carnap, R., 1928, *Der Logische Aufbau der Welt*, Felix Meiner Verlag, Leipzig. Traslated by Rolf A. George, 1967, *The Logical Structure of the World: Pseudoproblems in Philosophy*, University of California Press, California.
14. Cartwright, N., 1989, *Nature's Capacities and their Measurement*, Oxford University Press, Oxford.
15. Cartwright, N., 2007, *Causal Powers: What Are They? Why Do We Need Them? What Can and Cannot be Done with Them?*, Contingency and Dissent in Science Series, LSE, London.
16. Cohen, S.M., "Aristotle's Metaphysics", *The Stanford Encyclopedia of Philosophy (Spring 2009 Edition), E. N. Zalta (ed.)*, URL: http://plato.stanford.edu/archives/spr2009/entries/aristotle-metaphysics/.
17. Cordero, N.L., 2014, *Cuando la realidad palpitaba*, Biblos, Buenos Aires.
18. Curd, M. and Cover, J. A., 1998, *Philosophy of Science. The central issues*, Norton and Company (Eds.), Cambridge University Press, Cambridge.
19. da Costa, N. and de Ronde, C., 2013, "The Paraconsistent Logic of Quantum Superpositions", *Foundations of Physics*, **43**, 845-858.
20. da Costa, N. and de Ronde, C., 2016, "Revisiting the Applicability of Metaphysical Identity in Quantum Mechanics", preprint. (quant-ph:1609.05361)
21. de Ronde, C., 2011, *The Contextual and Modal Character of Quantum Mechanics: A Formal and Philosophical Analysis in the Foundations of Physics*, PhD dissertation, Utrecht University, The Netherlands.
22. de Ronde, C., 2013, "Representing Quantum Superpositions: Powers, Potentia and Potential Effectuations", preprint. (quant-ph:1312.7322)
23. de Ronde, C., 2014, "The Problem of Representation and Experience in Quantum Mechanics", in *Probing the Meaning of Quantum Mechanics: Physical, Philosophical and Logical Perspectives*, pp. 91-111, D. Aerts, S. Aerts and C. de Ronde (Eds.), World Scientific, Singapore.

24. de Ronde, C., 2015, "Modality, Potentiality and Contradiction in Quantum Mechanics", *New Directions in Paraconsistent Logic*, pp. 249-265, J.-Y. Beziau, M. Chakraborty and S. Dutta (Eds.), Springer.
25. de Ronde, C., 2016, *"Probabilistic Knowledge* as *Objective Knowledge* in Quantum Mechanics: *Potential Powers* Instead of *Actual Properties"*, in *Probing the Meaning and Structure of Quantum Mechanics: Superpositions, Semantics, Dynamics and Identity*, pp. 141-178, D. Aerts, C. de Ronde, H. Freytes and R. Giuntini (Eds.), World Scientific, Singapore.
26. de Ronde, C., 2016, "Representational Realism, Closed Theories and the Quantum to Classical Limit", in *Quantum Structural Studies*, pp. 105-136, R. E. Kastner, J. Jeknic-Dugic and G. Jaroszkiewicz (Eds.), World Scientific, Singapore.
27. de Ronde, C., 2016, "QBism, FAPP and the Quantum Omelette. (Or, Unscrambling Ontological Problems from Epistemological Solutions in QM)", preprint. (quant-ph:1608.00548)
28. de Ronde, C., 2016, "Unscrambling the Omelette of Quantum Contextuality (PART I): Preexistent Properties or Measurement Outcomes?", preprint. (quant-ph:1606.03967)
29. de Ronde, C., 2017, "Quantum Superpositions and the Representation of Physical Reality Beyond Measurement Outcomes and Mathematical Structures", *Foundations of Science*, https://doi.org/10.1007/s10699-017-9541-z. (quant-ph:1603.06112)
30. de Ronde, C., 2017, "Causality and the Modeling of the Measurement Process in Quantum Theory", *Disputatio*, forthcoming. (quant-ph:1310.4534)
31. de Ronde, C., 2017, "Hilbert space quantum mechanics is contextual. (Reply to R. B. Griffiths)", *Cadernos de Filosofia*, forthcoming. (quant-ph:1502.05396)
32. de Ronde, C. and Bontems, V., 2011, "La notion d'entité en tant qu'obstacle épistémologique. Bachelard, la mécanique quantique et la logique.", *Bulletin des Amis de Gaston Bachelard*, **13**, 12-38.
33. de Ronde, C., Freytes, H. and Domenech, G., 2014, "Interpreting the Modal Kochen-Specker Theorem: Possibility and Many Worlds in Quantum Mechanics", *Studies in History and Philosophy of Modern Physics*, **45**, 11-18.
34. de Ronde, C. and Fernandez Moujan, R., 2017, "Epistemological versus Ontological Relationalism in Quantum Mechanics", in *Probing the Meaning and Structure of Quantum Mechanics*, D. Aerts, M.L. Dalla Chiara, C. de Ronde and D. Krause (Eds.), World Scientific, Singapore, forthcoming.
35. Dieks, D., 1988, "The Formalism of Quantum Theory: An Objective Description of Reality", *Annalen der Physik*, **7**, 174-190.
36. Dieks, D., 2007, "Probability in the modal interpretation of quantum mechanics", *Studies in History and Philosophy of Modern Physics*, **38**, 292-310.
37. Dieks, D., 2010, "Quantum Mechanics, Chance and Modality", *Philosophica*, **82**, 117-137.
38. Dorato, M., 2006, "Properties and Dispositions: Some Metaphysical Remarks on Quantum Ontology", *Proceedings of the AIP*, **844**, 139-157.
39. Dorato, M., 2010, "Physics and metaphysics: interaction or autonomy?", *HumanaMente*, forthcoming.

40. Dorato, M., 2011, "Do Dispositions and Propensities have a role in the Ontology of Quantum Mechanics? Some Critical Remarks", In *Probabilities, Causes, and Propensities in Physics*, 197-218, M. Suárez (Ed.), Synthese Library, Springer, Dordrecht.
41. Dorato, M., 2015, "Events and the Ontology of Quantum Mechanics", *Topoi*, **34**, 369-378.
42. Dorato, M. and Esfeld, M., "GRW as an Ontology of Dispositions", *Studies in History and Philosophy of Modern Physics*, forthcoming.
43. Einstein, A., Podolsky, B. and Rosen, N., 1935, "Can Quantum-Mechanical Description be Considered Complete?", *Physical Review*, **47**, 777-780.
44. Esfeld, M., 2011, "Causal realism", In *Probabilities, laws, and structures*, D. Dieks, W. Gonzalez, S. Hartmann, M. Stöltzner and M. Weber (Eds.), Springer, Dordrecht.
45. Ghirardi, G. C. Rimini A. and Weber, T., 1986, "Unified Dynamics for Microscopic and Macroscopic Systems", *Physical Review D*, **34**, 470-491.
46. Heisenberg, W., 1958, *Physics and Philosophy*, George Allen and Unwin Ltd., London.
47. Heisenberg, W., 1971, *Physics and Beyond*, Harper & Row.
48. Heisenberg, W., 1973, "Development of Concepts in the History of Quantum Theory", In *The Physicist's Conception of Nature*, 264-275, J. Mehra (Ed.), Reidel, Dordrecht.
49. Kastner, R., 2012, *The Transactional Interpretation of Quantum Mechanics: The Reality of Possibility*, Cambridge University Press, Cambridge.
50. Karakostas, V., 2007, "Nonseparability, Potentiality and the Context-Dependence of Quantum Objects", *Journal for General Philosophy of Science*, **38**, 279-297.
51. Karakostas, V. and Hadzidaki, P., 2005, "Realism vs. Constructivism in Contemporary Physics: The Impact of the Debate on the Understanding of Quantum Theory and its Instructional Process", *Science & Education*, **14**, 607-629.
52. Kochen, S. and Specker, E., 1967, "On the problem of Hidden Variables in Quantum Mechanics", *Journal of Mathematics and Mechanics*, **17**, 59-87.
53. Laplace, P.S., 1902, *A Philosophical Essay*, New York.
54. Laurikainen, K. V., 1998, *The Message of the Atoms, Essays on Wolfgang Pauli and the Unspeakable*, Springer Verlag, Berlin.
55. Margenau, H., 1954, "Advantages and disadvantages of various interpretations of the quantum theory", *Physics Today*, **7**, 6-13.
56. Maxwell, N., 1988, "Quantum propensiton theory: A testable resolution to the wave/particle dilemma", *British Journal for the Philosophy of Science*, **39**, 1-50.
57. Melamed, Y., 2013, *Spinoza's Metaphysics and Thought*, Oxford University Press, Oxford.
58. Nagel E., 1961, *The structure of science: problems in the logic of scientific explanation*, Hackett publishing company, New York.
59. Pauli, W. and Jung, C. G., 2001, *Atom and Archetype, The Pauli/Jung Letters 1932-1958*, Princeton University Press, New Jersey.
60. Piron, C., 1976, *Foundations of Quantum Physics*, W.A. Benjamin Inc., Massachusetts.
61. Piron, C., 1981, "Ideal Measurements and Probability in Qunatum Mechanics", *Erekenntnis*, **16**, 397-401.

62. Popper, K.R., 1935, *Logik der Forschung*, Verlag von Julius Springer, Vienna, Austria. Translated as *The Logic of Scientific Discovery*, Hutchinson and Co., 1959.
63. Popper, K.R., 1959, "The propensity interpretation of probability", *British Journal for the Philosophy of Science*, **10**, 25-42.
64. Popper, K.R., 1982, *Quantum Theory and the Schism in Physics*, Rowman and Littlefield, New Jersey.
65. Rédei, M., 2012, "Some Historical and Philosophical Aspects of Quantum Probability Theory and its Interpretation", in *Probabilities, Laws, and Structures*, pp. 497-506, D. Dieks et al. (Eds.), Springer.
66. Sambursky, S., 1988, *The Physical World of the Greeks*, Princeton University Press, Princeton.
67. Schrödinger, E., 1935, "The Present Situation in Quantum Mechanics", *Naturwiss*, **23**, 807. Translated to english in *Quantum Theory and Measurement*, J. A. Wheeler and W. H. Zurek (Eds.), 1983, Princeton University Press, Princeton.
68. Simondon, G., 2005, *L'Individuation à la lumière des notions de forme et d'information*, Jérôme Millon, Paris.
69. Smets, S., 2005, "The Modes of Physical Properties in the Logical Foundations of Physics", *Logic and Logical Philosophy*, **14**, 37-53.
70. Stapp, H.P., 2009, "Quantum Collapse and the Emergence of Actuality from Potentiality", *Process Studies*, **38**, 319-39.
71. Suárez, M., 2003, "Scientific representation: Against similarity and isomorphism", *International Studies in the Philosophy of Science*, **17**, 225-244.
72. Suárez, M., 2004, "On Quantum Propensities: Two Arguments Revisited", *Erkenntnis*, **61**, 1-16.
73. Suárez, M., 2004, "Quantum Selections, Propensities, and the Problem of Measurement", *British Journal for the Philosophy of Science*, **55**, 219-255.
74. Suárez, M., 2007, "Quantum propensities", *Studies in History and Philosophy of Modern Physics*, **38**, 418-438.
75. Sudbery, A., 2016, "Time, Chance and Quantum Theory", in *Probing the Meaning and Structure of Quantum Mechanics: Superpositions, Semantics, Dynamics and Identity*, pp. 324-339, D. Aerts, C. de Ronde, H. Freytes and R. Giuntini (Eds.), World Scientific, Singapore.
76. Van Fraassen, B.C., 1980, *The Scientific Image*, Clarendon, Oxford.
77. Verelst, K. and Coecke, B., 1999, "Early Greek Thought and perspectives for the Interpretation of Quantum Mechanics: Preliminaries to an Ontological Approach", in *The Blue Book of Einstein Meets Magritte*, 163-196, D. Aerts (Ed.), Kluwer Academic Publishers, Dordrecht.
78. Wallace, D., 2007, "Quantum Probability from Subjective Likelihood: improving on Deutsch's proof of the probability rule", *Studies in History and Philosophie of Modern Physics*, **38**, 311-332. (quant-ph:0312157)
79. Wheeler, J. and Zurek, W. (Eds.) 1983, *Theory and Measurement*, Princeton University Press, Princeton.

OUTLINES FOR A PHENOMENOLOGICAL FOUNDATION FOR DE RONDE'S THEORY OF POWERS AND POTENTIA

MATÍAS GRAFFIGNA

Philosophisches Seminar, Georg-August-Universität Göttingen - Germany
Graduiertenschule für Geisteswissenschaften Göttingen
E-mail: m.graffignacostas@stud.uni-goettingen.de

Starting with the claim that Quantum Mechanics (QM) is in need of a new interpretation that would allow us to understand the phenomena of this realm, I wish to analyse in this paper de Ronde's theory of power and potentia from a phenomenological perspective. De Ronde's claim is that the reason for the lack of success in the foundations of QM is due to the reluctance of both physicists and philosophers to explore the possibility of finding a new ontology, new concepts for the physical theory. De Ronde proposes such new ontology and the question I wish to address here, is whether his ontology is conceptually plausible. I will, for this purpose, recur to Edmund Husserl's phenomenology. After presenting some of the basic concepts and methodological tools of this theory, I shall apply them to de Ronde's ontology to determine the viability of his theory.

Keywords: Quantum mechanics; phenomenology; immanent powers; potentia.

Introduction

In this article, I wish to analyze from a philosophical perspective —the phenomenological in particular— one possible approach to the foundations of Quantum Mechanics (henceforth QM). Taking as a starting point the premise that QM is in need of an interpretation that would allow us not only to carry out successful experiments, but also that will grant us the possibility of understanding comprehensively the phenomena the theory describes and explains,[a] it is that I wish to examine one particular approach to this problem: de Ronde's theory of *powers* and *potentia*. De Ronde's claim, to anticipate, is essentially that most of the theoretical problems QM faces nowadays are due to misunderstandings at the basic conceptual

---

[a]For this discussion, see [3, pp. 54-55], [24, p. 230], [17, p. 2], [6, p. 9]; and my own [18, pp. 3-5]

level, and the reluctance of physicists, since the early days of the theory, to adopt new metaphysical/ontological principles that suit the new theory, rather than keep insisting in trying to adapt QM to the 'classical picture' of the world that stems from Aristotelian metaphysics and Newtonian classical mechanics.

In order to do this, I will divide this work in four parts: first, I will discuss the situation concerning the relation between ontology/metaphysics and physics, to elucidate the notion of an alternative ontology, and will offer some examples of proposals that go in this line; I will then introduce de Ronde's alternative ontology for QM, mainly his concepts of *power* and *potentia* and how they stand in reference to the classical metaphysical concepts; in the third section I shall offer a very schematic and introductory presentation of some basic phenomenological concepts; and, finally, I shall try to apply these phenomenological concepts to de Ronde's ontology. The thesis I will try to defend is humble in its reach: it is my claim that de Ronde's ontology is *conceptually plausible* from a phenomenological perspective. Whether or not it is the right ontology for QM is something that exceeds my present purpose. Yet, I believe the claim is not trivial, inasmuch de Ronde's proposal for an alternative ontology is highly controversial, due to the fact that it contradicts several basic intuitions regarding our understanding of physical reality.

1. What is an alternative ontology?

De Ronde's theory is presented as an 'alternative ontology'. We could begin by asking ourselves, alternative to what? The first answer to this question would be one of a very general nature: to classical ontology. Classical ontology, or even classical metaphysics, is not a theory presented and defended by one author, but rather a series of principles, beliefs and suppositions that have settled through time in western thought, philosophy and science. There are multiple ways to understand this history and multiple ways to reconstruct what a classical ontology is. Since the discussion we are facing is one concerning the philosophy of physics, of QM in particular, it is not of the utmost importance to unequivocally determine what these principles are and where do they stem from. What is important, is to understand and determine what it is that physics has taken to be classical ontology.

If we then take classical ontology to be the set of basic metaphysical principles that determine a basic worldview, upon which classical mechanics will be based in order to be developed as the standard theory that we know today, it would seem we need to go back to Aristotle. But let us begin from the 'end': classical Newtonian mechanics is based upon a series of metaphysical presuppositions that are well known by both philosophers and physicists: time and space are absolute; that which is *real*, which the theory describes, explains and predicts are the properties and states of *bodies* that inhabit this space-time, and whose objective properties can be given at any time (mass, acceleration, speed, position, etc.). In other words: given a body, a real being, in classical Euclidean space, and given some initial conditions, Newtonian mechanics is able to calculate and predict the objective properties of said body. This is nowadays so obvious, that even a school child can grasp it. Now, in what sense do we claim that these basic principles of physics are based upon a classical ontology?

"One of the first such metaphysical systems, which still today plays a major role in our understanding of the world around us is that proposed by Aristotle through his logical and ontological principles: the Principle of Existence (PE), the Principle of Non-Contradiction (PNC) and the Principle of Identity (PI)" [8, p. 2]. These principles, as formulated by Aristotle, constitute the building blocks of classical metaphysical thinking, and hence, the basis upon which classical mechanics is grounded. What they do, is basically determine the notion of entity, which we can understand as: a real existent in space-time, with definite non-contradictory properties that can be known, that has an identity with itself, in the sense that it is the same entity throughout time. But these principles do not only determine an entity in a positive sense, they also exclude from the ontology everything else: something is either an entity, a property of an entity, or it doesn't belong to the realm of the physical, i.e., real existence, susceptible of physical scientific study. It would, of course, be an overstatement to say that all western metaphysical thinking can be reduced to these Aristotelian principles; that is not the point. The claim is that these are the ontological foundations upon which Newton developed his physical theory and since then, they have become the basic ontological presuppositions for physicists: "It was Isaac Newton who was able to translate into a closed mathematical formalism both, the ontological presuppositions present in Aristotelian (Eleatic) logic and the materialistic ideal of *res extensa*, with actuality as its mode of existence." [*Op. cit.*, p. 4]

It was Kant who would later offer what he understood as the transcendental justification of the possibility of scientific knowledge [23]. Put very simply, an object of possible experience is that from which we receive sensible impressions under the pure forms of sensibility (space-time) and that falls under the transcendental categories of understanding. Everything that does not meet these two conditions is not an object of possible experience and, hence, is not an object of science, but of speculative metaphysics. We can see how with his philosophy, what we are presenting as classical ontology obtains a transcendental foundation.

Some years later, Logical Positivism, as the heir of British empiricism and German logicism, would determine the basic principles upon which our contemporary notion of science is based: what is real and scientifically knowable are entities, objects of possible experience, and these are *observable*, inasmuch as there is a definite process by which they can be empirically determined. This is what is known as the *operationalist* definition of concepts, or methodology: no concept can claim scientific validity, unless there is a definite process that allows to empirically measure the concept in a series of finite steps. "The notion of operational definition thus lies at the heart of contemporary physics" [5, p. 26] and thus limits the realm of the physically real to a materialist-empiricist ontology. Everything else, once again, is outside the realm of the real, of the scientific and, thus, is nothing more than metaphysical speculation, only this time, "metaphysical" is synonym of an illegitimate form of knowledge.

Now, beyond the criticisms and discussions that could arise from any of these theories in particular, the problem that we face at present is the one brought about by the appearance of QM in the early 20th century. It is the basic principles of this theory, in its mathematical formalism and its unprecedented empirical success, that challenge our very understanding of what it means for something to be real, in particular, to be physically real.

Heisenberg's principle of indetermination, superposition states, entanglement, contextuality and, in general, all the phenomena belonging to the domain of QM, seem to contradict this classical picture of the world which was so obvious and unquestionable until not so long ago. The peculiar situation with QM is that the theory was developed first in its mathematical formalism, it was successfully taken to the labs, but until today lacks a conceptual framework: One that would allows us to understand the domain described by the theory beyond its abstract mathematical formalism and more comprehensively than in its isolated experimental results —which, as it's always the case with experimental data, need a conceptual interpretation and understanding. In the words of Griffiths:

> Scientific advances can significantly change our view of what the world is like, and one of the tasks of the philosophy of science is to take successful theories and tease out of them their broader implications for the nature of reality. Quantum mechanics, one of

the most significant advances of twentieth century physics, is an obvious candidate for this task, but up till now efforts to understand its broader implications have been less successful than might have been hoped. The interpretation of quantum theory found in textbooks, which comes as close as anything to defining "standard" quantum mechanics, is widely regarded as quite unsatisfactory. Among philosophers of science this opinion is almost universal, and among practicing physicists it is widespread. It is but a slight exaggeration to say that the only physicists who are content with quantum theory as found in current textbooks are those who have never given the matter much thought, or at least have never had to teach the introductory course to questioning students who have not yet learned to "shut up and calculate!" [17, p. 2]

The lack of a commonly accepted interpretation of the Quantum Theory is also remarked by Dorato, who also strives to find an alternative understanding of the ontological principles for QM:

One of the most frequent points of misunderstanding between physicists and philosophers of physics or metaphysicians is not only caused by differences in language but also by the fact that philosophers worry much more than physicists about ontological issues, namely interpretive questions involving what (typically a poorly understood) physical theory tells us about the world. In the case of quantum mechanics, however, interpretive questions calling for ontological analyses ("how could the world be like if quantum theory is true?") become murky since, at least according to philosophers, it is still controversial how quantum theory should be formulated, given that in the market there are various proposals. [13, p. 370]

Now, where does the cause for this failure lie? How is it that after one century physics has been unable to find meaning for these discoveries? As always, there are multiple possible answers to this question. One of them is, in the line of instrumentalists such as Fuchs & Peres (Cf. [16]), to renounce to the question altogether and understand QM as a mere tool to produce calculations and predictions. I believe the best answer to this question, following de Ronde's line of research (see specially [6,8–10]), is to say that all attempts to offer an interpretation of QM have the same common ground: they all try to do so limiting themselves to a classical ontology. Ever since Bohr claimed that "the unambiguous interpretation of any measurement must be essentially framed in terms of classical physical theories, and we may say that in this sense the language of Newton and Maxwell will remain the language of physicists for all time" [27, p. 7], it has become a dogma for physicists and philosophers to constraint any attempt to bring meaning to QM to the very closed limits of classical metaphysical principles. This is so up to the point that physicists have decided to modify the mathematical formalism of the theory, like in Bohm's case, rather than trying to search for a different ontology. Moreover, the triumph of the scientific program of logical positivism has settled that ontology as the only possible one and precluded the possibility of having metaphysical discussions, naively believing that materialistic experimental science is free of metaphysics, just because they chose a specific set of ontological principles instead of some other. So, the situation is as tricky as it can be historically appreciated: there's only the classical ontology to serve as basis for physical theories (Bohrian lines of inquiry); discussing ontology/metaphysics is an unscientific enterprise and should be avoided (logical positivism). The result: a century of failed attempts to comprehend QM.

It seems, then, that both these constraints need to be overcome in order to have a chance to move forwards. And hence, the idea of an alternative ontology for QM. Now, of course, there is no guarantee that this path shall bring us success at the end, but there seems to be already enough guarantee that the other one has failed. Developing an alternative ontology has problems of its own, some could even be considered more challenging than the ones faced up to now, and this shall be the focus of the rest of this article.

I wish to conclude this section by mentioning three examples of proposals that go in this direction. Unfortunately, I will not be discussing them here, but they shall serve nonetheless to illustrate the general project of developing an alternative understanding for the ontological foundations of QM. In the first place, the dispositionalist account of powers in QM is worth mentioning. French offers such an account and acknowledges as well the need for a metaphysical discussion regarding physical theories. Yet, there is a crucial difference between his approach and my understanding of de Ronde's, as we shall see, which lies in the role itself that metaphysics are called to play: "I have tried to articulate what we have called the 'toolbox' approach to metaphysics, according to which metaphysics can be viewed as providing a set of tools that philosophers in other sub-disciplines, and particularly, the philosophy of science, can use for their own purposes" [15, p. 1]. The concept of *disposition* is one of such 'tools' French sets out to implement. Such a conception is, in my opinion, not free of some serious risks. We could very well call this position an instrumentalist approach to ontology, inasmuch as ontology is not the basis upon which scientific concepts are developed, but the source of *ad hoc* utilizations of different metaphysical concepts.

In a closer line to our own, Esfeld actually proposes a different ontological grounding for QM, with a realist aim:

> Ontic structural realism is a current in contemporary metaphysics of science that maintains that in the domain of fundamental physics, there are structures in the first place rather than objects with an intrinsic identity. Its main motivation is to develop a tenable version of scientific realism in form of an ontology that meets the challenges of modern physics, giving an account of entanglement in quantum physics and of space-time in the theory of general relativity. The claim is that there are structures of entanglement instead of objects with an intrinsic identity in the domain of quantum physics (Ladyman 1998, French & Ladyman 2003, Esfeld 2004) and metrical structures, which include the gravitational energy, instead of space-time points with an intrinsic identity in the domain of the theory of general relativity. [14, p. 1]

The last attempt I wish to mention is Dorato's proposal for an event ontology, that would replace the ontology of objects used in order to interpret the quantum formalism. This ontology also seeks to be a foundation for a realist physical science, the conclusions of which are of high interest. After presenting the different alternatives for an event ontology and their connection to QM, Dorato offers, among others, the following conclusion: "The point I want to urge is that it is possible to claim that the individual particle has no definite spin in a given direction, but that it manifests an intrinsic disposition or has a concrete power to display spin up or spin down by interacting with the Stern-Gerlach apparatus (the stimulus of the disposition)" [13, p. 376]. These two quotes exemplify what it means to search for an alternative ontology: a fundamentally different metaphysical basis for physics should have as a result a fundamentally different understanding of the quantum phenomena. It concerns the question of what it is that we describe and explain through QM, and has the explicit aim of achieving a deeper understanding that could have actual consequences for the way in which physics is done today.

2. De Ronde's theory of immanent powers and potentia

In this section I will present de Ronde's alternative ontology for the interpretation of QM, which is, as must be warned, a work in progress. The purpose of the present work is not to determine whether or not this ontology is, as a matter of fact, the "correct" ontology for QM, but rather to analyze its plausibility from a purely philosophical perspective. In this sense, it might well be the case that even if this proposal turned out to be philosophically solid and acceptable, it could still not be the appropriate interpretation for QM. This last item, of no minor importance, will not be considered here. Yet, I consider the philosophical plausibility of the theory to be logically prior to its adequacy, and hence the relevance of the present work.

De Ronde presents the basic tenants of his theory in direct contraposition to Aristotle's basic metaphysical principles that serve as the underlying base for classical mechanics:

> The first important point according to our stance is to recall the fact that Aristotle grounded the notion of entity in the logical and ontological principles of existence, non-contradiction and identity. Our proposal is that in fact there exist analogous principles in QM which can allow us to develop new concepts. The principles of indetermination, superposition and difference could be considered as providing the logical and ontological foundation of that of which QM is talking about. [7, p. 11].

As Aristotle's metaphysical principles determined the concept of entity, which is central both to his philosophy and to the following history of western thought, de Ronde proposes three new different principles that basically stem from the quantum theory itself. We shall focus on each one of them later, the point to be made now is that from these principles, that are 'no more' than the ontological understanding of what is already present in QM, a new concept can be developed: "We claim that just like the logical and ontological principles of existence, non-contradiction and identity provide the constraints for a proper understanding of the concept of entity; the principles of indetermination, superposition and difference are able to determine the notion of power" (ibid). The concept of immanent[b] power, then, shall be the central core of this new ontology, just as Aristotle's

---

[b]The "immanence" is related to the understanding of the measurement process in terms of the *immanent cause*. This is explained in detail in: [10].

concept of entity was for his theory. Within the characterization of this concept we shall find the essence of this new ontology.

The first element of a Derondian power is its *indetermination*. This principle obviously stems from Heisenberg's relations, central core of QM, and seeks to understand that which is referred to by this postulate in its being indeterminate, rather than making of that indetermination an uncertainty, i.e., an epistemological/subjective problem. Indetermination is then an ontological characterization of a specific type of being. Powers are not entities: this means that they are not actual, they are not determined and they are not the substratum of classical properties. "The mode of being of a power is potentiality" [7, p. 12], it is probably in this statement where the true ontological innovation lies. A power is a potential being, as opposed to actual beings, and yet this potentiality is not to be understood in reference to actuality. In Aristotelian metaphysics potentiality, though a mode of being in its own right, cannot be without the actual. This is ultimately expressed in Aristotle's postulation of the first unmoved motor, which in its being perfect is pure act, no potentiality. So, a power must be understood as a being in its own right, which is in the mode of the potential and which is, in that potentiality, ontologically independent (from actuality). This means that both the realm of the potential, as its inhabitants, the powers, are ontologically separate and independent from the actual realm and actual entities. In the same way that entities inhabit actual space-time and in their being related to other entities and properties are part of what is called 'state of affairs', powers in their being related to other powers (and entities as well, since ontological independence does not mean absence of relations) belong to what de Ronde calls "potential states of affairs".[c]

The second principle that determines a power is that of *superposition*, another known tenant of QM. Superposed states are a normal currency in the quantum domain and yet, as famously shown by Schrödinger and his cat, the notion of having two mutually contradictory states superposed with one another is, from the perspective of a classical ontology, simply senseless. Superpositions seem to violate the principle of non-contradiction and I say 'seem', because the only way to violate the principle is if we accept it as operating. De Ronde's claim is that this principle is not valid in the quantum domain, since powers as being existents in the mode of the potential do not 'collapse' when 'supporting' mutually contradictory properties.

---

[c]In order to make the distinction more explicit de Ronde refers to what is usually known as simply 'state of affairs' as 'actual state of affairs'.

The analog for a classical binary truth valuation to the elements $\{0, 1\}$ is in this new ontology a *potentia*, that is, a measure for the intensity of a power: powers have definite potentia and they can be, from the perspective of actuality, contradictory, but from the perspective of the quantum we say they are simply superposed.

The third and last fundamental principle is that of *difference* as opposed to the Aristotelian identity. A power cannot be ascribed identity through time, inasmuch as it is not a temporal being and hence its unity, both conceptual and ontological, cannot rely in a spatio-temporal identification nor in a property-bundle theory. In this sense, we should think of a power as a being that does possess a unity that can be scientifically described, but not a strict identity as an actual entity would. The unity of the power is then found in difference, in that they change from a temporal perspective or are superposed. It should be noted, however, that the issue of identity is an extended metaphysical problem that applies as well to actual entities. Spatio-temporal beings also change through time, and the notion of identity through time requires some kind of metaphysical commitment: essence, material constitution, etc.

From this it follows, as anticipated, that powers, being in the mode of the potential, do not belong to the realm of the spatio-temporal: "A power cannot be thought as existing in space-time. It is only in the process, through which the power is exposed, that space-time enters the scene. The process builds a bridge to bring the power from its potential existence into its space-time actual effectuation" [7, p. 1]. Powers are not space-time beings, but we are, and so are the experiments we perform. For this reason, even if we speak about beings outside space-time, we do it from inside. We might occasionally evade space (in abstract mathematical reasoning, for example), but time is for us constitutive of our conscious existence. In this sense, we should not take our human limitation as an objective property of all being. Powers are non spatio-temporal, even if our means of experimentally accessing them are, and even if in our getting to know these powers empirically we 'change' them and submit them to our type of actual being. When we do so, we must understand that we access only a very partial aspect of these beings that are of a very different nature. 'Just as' we can only see an entity from one perspective at a time, never being able to stand from God's vantage point, our access to powers is an access to just one perspective of them, in this case though, probably much more limited than when we only see a face of a body.

As a way of summing up, I quote de Ronde's own self-valuation: '

> We would like to remark the fact that our notion of immanent power is maybe the first physical notion to be characterized ontologically in terms of an objective probability measure. This concept escapes the ruling of actuality since it is founded on a different set of metaphysical principles to that of classical entities. Indeed, powers are indetermined, paraconsistent and contextual existents. Powers can be superposed and entangled with different —even contradictory— powers. A power, contrary to a property which can be only true or false possesses an intrinsic probabilistic measure, namely, its potentia. A potentia is intrinsically statistical, but this statistical aspect has nothing to do with ignorance. It is instead an objective feature of quantum physical reality itself. [8, p. 34]

Now, there are several evidently controversial aspects about this proposal and hence the present work, in the hope of elucidating them. Here, I shall only enumerate these aspects, the analysis of which will be left for the final section:

I. **Non-spatio-temporal physical beings:** De Ronde's power is obviously not the first being outside of space-time that has been proposed. Plato's ideas could be considered as such. In the context of QM, also Aerts has recently proposed to consider non-spatial existents [1]. Even if it turns into a matter of beliefs and acceptance, we cannot simply discredit this possibility by assuming a materialist ontology. There is nothing particularly controversial about beings outside space-time, even if one would rather not accept their existence. What is highly controversial is considering beings outside space-time as physical existents that can be studied and explained by an empirical science such as physics, which is exactly what Derondian powers strive to be: the ontological concept of what occurs in the quantum domain, i.e., a domain of natural being studied by the physical science, formulated as QM.

II. **Contradictory beings:** even if we speak of powers instead of entities and of potentia instead of property, the possibility of something we understand as a comprehensible unitary being, being contradictory requires more explanation. De Ronde's thesis is undoubtedly controversial and we require more than new names in order to accept the idea that we can think, explain and describe beings that are in their nature indeterminate and contradictory.[d]

III. Unity without identity: if we are to accept the claim that powers *are* indeed existent beings to be known and yet they do not possess self-identity, we need a clear substitute, for the concept to be meaningful. If the claim is that these beings possess some sort of unity that allows us to speak about *powers* as beings, and yet there is no possible identity to be ascribed to them, we need something that will stand in its place and offer an ontological unity for these powers. Even the concept of event which is clearly not an entity, has clear conditions of identity. This point violates Quine's famous dogma of "no entity without identity", and if we simply answer that we are not discussing entities then we are back to point **I.** of this list.

A first reaction to this controversial theory is to deny it as implausible. Doing so is not hard, there are plenty of philosophers that could help in that task. The one that would do it better is probably Kant, as we can appreciate in Pringe's very thorough Kantian reading of Bohr's interpretation of QM:

> If the quantum postulate is assumed, all pretension of reaching a spatial-temporal representation, which is at the same time causal, of an object subject to the postulate, must be abandoned. That is, if an object is within the domain of validity of the postulate, it won't be possible —as it is in classical physics— to synthesize the set of contingent data of a measurement, according to the concept of cause, as the effect of said object, representing this in space and time, in such a way that its states modify each other causally. [25, p. 183]

As it was mentioned in the previous section, Kant's result in the *Critique of Pure Reason* [23] is that an object of possible experience is that which is given in impressions under the pure forms of space and time, and under the *a priori* concepts, called categories, that synthesize those impressions into coherent objects. Anything outside these conditions cannot be legitimately called an object and therefore, the idea of it being studied by an empirical science is completely out of the question. Kant's result applied to QM, as Pringe's reading of Bohr proposes, is that we need to understand

---

[d]Da Costa and de Ronde have proposed to consider quantum superpositions in terms of a paraconsitent logic in [4]. This has lead to an interesting debate with Krause and Arenhart. See e.g.: [2,11].

the quantum domain under symbolic analogies and regulative principles grounded in our experience of the actual realm, that is the only realm we can really experience, the scientific correlate of which is classical mechanics. Hence, this alternative ontology, as any other which goes against Kant's conclusions, is mere nonsense. My only answer to this position is a very simple question: why settle with Kant?

3. Brief introduction to phenomenology

Indeed, why should we settle with an 18th century philosophy, when other theories are available to us, which could help us to better understand the peculiar situation brought about by QM. In this section, I face a difficult challenge, that of schematically presenting some of the basic tenants of one of the most important theories of the 20th century —one of the richest in production and bibliography— to readers most likely not familiar with any of it. I shall try to steer clear of technicalities, discussions and disagreements, in the hope of being able to present some basic notions of the theory, that I can later apply to de Ronde's alternative ontology. I shall do this, then, in a way that would most probably upset most phenomenologists, but that I hope could serve as an introduction to those unfamiliar with the theory, who could, if so they wished, continue reading on their own.

Phenomenology was founded by Edmund Husserl in the turn of the 19th century. It has many stages, already within Husserl's work, and many more if we consider his disciples, direct and indirect, and all the later developments through the 20th century up until today. Husserl was a mathematician, whose original concern was that of explaining how it is that we come to have numerical concepts and mathematical thinking in general. After his first works on the mathematical-psychological issues, the *Logical Investigations*, the most important work of the so called 'early phenomenology', are mainly concerned with discrediting psychologism, the position that sustains that the laws of logic and mathematics are dependent on the psychological laws of human thinking. Around 1907 Husserl is said to have taken a 'transcendental turn' in his philosophy, moving away from his early realism-empiricism to a transcendental philosophy. *Ideas pertaining to a pure Phenomenology and to a phenomenological Philosophy* [21] or simply *Ideas I* as it is usually referred to, is the work in which we find Husserl's transcendental static phenomenology explained. This we could call, not without some controversy, 'classic' or 'standard' phenomenology. It is mainly in this theory, with the backing of some later texts —mainly the *Cartesian Meditations* [20]— that I will be basing myself for this succinct presentation.

To begin with, let us characterize phenomenology in the most general terms. We can say phenomenology is a method that consists of describing experience such as it is experienced by the living subject. The first step in this method is to abandon the 'natural attitude' in which we live, and enter a critical attitude that will allow us to perform these descriptions from an objective standpoint. The phenomenological method consists in performing what Husserl called 'epoché', taken from the Greek word that means 'suspension of judgement'. Phenomenological epoché consists, first, in abandoning all our presuppositions, be them naive or coming from other theories or from science; second in suspending the 'effectivity thesis of the existence of the world', which means that, when we enter a reflexive attitude in order to describe our experience, we must ignore the actual existence of the world and its objects. In doing so, we access our own experience as such, without concerning ourselves with actuality, truth or correspondence. The basic, most important property of consciousness we immediately discover is that consciousness is always consciousness of something. This property is called *intentionality*. Consciousness is directed towards an object that is given to it as a correlate of experience.[e] The results of this method claim *a priori* transcendental validity: "The epoché can also be said to be the radical and universal method by which I apprehend myself purely: as Ego, and with my own pure conscious life, in and by which the entire objective world exists for me and is precisely as it is for me" [19, p. 21]. Transcendental phenomenology claims that the world and its objects (or phenomena) are constituted by the subject, but this constitution is not, so to say, arbitrary or relative, but objectively guided by the worldly objects themselves. The meaning of this is, that we cannot experience the world by any other way than that in which we constitute it,[f] but the fact that we do constitute does not mean that there is no world, or that the world is relative to each constituting individual, since, precisely, in the phenomenological epoché we are describing transcendental properties of consciousness and not psychological properties of empirical individuals.

Next, I would like to present three phenomenological themes which are of great importance for the theory, and which I will be applying to de Ronde's theory of powers.

The first of these is the so called *noetic-noematic* structure of consciousness, also known as the a priori correlation. What this means is not so obscure as it might seem: by 'noesis' we understand what pertains to the

---

[e] "In general, it belongs to the essence of each actual cogito to be consciousness of something" [21, p. 73]

[f] "All real unities are unities of sense" [21, p. 120]

mental acts (intentional acts) that consciousness performs. By 'noema' we understand the objective correlate of said acts, such as they are constituted by the intentional acts. The thesis of the correlations runs as follows: "Thus the eidetic law, confirmed in every case, states that there can be no noetic moment without a noematic moment specifically belonging to it" [22, p. 226]. The basic idea, as anticipated, is that constitution involves attributing sense to the objects we experience, and this phenomenon, rather than make our experience relative, it makes it possible:

> Like perception, every intentive mental process —precisely this makes up the fundamental part of intentionality— has its 'intentional object,' i.e., its objective sense. Or, in other words: to have sense or "to intend to" something is the fundamental characteristic of all consciousness which, therefore, is not just any mental living whatever, but is rather a <mental living> having sense, which is 'noetic'. [22, p. 217]

Noetically speaking, an act has different possible modes: one can perceive an object, remember it, desire it, fantasize it, imagine it, etc. For any act that consciousness performs, there *must* be a noematic correlate. The noema is an object such as it is constituted by the I, with the meaning invested upon the sensible impressions that we passively receive. We can never perceive all faces of a given object at the same time, nor can we perceive an object forever, without interruption. The perceived sides are said to be given, and the non-given sides are "apperceived" and intuited through acts of presentification. Both the given and not given "adumbrations" (*Abschattungen*) of the object are synthesized into unity. Perception is the paradigmatic act of consciousness, which gives us the object 'in the flesh', such as it is. Upon perception are founded all other acts, ideally speaking, which Husserl calls 'presentifications'. These are intuitive acts that give us the object not in the flesh, but in some other form (memory, imagination, etc.).

The second element I wish to present is merely an aspect of the general thesis of the correlation and it is that of the correlation between belief-characteristics and being-characteristics. In any intuitive act, noetically speaking, our act has a belief-characteristic that depends on the conditions under which we are intuiting, how well in particular we can intuit the given object, on attention and other noetic aspects. Noematically, to each corresponds a being-characteristic, as that with which the object is given to us. We can see this correlation in figure 1.

Belief (noetic)	Being (noematic)
Certainty	Truth – the object 'pure and simple'
Suspicion	Probability
Conjecture	Possibility
Question	Questionability
Doubt	Dubious

Fig. 1.

Certainty and *truth* are, respectively, the mother-forms of all character-istics, which also serve as ideals in guiding the different acts. If we take as example any simple case of perception, we can see how this works: in the middle of the night, on a dim-lighted street, we see what appears to be 'something' moving in the distance. We doubt whether or not there's something really there, and the 'something there' is for us of a dubious nature. Some moments later, a figure forms and we can no longer doubt whether there is or isn't something really there. We conjecture what it is. It is the possibility of a person approaching, as it would seem due to the way it moves. As it gets even closer and passes under a lamppost we reach the strong suspicion that this is *most probably* a human being, who finally stands before us, asks us for a cigarette, and we reach the certainty that, truly, there is a human being in the flesh standing right there.[g]

The last point I wish to present is that of empathy. Empathy is a specific form of constitution that differs from the constitution of other objects of our experience. Empathy is the intentional act through which we constitute 'others as myself', other human beings, conscious subjects. The main differ-ence is that, in this case, we cannot access that which defines the thing as what the thing is: we cannot perceive the stream of experiences of others, it is never given to us in the authentic intuition of perception. Therefore, we must do something else in order to know that these special objects that stand before us in our life are not mere bodies, like other physical objects, but are in fact consciousnesses that have experience of the world just as we do. In order to reach this knowledge, we ground our acts in the perception of the body of the other. The body is given to us, just as any other body is. When we perceive any physical body, say a tree, we see one side of it. The other sides, not actually being perceived, are said to be *appresented*,

---

[g]Cf. [22, p.249-252].

which means that they are not being perceived at the moment, and could be either presentified (i.e. imagined) or empty, but could be eventually perceived, if one just moves around the tree and sees it from the opposite side. In the case of empathy, though, the stream of consciousness of the other is appresented, but can never be perceived. That appresentation can only be fulfilled in a presentification. So, in order to reach this level we need something other than the mere perception of a body. We see, for example, how that peculiar body is not just there, like it happens with tables and stones, but it also moves. Not only it moves, but it seems to move in a very specific way, i.e. like I move, with purpose. We see that body interacting within nature, with other bodies in a way we can consider analogic to our own way of moving and interacting (what Husserl calls 'parification'). We also see that this body speaks, and in doing so, expresses beliefs, desires, goals, fears, etc. The other is constituted as an alter-ego, when we appresent her conscious life based on the perception of the body, and thus understand that there is someone who has experience of the world, in the same sense as I do. In superior forms of empathy, which require the intervention of active reasoning, we can presentify the specific mental experiences of a concrete individual, what this particular person wants, believes, loves, etc.[h]

With this, we move on to the analysis of de Ronde's theory, under the phenomenological perspective.

4. Phenomenology and immanent powers

I would like to begin with a general consideration that stems from the phenomenological thesis of the noetic-noematic correlation. De Ronde offers in his theory the rudimentary principles to build a new ontology that shall serve as the basis for QM. In this line, we could say that he is describing a specific ontological region,[i] inhabited by certain type of noemata: immanent powers with a definite potentia. We find in his theory (leaving aside how

---

[h]For Husserl's most systematic presentation of *empathy*, see the *V Cartesian Meditation* [19,20]. An excellent work (in Spanish) on the subject is [26].

[i]The concept of *ontological region* is presented by Husserl in Chapter 1 of *Ideas I*. It is a complicated concept, as the whole chapter is. The concept is presented as being defined by a kind of object, whose ontological properties are unique. The mother-forms of ontological regions are Nature and Consciousness. Nature is defined by being spatio-temporal and Consciousness by being only temporal. Thus, another form to present de Ronde's claim is precisely the idea that immanent powers constitute a unique type of object (the concept of object is not to be confused with that of entity, object in phenomenology is the most general category to speak about "something") and therefore belong to a distinct ontological region, different from that of the macroscopic objects described by classical physics and experienced in the natural attitude.

developed or in need of development it might stand) the formal principles to describe such an ontological realm. We could say, then, that they stand on the noematic side of the correlation. What we are still missing is a noetic description of what acts would be involved in the constitution of said noemata. Which acts must consciousness perform in order to constitute meaningful noemata, such as the Derondian powers?

A first answer to this problem, in strong contraposition to a Kantian approach, is that it is possible to find this noetic description. For Kant, objective experience is limited by impressions and concepts. In the absence of one we have no object, and hence no possibility of science. Phenomenology's concept of what can be objective experience is much, much broader than that. It includes the possibility of everything that consciousness experiences, lives, of what is given to it. Perceptual-physical objects are one domain of experience. Mathematical, abstract thinking is another, art and society yet others. As long as we can find the noetic acts involved in the specific type of constitution, we can legitimately speak about experience. If we can't, it would mean that the discourse about powers is indeed 'empty talk', just words with no reference, with no kind of intuitive comprehension whatsoever. So, the first point to claim from the phenomenological perspective would be: there is no *a priori* impediment against the notion of a Derondian power —as there would be from a Kantian perspective. Of course, it is yet to be seen whether it is in fact phenomenologically plausible to constitute the proposed beings.

Now the second question to answer phenomenologically is: is it possible to constitute these contextual, superposed, different, outside space-time, even contradictory beings, which we cannot even perceive? Again, a first phenomenological answer would be: it is possible. Here is where empathy comes in. I have brought this theme from phenomenology precisely for this particular point. In the case of empathy, we constitute very specific beings, under quite peculiar conditions. As it was explained, the consciousness of the other is never given to perception. Yet, at least from the phenomenological perspective, its reality cannot be denied. We *know* these funny bodies that walk about are conscious entities such as myself, even if I can *never* access their consciousness directly, and even if I sometimes wonder whether I am alone in the universe.

What I want to state is that empathy is not only an example of successful constitution without direct perception, but it is also an example of constitutions that involve contradictory (superposed) beings. When we come to the higher forms of empathy and reflect upon the specific conscious life of a particular individual, we are faced with something quite similar to a superposition state. What does this person want? Does she love me? Will he accept my offer? We could consider this mere ignorance regarding our knowledge of the other person's state of mind, and in many situations, this might very well be the case. But if we reflect upon our own beliefs, desires, projects, ambitions, we can very easily verify that in many other cases the problem is not one of ignorance, but of actual indetermination. We do not know what we want. We want something, but don't want the means it implies, or some yes, but not all, and we are afraid of the consequences, but we would really like it to turn out this way... So? What is the answer, in a specific situation, to the question "Does S want X or Y?". Funny as it might sound, one way of expressing it could be: $X = 60\%$; $Y = 40\%$. And then we ask, and find in actuality only one answer, because that is all that actuality tolerates, one definite state. But our consciousness, though invariably and inexorably bound by temporality, does not inhabit physical space and is not susceptible to the deterministic laws of classical mechanics.[j] We can be, as we so often are, in a state of indetermination, superposed by contradictory desires. And being this the way it is, we can still successfully constitute alter-egos as coherent, unitary beings and say: this is John, he has experience of the world just like I do, even if I don't see his experience. He is a catholic, but he also believes that abortion is right (contradiction), he says he wants to come to the party on Saturday, but he also says he might not come; I know he is my friend, even though sometimes he is bad to me, etc., etc. John is a coherent unitary being, made of contradictions and superpositions which I can only presentify and never perceive.

It is not my claim, under no circumstance and in no sense, that human consciousness *are* quantum superposition states. My only claim is that we cannot rule out the possibility of constituting something like a Derondian power, 'just because' it is outside normal macroscopic perception, because it involves the superposition of mutually contradictory states. We do this already in the case of the empathic constitution of the alter-ego, and we do it successfully. It is my claim, then, that if it is *possible* in one domain, it

---

[j]It is worth noting that the attempts to clarify the phenomenon of empathy (or theory of mind) in other philosophical theories and other disciplines, such as cognitive sciences, would state something quite different, if not the exact opposite. It is therefore also worthy of attention, that these programs that accept naturalistic principles for the study of the human "mind" incur in the same attitude as the philosophers of physics that pretend to limit QM to the classical ontology: they accept only one realm of the real (actual, material existence) and only one method to study it (logical-empiricism, observability).

might be *possible* in another, such as the quantum realm.

Two problems in this direction still stand. In the case of empathy, as we saw, the constitution of the other is founded upon the perception of the physical body. There is a perception that serves as basis for the presentification of the not-given aspects, in this case, the conscious life itself. What is the founding perception in the case of powers? This question remains unanswered. We need something that serves as basis and guide for the constitution of other elements. We cannot simply do everything in presentification, because that would be mere imagination and we can hardly call that science. A possible candidate could be the actual, already performed, experimental arrangements. The problem is that they are incompatible with de Ronde's pretentions. According to him, in experimental arrangements we only access the actualized power, an actual effectuation, but not the power itself, nor the power in its most important form, which is that which is not actualized. Still, they could be a candidate for an actual perception, upon which we would be able to presentify the other, most important, elements of the power (just like in empathy, "the most important" aspect, which is the consciousness itself, is only appresented and not perceived).

The second associated problem, much harder to solve, is that which I listed as number 1: powers are not actual, but they are physical. In phenomenological terms, real means space-time bound, while ideal means outside space-time. Ideal entities are obviously much more tricky than real ones, but they are nothing to discourage us: mathematical entities, meanings, essences, species, are all more or less familiar ideal entities. This poses, however, a huge ontological problem: we are intersecting what up to now were two completely separated realms of being, the 'ideal' (in the Husserlian sense) and the physical. The solution to this problem is, I believe, intimately bound to the previous point: if something, say a Derondian power, belongs to physical nature, no matter what other peculiarities it has (being outside space-time, being contradictory, etc.) there *must* be some sort of perception of it, or of one of it sides. Otherwise, all concepts lose meaning. I represent this problem in Figure 2.

The remaining point concerns the phenomenological notion of belief- and being-characteristics, and how this can be applied to the issue of identity and truth in regards to powers. As we saw, de Ronde claims that the information we obtain when we perform a measurement is only very partial regarding the ontological existent beneath it, which is the power. Yet, due to contextuality, our knowledge of the power itself can never be expressed in terms of a 'truth or false' '1 or 0' valuation: "Quantum contextuality,

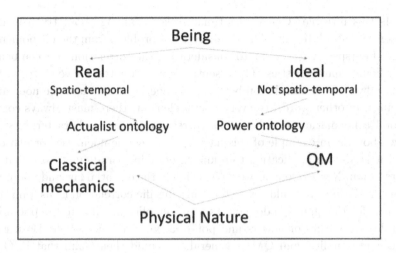

Fig. 2.

which was most explicitly recognized through the Kochen-Specker theorem, asserts that a value ascribed to a physical quantity A cannot be part of a global assignment of values but must, instead, depend on some specific context from which A is to be considered" [12, p. 5]. In this sense, as the problem was listed under number 3, we need new criteria to determine the identity of a power and a new corresponding notion of truth to value our statements about powers:

> Our proposed representation of quantum physical reality in terms of powers with definite potentia opens the possibility of considering a reference which is not exclusively defined in terms of 0 or 1 values (true or false). Instead, the power can be understood as possessing an intensity, or in other words, a potentia which pertains to the closed interval $[0; 1]$. According to our approach, given a Ψ, there is a set of powers which can be defined as being true in the ontic level but, contrary to the classical case, their relation to actual effectuations is not that of a one-to-one correspondence. Indeed, due to the fact that a power has assigned a potentia in the closed interval $[0,1]$, statistics is required in order to acquire complete knowledge of each power. In order to gain knowledge about a PSA we require the (contextual) measurement of each (non-contextual) power and its potentia. Thus, while classical physics has a binary valuation of properties, QM presents instead an intensive valuation of powers. [12, p. 18]

Is this possible? Can we legitimately keep talking about truth in the absence of absolute values? Let's analyze the problem from the phenomenological perspective, taking into consideration the correlation between belief and being characteristics. The essential law we quoted above states there must always be a correlation between the noetic (belief) and the noematic (being), in other words, to every noetic element, there must always correspond a noematic one. The *actual* correlation between characteristics we saw above is an example of this more general noetic-noematic correlation, and it is also an application to a known domain, paradigmatically that of perception. Now, as long as we do not violate the essential law and maintain the correlation, we could revise how this specific correlation is, as a matter of fact, for the quantum domain. I mean the following: due to the potential ontological nature of powers and potentia, known to us by the Born rule and contextuality, and QM in general, we could then claim that in QM, what corresponds to certainty is not actual truth, but a probability value. And what corresponds to truth, in the sense of something fully given in actuality, is not certainty, but a conjecture regarding the full nature of the power that can only be partially known in an actual measurement; changing some of the correspondences in the table, so as to have:

Fig. 3.

The idea would then be that we have certainty, certain knowledge of a quantum power, when we know it in its being probabilistic and indeterminate and not when it is given in a determined value in the actuality of a measurement process. Through an experimental result, we access only a partial aspect of the power, and therefore can only suspect what the true nature of the power really is. A power can then be identified by its intensive probability value, which is a specific value that is assigned to the power, even though it is not a determinate value.

Of course, this is not enough to fully determine the identity of a power, and neither is the correlation a definitive notion of truth. But again, we can appreciate the plausibility of utilizing new notions of truth and identity to describe a domain that is, in itself, radically new. Regarding the violation of Quine's dogma, all we can say is: of course! We are not talking about entities, but about powers!

Conclusions

In this article, I have surveyed the notion of an alternative ontology to QM, a necessity that stems from the impossibility of describing what QM talks about. Most attempts to do so, have done it in the line of maintaining themselves within the limits of classical ontology and classical mechanics, even at the expense of the formalism itself. When it comes to a new ontology, alternative to the classical one, we are immediately faced with a myriad of problems. I have presented de Ronde's alternative ontology of powers and potentia, and I have analyzed it from the phenomenological perspective. My conclusion is, then, quite simple: I do not claim that de Ronde's ontology is the correct one for QM (that question falls out of the scope of my work). Neither do I claim that de Ronde's ontology is finished or without problems to be revised. What I do claim is that there is no a priori restriction for the possibility of said ontology, that it does not violate any phenomenological restriction. This not only clears the way to keep working on this project, but it also helps to point out what the problems to be solved are. In particular, I would like to highlight two: the need to find an element in the order of the intuitive (noetic) that would allow and guide the proper meaningful constitution of the power itself; and the huge problem that implies considering an ontological realm that is not spatio-temporal, and yet belongs to physical nature. I believe that, in what concerns the philosophical development of this ontology, these are the major challenges for the theory.

Acknowledgements

This work was completed under a DAAD grant. I would like to thank Christian de Ronde for the discussion and valuable comments on the manuscript.

182

References

1. Aerts D., 2010, "A potentiality and Conceptuality Interpretation of Quantum Physics", *Philosophica*, **83**, 15-52.
2. Arenhart, J. R. and Krause, D., 2016, "Contradiction, Quantum Mechanics, and the Square of Opposition", *Logique et Analyse*, **59**, 273-281.
3. Bitbol, M., 2010, "Reflective Metaphysics: Understanding Quantum Mechanics from a Kantian Standpoint", *Philosophica*, **83**, 53-83.
4. da Costa, N. and de Ronde, C., 2013, "The Paraconsistent Logic of Quantum Superpositions", *Foundations of Physics*, **43**, 845-858.
5. D'Espagnat, B., 1989, *Reality and the Physicist: Knowledge, duration and the quantum world*, Cambridge University Press.
6. de Ronde, C., 2011, *The Contextual and Modal Character of Quantum Mechanics: A Formal and Philosophical Analysis in the Foundations of Physics*, PhD dissertation, Utrecht University, The Netherlands.
7. de Ronde, C., 2013, "Representing Quantum Superpositions: Powers, Potentia and Potential Effectuations", preprint. (quant-ph:1312.7322)
8. de Ronde, C., 2016, "Probabilistic Knowledge as Objective Knowledge in Quantum Mechanics: Potential Powers Instead of Actual Properties", in *Probing the Meaning and Structure of Quantum Mechanics: Superpositions, Semantics, Dynamics and Identity*, pp. 141-178, D. Aerts, C. de Ronde, H. Freytes and R. Giuntini (Eds.), World Scientific, Singapore.
9. de Ronde, C., 2017, "Quantum Superpositions and the Representation of Physical Reality Beyond Measurement Outcomes and Mathematical Structures", *Foundations of Science*, https://doi.org/10.1007/s10699-017-9541-z. (quant-ph:1603.06112)
10. de Ronde, C., 2017, "Causality and the Modeling of the Measurement Process in Quantum Theory", *Disputatio*, forthcoming. (quant-ph:1310.4534)
11. de Ronde, C., 2018, "A Defense of the Paraconsistent Approach to Quantum Superpositions (Reply to Arenhart and Krause)", *Metatheoria*, forthcoming. (quant-ph:1404.5186)
12. de Ronde, C., 2018, "Potential Truth in Quantum Mechanics", preprint.
13. Dorato, M., 2015, "Events and the Ontology of Quantum Mechanics", *Topoi*, **34**, 369-378.
14. Esfeld, M., 2009, "The modal nature of structures in ontic structural realism", *International Studies in the Philosophy of Science*, **23**, 179-194.
15. French, S., 2015, "Doing Away with Dispositions: Powers in the Context of Modern Physics", in *Dispositionalism: Perspectives from Metaphysics and the Philosophy of Science* A. S. Meincke-Spann (Eds.), Springer Synthese Library, Springer.
16. Fuchs, C. and Peres, A., 2000, "Quantum theory needs no 'interpretation'", *Physics Today*, **53**, 70.
17. Griffiths, R., 2013, "A Consistent Quantum Ontology", *Studies in History and Philosophy of Science Part B*, **44**, 93-114.
18. Graffigna, M., 2016, "The Possibility of a New Metaphysics for Quantum Mechanics from Meinong's Theory of Objects", in *Probing the Meaning and Structure of Quantum Mechanics: Superpositions, Semantics, Dynamics and Identity*, pp. 280-307, D. Aerts, C. de Ronde, H. Freytes and R. Giuntini (Eds.), World Scientific, Singapore.

19. Husserl, E., 1960, *Cartesian Meditations: An Introduction to Phenomenology*. Martinus Nijhoff, The Hague.
20. Husserl, E., 1973, *HUA I: Cartesianische Meditationen und Pariser Vorträge*. S. Strasser (Eds.), Martinus Nijhoff, The Hague.
21. Husserl, E., 1977, *HUA III-I: Ideen zu einer reinen Phänomenologie und phänomenologischen Philosophie. Erstes Buch: Allgemeine Einführung in die reine Phänomenologie 1. Halbband: Text der 1.-3. Auflage - Nachdruck*, Karl Schuhmann (Eds.), Martinus Nijhoff, The Hague.
22. Husserl, E., 1983, *Ideas 1*, Martinus Nijhoff, The Hague. (Originally published in 1913)
23. Kant, I., 1998, *Kritik der reinen Vernunft*, Akademie Verlag, Berlin.
24. Lurat, F., 2007, "Understanding Quantum Mechanics", in *Rediscovering Phenomenology*, L. Boi, P. Kerszberg, P. and F. Patras (Eds), Springer, Dodrecht.
25. Pringe, H., 2012, "La filosofía trascendental y la interpretación de Bohr de la teorá cuántica", *Scientiae Studia*, **10**, 179-194.
26. Walton, R., 2001, "Fenomenología de la empatía", *Philosophica*, **24**, 25.
27. Wheeler, J. A. and Zurek, W. H. (Eds.) 1983, *Theory and Measurement*, Princeton University Press, Princeton.

ON AERTS' OVERLOOKED SOLUTION TO THE
EINSTEIN-PODOLSKY-ROSEN PARADOX

MASSIMILIANO SASSOLI DE BIANCHI

*Center Leo Apostel for Interdisciplinary Studies, Brussels Free University, 1050
Brussels, Belgium and Laboratorio di Autoricerca di Base, via Cadepiano 18, 6917
Barbengo, Switzerland.*
E-mail: msassoli@vub. ac. be

The Einstein-Podolsky-Rosen (EPR) paradox was enunciated in 1935 and since
then it has made a lot of ink flow. Being a subtle result, it has also been largely
misunderstood. Indeed, if questioned about its solution, many physicists will
still affirm today that the paradox has been solved by the Bell-test experimental results, which have shown that entangled states are real. However, this
remains a wrong view, as the validity of the EPR *ex-absurdum* reasoning is
independent from the Bell-test experiments, and the possible structural short-
comings it evidenced cannot be eliminated. These were correctly identified by
the Belgian physicist Diederik Aerts, in the eighties of last century, and are
about the inability of the quantum formalism to describe separate physical
systems. The purpose of the present article is to bring Aerts' overlooked result
to the attention again of the physics' community, explaining its content and
implications.

Keywords: EPR paradox; quantum structures; quantum entanglement.

1. Introduction

In 1935, Albert Einstein and his two collaborators, Boris Podolsky and
Nathan Rosen (abbreviated as EPR), devised a very subtle thought experiment to highlight possible inadequacies of the quantum mechanical formalism in the description of the physical reality, today known as the *EPR
paradox* [1]. The reason for the "paradox" qualifier is that the predictions
of quantum theory, regarding the outcome of their proposed experiment,
differed from those obtained by means of a reasoning using a very general
reality criterion.

Despite the fact that the EPR objection to quantum mechanics has
been the subject of countless discussions in the literature, many physicists
still believe today that the EPR paradox has been solved by the celebrated

coincidence experiments on pair of entangled photons in singlet states, realized by Alain Aspect and his group in 1982 [2], which were later reproduced under always better controlled experimental situations [3], closing one by one all potential experimental loopholes [4]. More precisely, the belief is that these experiments would have invalidated EPR's reasoning by confirming the exactness of the quantum mechanical predictions.

This conclusion, however, is the fruit of a misconception regarding the true nature of the EPR paradox, which was not solved by experiments like those conducted by Aspect et al., but by a constructive proof presented almost forty years ago by Diederik Aerts, in his doctoral dissertation [5–10]. Contrary to what is generally believed, Aerts' solution says that the quantum mechanical description of reality is indeed incomplete, because, as we are going to explain, it cannot describe separate physical systems. Aerts' result remains to date largely unknown, and the main purpose of the present article is to bring it back to the attention of the scientific community. I will do so by trying to explain it in the simplest possible terms, also indicating its consequences for our understanding of classical and quantum theories.

2. Correlations

We start by observing that quantum entanglement, which was firstly discussed by EPR [1] and Schrödinger [11,12], is incompatible with a classical spatial representation of the physical reality. Indeed, in this representation a spatial distance also expresses a condition of *experimental separation* between two physical entities, in the sense that the greater the spatial distance Δx between two entities A and B, and the better A and B will be experimentally separated. To be experimentally separated means that when we test a property on entity A, the outcome of the test will not depend on other tests we may want to perform, simultaneously or in different moments, on entity B, and vice versa.[a]

For two classical entities this will be the case if Δx and the time interval Δt between the different tests is such that no signal can propagate in time

---

[a]More precisely, it will not depend on them in an ontological sense, rather, possibly, in a dynamical sense, for instance because both entities may interact by means of a force field, such as the gravitational or electromagnetic fields. In other words, quoting from [8]: "In general there is an interaction between separate systems and by means of this interaction the dynamical change of the state of one system is influenced by the dynamical change of the state of the other system. In classical mechanics for example almost all two body problems are problems of separate bodies (e.g. the Kepler problem). Two systems are non-separate if an experiment on one system changes the state of the other system. For two classical bodies this is for example the case when they are connected by a rigid rod."

between the two entities to possibly influence the outcomes of the respective tests, which will be the case if $\frac{\Delta x}{\Delta t} > c$, with c the speed of light in vacuum. Of course, in the limit $\Delta t \to 0$, where the two tests are performed in a perfectly simultaneous way, any finite distance Δx will be sufficient to guarantee that we are in a non-signaling condition, i.e., that we are in a situation of experimental separation. In other words, in classical physics the notions of *spatial separation* and *experimental separation* were considered to be intimately connected, in the sense that the former was considered to generally imply the latter.

Consider now that the two entities A and B are two bodies moving in space in opposite directions and assume that two experimenters decide to jointly measure their positions and velocities. Since the two entities are spatially separate, and therefore perfectly disconnected, no correlations between the outcomes of their measurements will in general be observed. However, if the two objects were connected in the past, the physical process that caused their disconnection may have created correlations that subsequently can be observed. As a paradigmatic example, consider a rock initially at rest, say at the origin of a laboratory's system of coordinates, and assume that at some moment it explodes into two fragments A and B, having exactly equal masses (see Fig. 1). The positions and velocities of these two flying apart fragments of rock will then be perfectly correlated, due to the conservation of momentum: if at a given instant the position and velocity of (the center of mass of) fragment A are $\mathbf{x}$ and $\mathbf{v}$, respectively, then the position and velocity at that same instant of fragment B will be $-\mathbf{x}$ and $-\mathbf{v}$. This situation of perfect correlation is clearly the consequence of how the two fragments were created in the past, out of a single whole entity, and is not the result of a connection that is maintained between the two fragments while moving apart in space.

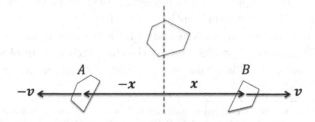

Fig. 1. A rock initially at rest explodes into two fragments which are here assumed to be of equal masses, flying apart in space with opposite velocities.

It is in fact important to distinguish between the correlations that can only be *discovered*, between the two components of a bipartite system, and which are due to previous processes of connections-disconnection, from the correlations that are literally *created* by the very process of their observation, i.e., which are created out of an actual connection between the two components of the bipartite system, when these two parts are subjected to a measurement. This fundamental distinction was made in the nineties by Aerts, who specifically named the correlations that are only discovered in a measurement *correlations of the first kind*, and those that are instead created in a measurement *correlations of the second kind* [13].

The key role played by Bell's inequalities [14,15] in identifying the presence of entanglement in composite physical systems can then be identified in their ability to demarcate between correlations of the first kind and correlations of the second kind, as only the latter can violate them. In that respect, it is important to note that the violation of Bell's inequalities is not a specificity of micro-physical systems: also classical macroscopic systems can violate them, as what is truly important for the violation is to have correlations that can be created during the very process of measurement, which will be generally the case when the two entities forming the bipartite system are connected in some way, for instance because they are in direct contact, or because of the presence of a third connecting element. So, to give examples, two vessels of water connected through a tube [10,16], or two dice connected through a rigid rod [17,18], can easily violate Bell's inequalities in specifically designed coincidence experiments.[b]

But then, if classical entities can also produce quantum-like correlations of the second kind, violating Bell's inequalities, why Einstein famously called the quantum correlations "spooky actions at a distance"? The answer is simple: a tube connecting two vessels of water, or a rod connecting two dice, are elements of reality that can be easily described in our three-dimensional Euclidean theater, so there is no mystery in their functioning, whereas what keeps two micro-physical entities connected in a genuine quantum entangled state apparently cannot. In other words, the "spookiness" of the quantum correlations comes from the fact that: (1) they are not correlations of the first kind and (2) the connectedness out of which the correlations are created is a *non-spatial* element of our physical reality.[c]

---

[b]To give another remarkable example, abstract conceptual entities, which are connected through meaning, are also able to violate Bell's inequalities similarly to quantum micro-entities, in specific psychological measurements [19,20].

[c]Interestingly, this "quantum connectedness element," characterizing the potential cor-

3. The paradox

Let me now explain the EPR reasoning in their celebrated article [1]. First of all, they introduced the important notion of *element of reality*, corresponding to the following definition [1]: *"If without in any way disturbing the state of a physical entity the outcome of a certain observable can be predicted with certainty, there exists an element of reality corresponding to this outcome and this observable."* Contrary to what is often stated, the important part in this definition is not the "without in any way disturbing" one, but the "can be predicted with certainty" one. Indeed, as observed by Aerts [7]: "it is possible for an entity to have an element of reality corresponding to a physical quantity even when this physical quantity cannot be measured without disturbing the entity."

Consider for instance the burnability property of a wooden cube [6]. We know it is an actual property, as we can *predict with certainty* that if we put the cube on fire it will burn with certainty (i.e., with probability equal to 1), but of course we cannot test the burnability property without deeply disturb the wooden entity. In their paper, EPR did not mention explicitly this subtle point, which however was later on integrated by Constantin Piron in his fundamental definition of an *actual property* [22,23] and used in his construction of an axiomatic operational-realistic approach to the foundations of quantum mechanics. According to this definition, a property is actual if and only if, should one decide to perform the experimental test that operationally defines it, the expected result would be certain in advance. If this is the case, the entity in question is said to have the property (i.e., to possess it in actual terms) even before the test is done, and in fact even before one has chosen to do it (independently on the fact that the test might be invasive or not). And this is the reason why one is allowed to say that the property is an element of reality, existing independently from our observation.

So, even if not fully expressed at the time, the EPR reasoning contained the deep insight that the (actual) properties of physical systems are "states of predictions." EPR then considered the situation of two quantum entities A and B that, after interacting, subsequently flew apart in space, becoming in this way spatially separate and, according to EPR's prejudice, also experimentally separate. The additional step taken by EPR in their

---

relations that can be actualized in a coincidence measurements with entangled entities, can be explicitly represented in the generalized quantum formalism called the Extended Bloch Representation (EBR) of quantum mechanics [21].

1935 paper is to consider the quantum mechanical formalization of this situation, in accordance with the notion of entanglement, from which they observe that the positions and velocities of the two quantum entities are strongly correlated.

More precisely, the EPR reasoning goes as follows. They consider the possibility of measuring the position of one of the quantum entities, say entity B that is flying to the right. Assuming that such measurement has been carried out, and that the position of entity B has been observed to be $\mathbf{x}$, then, according to the quantum description, the experimenter is in a position to predict that if a position measurement would be performed on entity A, the outcome $-\mathbf{x}$ would be obtained with certainty (considering a system of coordinates such that the place where the two entities interacted before flying apart corresponds to its origin). The subtle point here is that since A and B are separated by an arbitrarily large spatial distance, and that the assumption is that a spatial separation also implies an experimental separation, the previous measurement on B could not affect in whatsoever way the state of A. Hence, the prediction that the position of A is $-\mathbf{x}$ establishes the actuality of the property, and of course the same reasoning holds in case it is the velocity (or momentum) that is measured on B, as also in this case, if the outcome of the measurement was, say, $\mathbf{v}$, then the outcome $-\mathbf{v}$ could have been predicted with certainty for entity A.

Is the above sufficient to conclude that entity A has both a well-defined position and velocity? To clarify the situation, let me come back to Aerts' example of the wooden cube, which as we observed has the property of being burnable. We also know that it has other properties, like the property of floating on water. How do we know that? Again, because if we would perform the test of immersing the cube in water, the "floating on water" outcome would be obtained with certainty. But then, we can ask the following question: Does the wooden cube jointly possess the properties of burnability and floatability? Our common sense tells us that this has to be the case, but how do we test this *meet property* obtained by the combination of the burnability and floatability properties? Because a wet cube does not burn, and a burned cube does not float, so we cannot conjunctly or sequentially test these two properties.

In fact, we don't have to, because, as noted by Piron [22], the test for a meet property is a so-called *product test*, consisting in performing only one of the two tests, but chosen in a random (unpredictable) way. Indeed, the only way we can then predict the positive outcome of a procedure consisting in randomly selecting one of the two tests, then executing it, is to be able to predict the positive outcome of both tests, which precisely corresponds to the situation where both properties are simultaneously actual.

Having clarified that even when the experimental tests of two properties are mutually incompatible this does not imply that they cannot be jointly tested by means of a product test, and therefore the properties be simultaneously actual, we can now observe that the EPR reasoning precisely describes a situation where the outcome of a product test (or product measurement) for the position and velocity of entity A can be predicted with certainty. Indeed, in case it is the position measurement that is randomly selected, the experimenter can perform that same measurement on entity B and then predict with certainty the outcome of the position measurement on entity A, without the need to perform it. And in case it is the velocity (or momentum) measurement that is randomly selected, the experimenter can perform the velocity measurement on entity B and again predict with certainty the outcome of the same measurement on entity A, again without the need to perform it. In other words, we are exactly in a situation where the outcome of a product measurement of position and velocity observables can be predicted with certainty, hence, we are allowed to conclude, with EPR, that both position and velocity have simultaneous well-defined values for entity A. This is of course in flagrant contradiction with Heisenberg's uncertainty relations, hence the paradox and EPR's conclusion that quantum mechanics is an incomplete theory, as unable to represent all possible elements of reality associated with a physical entity.

Bohr's reaction to the EPR argument, that same year, was quite obscure [24]. Basically, the Danish physicist affirmed that one "is not allowed in quantum mechanics to make the type of reasoning proposed by EPR, and more specifically, the notion of element of reality does not make sense for quantum mechanical entities." With the exception of Schrödinger, Bohr's authority (and the influence of the Copenhagen interpretation) resulted in most leading quantum physicists simply accepting that there was not really a serious problem involved in the EPR reasoning and resulting paradox. Many years later though, perhaps because also of the influence of David Bohm, who certainly took the EPR argument seriously (inventing the entangled spin example as a more transparent description of the EPR situation), a small group of physicists, among whom was John Bell, believed that EPR highlighted a fundamental problem in quantum mechanics related to its possible incompleteness. However, different from what EPR, Bohm, Bell and others believed, the incompleteness in question was not an issue of "providing additional variables" to make it complete, or more complete, but a question of a shortcoming related to the impossibility for the quantum formalism to describe experimentally separate entities, as subsequently shown by Aerts [5–10].

4. The solution

To explain Aerts' solution, it is important to emphasize that EPR's reasoning is an *ex absurdum* one, that is, a reasoning which starts from certain premises and reaches a contradiction. What EPR have shown is that if their premises are assumed to be correct, then quantum theory has to be considered incomplete, as unable to describe all elements of reality of a physical system. Those who have taken seriously this conclusion thus tried to find remedies, for instance by supplementing the theory with additional variables for the quantum states, to allow position and velocity to have simultaneous definite values and escape the limitation of Heisenberg's uncertainty relations. This hidden variables program, however, subsequently met the obstacle of so-called *no-go theorems*, drastically limiting the class of admissible hidden-variable theories [25–30].

The premise that was part of the EPR reasoning, as we explained, is that for two quantum entities that have interacted and flown apart, it was natural to expect that their spatial separation was equivalent to an experimental separation. In addition to that, EPR applied the quantum formalism to describe the situation, which means they implicitly also assumed that quantum mechanics is able to describe a system formed by *separate physical entities*. But since this produced a contradiction, one is forced to conclude that the assumption is incorrect, that is, that quantum mechanics is unable to describe separate entities.

Now, one may object that this is a too strong conclusion, in the sense that the only mistake committed by EPR was to expect that spatial separation would also necessarily imply disconnection. This expectation, as we know today, has been overruled by numerous experiments, showing that by making sufficient efforts and taking all necessary precautions, experimental situations can indeed be created where microscopic entities, after having interacted, can remain interconnected, even when arbitrarily large spatial distances separate them. The mistake of EPR was therefore to think about a situation where there is no experimental separation between two entities, as a situation of actual experimental separation.

So, apparently problem solved: EPR-like experiments, like those performed by the group of Alain Aspect, have precisely shown that in the situation considered by EPR quantum mechanics does actually provide the correct description of two quantum entities flying apart, since Bell's inequalities are violated, in accordance with the quantum predictions. Thus, one would be tempted to conclude that EPR's reasoning is not valid. Well, yes and no. Yes, because at their time the possibility of producing these non-local/non-spatial states was a truly remarkable and totally unexpected possibility, based on classical prejudices, so the EPR *ex absurdum* reasoning was indeed applied to a wrong experimental situation, if such situation is considered to be correctly described by an entangled state. No, because the possibility of producing and preserving entangled states has very little to do with EPR's reasoning per se. Indeed, one can in principle also assume that experiments could be performed where instead of making efforts to preserve the quantum connectedness of the two flying apart entities, an effort is made instead to obtain the opposite situation of two flying apart entities eventually becoming perfectly disconnected, i.e., separated.

Experiments of this kind have never been worked out consciously, but these would indeed correspond to situations leading to the EPR paradox. In other words, the incompleteness of quantum mechanics is not revealed in the physical situation of quantum entities flying apart and remaining non-separate, as these are the situations which are perfectly well described by the quantum formalism (as the violation of Bell's inequalities proves), and there is no contradiction/paradox in this case, but by the experimental situations that can produce a disconnection, and which in the setting of EPR-like experiments would be interpreted as "badly performed experiments." These are precisely the situations that quantum mechanics would be unable to describe, certainly not by means of entangled states, as if we assume it can, then we reach a contradiction.

Having clarified that the logical reasoning of EPR is not directly affected by the experimental discovery of entangled states, the question thus remains about the completeness of the quantum formalism, in relation to its ability to describe separate physical entities. It is here that Aerts' work join the game. Indeed, among the topics of his doctoral research there was that of elaborating a mathematical framework for the general description of separate quantum entities. Aerts approached the issue using Piron's axiomatic approach to quantum mechanics, a very general formalism which was precisely [31]: "obtained by taking seriously the realistic point of view of Einstein and describing a physical system in terms of 'elements of reality'." This allowed him to view the EPR work from a completely new angle. Indeed, while describing the situation of bipartite systems formed by separate quantum entities, he was able to prove, this time in a perfectly constructive way, that quantum mechanics is structurally unable to describe these situations.

5. Aerts' proof

EPR were thus right about the incompleteness of quantum mechanics, but not for the reason they believed: quantum mechanics is incomplete because unable to describe separate physical systems. Of course, depending on the viewpoint adopted, this can be seen as a weak or strong trait of the theory. If separate systems exist in nature, then it is a weak trait, if they don't, then it is a strong trait. We will come back on that in the conclusive section, but let us now sketch the content of Aerts' constructive proof, which is actually quite simple.

Note that despite the simplicity of the proof, it usually comes as a surprise that quantum mechanics would have this sort of shortcoming. Indeed, the first reaction I usually get, when discussing Aerts' result with colleagues, is that this cannot be true, as separate systems are perfectly well described in quantum mechanics by so-called product states, that is, states of the tensor product form $\psi \otimes \phi$, where $\psi \in \mathcal{H}_A$ and $\phi \in \mathcal{H}_B$, with $\mathcal{H}_A$ the Hilbert (state) space of entity A and $\mathcal{H}_B$ that of entity B, the Hilbert space $\mathcal{H}$ of the bipartite system formed by A and B being then isomorphic to $\mathcal{H}_A \otimes \mathcal{H}_B$. This is correct, and in fact the shortcoming of quantum theory in describing separate systems cannot be detected at the level of the states, as in a sense there is an overabundance of them, but at the level of the properties, which in the quantum formalism are described by orthogonal projection operators. In fact, it is precisely this overabundance of states that produces a deficiency of properties, in the sense that certain properties of a bipartite system formed by separate components cannot be represented by orthogonal projection operators.

Technically speaking, the only difficulty of Aerts' proof is that one needs to work it out in all generality, independently of specific representations, like the tensorial one, so that one can be certain that its conclusion is inescapable [5–8,10]. Without entering into all details, the demonstration goes as follows. First, one has to define what it means for two entities A and B to be experimentally separate. As we mentioned already, this means that measurements individually performed on them do not influence each other. In other words, separate entities are such that their measurements are *separate measurements*. More precisely, two measurements $\mathcal{M}_A$ and $\mathcal{M}_B$ are separate if they can be performed together without influencing each others. This means that, from them, one can define a combined measurement $\mathcal{M}_{AB}$

such that: (1) the execution of $\mathcal{M}_{AB}$ on the bipartite entity formed by A and B corresponds to the execution of $\mathcal{M}_A$ on A and of $\mathcal{M}_B$ on B, and (2) the outcomes of $\mathcal{M}_{AB}$ are given by all possible couples of outcomes obtained from $\mathcal{M}_A$ and $\mathcal{M}_B$.

What Aerts then shows is that there is no self-adjoint operator O_{AB} that can represent such measurement $\mathcal{M}_{AB}$. To do so, he considers two arbitrary projections P_A^I and P_B^J, in the spectral decomposition of the self-adjoint operators O_A and O_B associated with measurements $\mathcal{M}_A$ and $\mathcal{M}_B$, respectively. Here I and J are subsets of the outcome sets E and F of the two measurements, respectively. He also defines the spectral projection $P_{AB}^{I \times J}$ of O_{AB}, where $I \times J$ is the subset of the outcome sets of $\mathcal{M}_{AB}$ formed by all couples (x, y) of elements $x \in I$ and $y \in J$. Then he shows (we do not go into the details of this here), as one would expect, that $[P_A^I, P_B^J] = 0$, so that also $[O_A, O_B] = 0$, and that $P_{AB}^{I \times J} = P_A^I P_B^J$.

The next step is to consider a state $\psi \in \mathcal{H}$ which can be written as a superposition $\psi = \frac{1}{\sqrt{2}}(\phi + \chi)$, where ϕ belongs to the subspace $P_A^I(\mathbb{I} - P_B^J)\mathcal{H}$ and χ to the subspace $(\mathbb{I} - P_A^I)P_B^J\mathcal{H}$, orthogonal to the latter. It follows that:

$$P_A^I \psi = \frac{1}{\sqrt{2}}\phi, \quad (\mathbb{I} - P_A^I)\psi = \frac{1}{\sqrt{2}}\chi,$$

$$P_B^J \psi = \frac{1}{\sqrt{2}}\chi, \quad (\mathbb{I} - P_B^J)\psi = \frac{1}{\sqrt{2}}\phi. \tag{1}$$

This means that when the bipartite system is in state ψ, there is at least two possible outcomes $x_1 \in I$ and $x_2 \in E - I$, for measurement $\mathcal{M}_A$, and at least two possible outcomes $y_1 \in J$ and $y_2 \in F - J$, for measurement $\mathcal{M}_B$. This means that the four outcomes $(x_1, y_1) \in I \times J$, $(x_1, y_2) \in I \times (F - J)$, $(x_2, y_1) \in (E - I) \times J$ and $(x_2, y_2) \in (E - I)(F - J)$ should be all possible outcomes of measurement $\mathcal{M}_{AB}$, if $\mathcal{M}_A$ and $\mathcal{M}_B$ are assumed to be separate measurements. But although we have:

$$P_{AB}^{I \times (F-J)}\psi = P_A^I(\mathbb{I} - P_B^J)\psi = \phi,$$

$$P_{AB}^{(E-I) \times J}\psi = (\mathbb{I} - P_A^I)P_B^J\psi = \chi, \tag{2}$$

so that (x_1, y_2) and (x_2, y_1) are possible outcomes of $\mathcal{M}_{AB}$, we also have that:

$$P_{AB}^{I \times J}\psi = P_A^I P_B^J\psi = 0,$$

$$P_{AB}^{(E-I) \times (F-J)}\psi = (\mathbb{I} - P_A^I)(\mathbb{I} - P_B^J)\psi = 0. \tag{3}$$

Hence, (x_1, y_1) and (x_2, y_2) are not possible outcomes of $\mathcal{M}_{AB}$, which means that $\mathcal{M}_A$ and $\mathcal{M}_B$ are not separate measurements.

In other words, because of the superposition principle, a joint measurement $\mathcal{M}_{AB}$ formed by two separate measurements $\mathcal{M}_A$ and $\mathcal{M}_B$ cannot be consistently described in quantum mechanics, which means that quantum mechanics, for structural reasons related to its vector space structure, cannot handle separate measurements.

Note that when one introduces the more specific tensorial representation $\mathcal{H} = \mathcal{H}_A \otimes \mathcal{H}_B$, the request for the self-adjoint operators associated with measurements $\mathcal{M}_A$ and $\mathcal{M}_B$ to commute is automatically implemented by writing them in the tensorial form $O_A \otimes \mathbb{I}_B$ and $\mathbb{I}_A \otimes O_B$, respectively, so that we also have in this case $P_{AB}^{I \times J} = P_A^I \otimes P_B^J$, and the superposition state ψ can for instance be written as an entangled state $\psi \frac{1}{\sqrt{2}}(\phi_A \otimes \phi_B + \chi_A \otimes \chi_B)$, with $\phi_A \in P_A^I \mathcal{H}_A$, $\phi_B \in (\mathbb{I} - P_B^J)\mathcal{H}_B$, $\chi_A \in (\mathbb{I} - P_A^I)\mathcal{H}_A$ and $\chi_B \in P_B^J \mathcal{H}_B$, thus making explicit the connection of Aerts' proof with EPR-like situations.

6. Discussion

Having provided the gist of Aerts' demonstration, I can conclude with a few important comments. First of all, I would like to highlight once more the importance of distinguishing the logic of the EPR reasoning, leading to a paradox (contradiction), from the subsequent Bell-test experiments, the validity and interest of EPR's *ex absurdum* reasoning being independent of the experimental violations of Bell's inequalities. To make this point even clearer, let me describe a different paradox, that Einstein and collaborators could have worked out at the time as an alternative reasoning to point to a possible incompleteness of quantum theory. For this, let me come back to the wooden cube and its properties of burnability and floatability. People confronted with the problem of designing an experiment able to test the joint actuality of these two properties, despite the experimental incompatibility of their individual tests, after some moments of reflection might come to the following proposal: take two additional cubes, identical to the one in question, then test the burnability on one and the floatability on the other. If both tests are successful, one can affirm that the cube under consideration jointly possess these two properties (i.e., that the meet property of "burnability and floatability" is actual for it).

This is of course a possible way out to the problem of having to deal with procedures that are experimentally incompatible, so EPR could also have considered this line of reasoning to try to make their point. More precisely, they could have considered the possibility to make two identical copies of the quantum entity under investigation, measure the position on the first copy and the momentum on the second one, then present the argument that they can predict in this way, with certainty, these same values for the entity under consideration (the one that was perfectly copied), thus showing again a contradiction with Heisenberg's uncertainty principle.[d] Of course, since quantum measurements appear to be non-deterministic, this argument, to be valid, requires the duplication process to be "dispersion free," that is, such that possible hidden variables determining the measurement outcomes are also assumed to be faithfully copied in the process.

The reader may object that this is an invalid reasoning because of the celebrated *quantum no-cloning theorem* [32,33], establishing the impossibility of making a perfect copy of a quantum state.[e] The no-cloning theorem, however, only concerns universal copying machines, working independently of any a priori knowledge of the state to be cloned, and if we relax this condition, which we do not need for the argument, then the cloning can always in principle be worked out [36]. So, EPR could also have concluded in this case that quantum mechanics is incomplete, and once more the incompleteness cannot be associated with its inability to jointly attach position and velocity elements of reality to a micro-entity, but with its inability here of describing a perfect cloning process, when the (hypothetical) hidden variables associated with the state to be copied are unknown.

In other words, from the above reasoning one can deduce a hidden variables variant of the no-cloning theorem: no machine can copy unknown hidden variables. Is this to be understood as an additional shortcoming of the quantum formalism? Not really, because we understand today the reason for this impossibility: hidden variables of this kind (delivering a deeper description of the reality of a physical entity) simply do not exist [25–30], so, they cannot be copied, as of course we cannot copy what does not exist.[f] *Mutatis mutandis*, Aerts' result can be understood as a *quantum no-separating theorem*, establishing the impossibility of separating two physical entities, or more generally of separating the measurements (or experimental tests) associated with two physical entities. Again,

---

[d]The reasoning using identical copies becomes of course more convincing if expressed in relation to Bohm's version of the EPR-type situation.

[e]Interestingly, the no-cloning theorem was proven *ante litteram* by Park in 1970 [34], when precisely investigating the possibility of achieving a universal non-disturbing measurement scheme [35].

[f]Note that although the no-go theorems tell us that there are no hidden variables associated with a quantum state, hidden variables can nevertheless be attached to the measurement interactions, in what was called the hidden-measurement interpretation of quantum mechanics, an approach initiated by Aerts in the eighties of last century which remains today a viable line of investigation (see [21,37] and the references cited therein).

we can ask: Should this be considered as a shortcoming of the quantum formalism?

Well, maybe, as to consistently talk about a physical entity, and do physics, one must be able to consider it as a phenomenon that is separate from the rest of the universe [6]. Consequently, any physical entity which belongs to "the rest of the universe" of that physical entity, will also have to be considered to be separate from it. But this is precisely a situation that cannot be consistently described by standard quantum mechanics. The *quantum measurement problem* could also be related to this limitation of the orthodox formalism, as in a measurement process the measured entity has to be initially separated from the measurement apparatus, enter into contact and interact with it, thus connect with it, then finally be separated again from it. If this connection-separation process cannot be properly described, the only way out seems that of reverting to a many-worlds picture/interpretation [37,38], where separations are introduced at the level of the universes (superposition states being then described as collections of collapsed states in different universes), a move that surely would not have pleased friar Occam.

Another difficulty one can consider, consequence of the limitations expressed by this structural impossibility of separating measurements and therefore entities, is in relation to the study of (mesoscopic) structures that are in-between the quantum and classical regimes, and the quantum-classical limit. Indeed, one would need for this a more general mathematical structure for the lattice of properties than that inherited from Hilbert space and the Born rule, able to integrate both classical and quantum features. This in turn means dispensing with two of the axioms of orthodox quantum mechanics, in its lattice approach, called *weak modularity* and *covering law* [6,22,23]. Quoting from [39]: "A new theory dispensing with these two axioms would allow for the description not only of structures which are quantum, classical, mixed quantum-classical, but also of intermediate structures, which are neither quantum nor classical. This is then a theory for the mesoscopic region of reality, and we can now understand why such a theory could not be built within the orthodox theories, quantum or classical."

One can of course object that what quantum mechanics has really shown us is that all in our physical reality is deeply interconnected, that is, entangled, and that separation would be an illusion or, better, something like an effect emerging from a fundamentally interconnected non-spatial substratum, described in a correct and complete way by quantum mechanics. This is of course a possibility, although not all physicists seem to be ready to accept all the consequences of it, like the one previously mentioned of resorting to parallel universes. We live surrounded by macroscopic entities which apparently do not show quantum effects, i.e., for which separate experimental tests can be defined. If we test a property of a wooden cube, this will not influence in whatsoever way a test we may want to perform on another wooden cube. But this cannot be generally true if the Hilbertian formalism and associated superposition principle is believed to be universal. Of course, to put two wooden cubes in a state such that experiments performed on them would not anymore be separate appear to be extremely difficult to achieve, but it remains a possibility if the standard quantum formalism is considered to be fundamental.

I personally believe that we do not know enough about our physical world to take a final stance on those difficult questions, so I think it is important to also have the possibility of studying the behavior of the different physical entities (and I stress again that the very notion of "physical entity" requires a notion of separability) in a theoretical framework which does not attach any a priori fundamental role to the linear Hilbert space structure and associated Born rule, particularly when addressing challenging scientific problems like the one of finding a full-fledged quantum gravity theory.

References

1. Einstein, A., Podolsky, B. and Rosen. N. (1935). Can quantum-mechanical description of physical reality be considered complete? *Phys. Rev.* 47, 777–780.

2. Aspect, A., Grangier, P. and Roger, G. (1982). Experimental Realization of Einstein-Podolsky-Rosen-Bohm Gedankenexperiment: A New Violation of Bell's Inequalities. *Phys. Rev. Lett.* 49, 91.

3. Aspect, A. (1999). Bell's inequality test: more ideal than ever. *Nature (London)* 398, 189–190.

4. Hensen B. et al (2015). Loophole-free Bell inequality violation using electron spins separated by 1.3 kilometres. *Nature* 526, 682–686.

5. Aerts, D. (1981). The One and the Many: Towards a Unification of the Quantum and Classical Description of One and Many Physical Entities. *Doctoral dissertation*, Brussels Free University.

6. Aerts, D. (1982). Description of many physical entities without the paradoxes encountered in quantum mechanics. *Found. Phys.* 12, 1131–1170.

7. Aerts, D. (1983). The description of one and many physical systems. In: *Foundations of Quantum Mechanics*, C. Gruber (Ed.), Lausanne: AVCP, 63–148.

8. Aerts, D. (1984). The missing elements of reality in the description of quantum mechanics of the EPR paradox situation. *Helv. Phys. Acta* 57, 421–428.

9. Aerts, D. (1984). How do we have to change quantum mechanics in order to describe separated systems? In: *The Wave-Particle Dualism*, S. Diner et al. (Eds.), D. Reidel Publishing Company, 419–431.

10. Aerts, D. (1984). The physical origin of the Einstein-Podolsky-Rosen paradox. In: *Open Questions in Quantum Physics*, G. Tarozzi and A. van der Merwe (Eds.), D. Reidel Publishing Company, 33–50.

11. Schrödinger, E. (1935). Discussion of probability relations between separated systems. *Mathematical Proceedings of the Cambridge Philosophical Society* *31*, 555–563.

12. Schrödinger, E. (1935). Die gegenwärtige Situation in der Quantenmechanik. *Die Naturwissenschaftern* *23*, 807–812, 823–828, 844-849. English translation: Trimmer, J. D. (1980). The present situation in quantum mechanics: a translation of Schrödinger's 'cat paradox' paper. *Proceedings of the American Philosophical Society* *124*, 323–328.

13. Aerts, D. (1990). An attempt to imagine parts of the reality of the microworld. In: *Problems in Quantum Physics II; Gdansk '89*, J. Mizerski, et al. (Eds.), World Scientific Publishing Company, Singapore, 3–25.

14. Bell, J. S. (1964). On the Einstein Podolsky Rosen paradox. *Physics 1*, 195–200. Reproduced as Ch. 2 of Bell, J. S. (1987). *Speakable and Unspeakable in Quantum Mechanics*, Cambridge University Press.

15. Bell, J. S. (1971). In: *Foundations of Quantum Mechanics*, Proceedings of the International School of Physics "Enrico Fermi," Course XLIX, B. d'Espagnat (Ed.), Academic Press, New York, 171–181; and Appendix B. *Speakable and Unspeakable in Quantum Mechanics* (Cambridge University Press, 1987).

16. Aerts, D. (1982). Example of a Macroscopical Classical Situation that Violates Bell Inequalities. *Lettere al Nuovo Cimento 34*, 107–111.

17. Sassoli de Bianchi, M. (2013). Quantum dice. *Ann. Phys. 336*, 56–75.

18. Sassoli de Bianchi, M. (2014). A remark on the role of indeterminism and non-locality in the violation of Bell's inequality. *Ann. Phys. 342*, 133–142.

19. Aerts, D., Arguëlles, J. A., Beltran, L., Geriente, S., Sassoli de Bianchi, M., Sozzo, S., Veloz, T. (2018). Spin and Wind Directions I: Identifying Entanglement in Nature and Cognition. *Found. Sci. 23*, 323–335.

20. Aerts, D., Arguëlles, J. A., Beltran, L., Geriente, S., Sassoli de Bianchi, M., Sozzo, S., Veloz, T. (2018). Spin and Wind Directions II: A Bell State Quantum Model. *Found. Sci. 23*, 337–365.

21. Aerts, D. and Sassoli de Bianchi, M. (2016). The Extended Bloch Representation of Quantum Mechanics. Explaining Superposition, Interference and Entanglement. *J. Math. Phys. 57*, 122110.

22. Piron, C. (1976). *Foundations of Quantum Physics*, W. A. Benjamin Inc., Massachusetts.

23. Piron, C. (1978). La Description d'un Système Physique et le Présupposé de la Théorie Classique. *Annales de la Fondation Louis de Broglie 3*, 131–152.

24. Bohr, N. (1935). Can quantum-mechanical description of physical reality be considered complete? *Phys. Rev. 48*, 696–702.

25. Von Neumann, J. (1932). Grundlehren. *Math. Wiss. XXXVIII*.

26. Bell, J. S. (1966). On the Problem of Hidden Variables in Quantum Mechanics. *Rev. Mod. Phys. 38*, 447–452.

27. Gleason, A. M. (1957). Measures on the closed subspaces of a Hilbert space. *J. Math. Mech. 6*, 885–893.
28. Jauch, J. M. and Piron, C. (1963). Can hidden variables be excluded in quantum mechanics? *Helv. Phys. Acta 36*, 827–837.
29. Kochen, S. and Specker, E. P. (1967). The problem of hidden variables in quantum mechanics. *J. Math. Mech. 17*, 59–87.
30. Gudder, S. P. (1970). On Hidden-Variable Theories. *J. Math. Phys 11*, 431–436.
31. Piron, C. (1975). Survey of General Quantum Physics. In: C. A. Hooker (Ed.), *The Logico-Algebraic Approach to Quantum Mechanics*, The University of Western Ontario Series in Philosophy of Science, vol. 5a. Springer, Dordrecht.
32. Wootters W. and Zurek, W. (1982). A Single Quantum Cannot be Cloned. *Nature 299*, 802–803.
33. Dieks, D. (1982). Communication by EPR devices. *Phys. Lett. A. 92* 271–272.
34. Park, J. L. (1970). The concept of transition in quantum mechanics. *Found. Phys. 1*, 23–33.
35. Ortigoso, J. (2018). Twelve years before the quantum no-cloning theorem. *Am. J. Phys. 86*, 201–205.
36. Buzek V. and Hillery, M. (1996). Quantum copying: beyond the no-cloning theorem. *Phys. Rev. A 54*, 1844–1852.
37. Aerts D. and Sassoli de Bianchi, M. (2014). The Extended Bloch Representation of Quantum Mechanics and the Hidden-Measurement Solution to the Measurement Problem. *Ann. Phys. (N. Y.) 351*, 975–102.
38. H. Everett, H. (1957). Relative State Formulation of Quantum Mechanics. *Review of Modern Physics 29*, 454–462.
39. Aerts, D. (1999): The Stuff the World is Made of: Physics and Reality. In: *The White Book of 'Einstein Meets Magritte'*, D. Aerts, J. Broekaert and E. Mathijs (Eds.), Kluwer Academic Publishers, Dordrecht, 129–183.

THE RELATIVISTIC TRANSACTIONAL INTERPRETATION: IMMUNE TO THE MAUDLIN CHALLENGE

R. E. KASTNER

Department of Philosophy, University of Maryland,
College Park, MD 20742, USA
* E-mail: ab_rkastner@umd.edu

The Transactional Interpretation has been subject at various times to a challenge based on a type of thought experiment first proposed by Maudlin. It has been argued by several authors that such experiments do not in fact constitute a significant problem for the transactional picture. The purpose of this work is to point out that when the relativistic level of the interpretation is considered, Maudlin-type challenges cannot even be mounted, since the putative 'slow-moving offer wave,' taken as subject to contingent confirmation, does not exist. This is a consequence of the Davies relativistic quantum-mechanical version of the direct-action theory together with the asymmetry between fermionic field sources and bosonic fields. The Maudlin challenge therefore evaporates completely when the relativistic level of the theory is taken into account.

Keywords: Transactional interpretation; Maudlin challenge; contingent absorber experiments.

1. The Basics: A brief review

The Transactional Interpretation (TI), first proposed by John Cramer [1], is based on the direct-action theory of electromagnetism by Wheeler and Feynman [2]. A relativistic extension of TI has been developed by the present author; that is based on Davies' direct-action theory of quantum electrodynamics [3]. Due to its possibilist ontology, that model has been termed 'PTI' [4], but the important feature is its relativistic nature, which provides further clarification of the conditions for emission and absorption. Therefore, for purposes of this discussion and going forward, I will refer to that model as the Relativistic Transactional Interpretation, RTI.

First, some terminology: in TI and RTI, the usual quantum state $|\Psi\rangle$ is called an 'offer wave' (OW), and the advanced response $\langle a|$ of an absorber A is called a 'confirmation wave' (CW). In general, many absorbers $A, B, C, \ldots$ respond to an OW, where each absorber responds to the component of the OW that reaches it. The OW component reaching an absorber X would be $\langle x|\Psi\rangle|x\rangle$, and it would respond with the adjoint (advanced) form $\langle x|\langle\Psi|x\rangle$. The product of these two amplitudes corresponds to the final amplitude of the 'echo' of the CW from X at the locus of the emitter (this was shown in [1] and reflects the Born Rule as a probabilistic weight of the 'circuit' from the emitter to absorber and back, the latter being called an *incipient transaction*. Meanwhile, the sum of the weighted outer products (projection operators) based on all CW responses–each representing an incipient transaction–constitutes the mixed state identified by von Neumann as resulting from the non-unitary process of measurement (cf [4], Chapter 3). Thus, TI provides a physical explanation for both the Born Rule and the measurement transition from a pure to a mixed state. The additional step from the mixed state to the 'collapse' to just one outcome is understood in RTI as an analog of spontaneous symmetry breaking; the 'winning' transaction, corresponding to the outcome of the measurement, is termed an *actualized transaction*. The absorber that actually receives the quantum is called the *receiving absorber*. This is to emphasize that other absorbers participate in the process but do not end up receiving the actualized quantum.

The other feature of this process, which gives it its possibilist ontology, is that the quantum entities (OW,CW, virtual quanta) are all pre-spacetime objects–Heisenbergian éxtitpotentiae. Spacetime events only occur as a final result of OW/CW negotiations, resulting in collapse to an actualized transaction. Thus, the collapse is not something that happens *within* spacetime; rather, collapse is the process of spacetime emergence. Specifically, what emerges as a result of collapse is the emission event, the absorption event, and their connection via the exchanged quantum (see [5]). (This point will be relevant later on.) It is only upon actualization of the transaction that a real quantum is emitted and absorbed at the receiving absorber.[a]

Now let us briefly review the Maudlin thought experiment ([6, p. 200];

---

[a]Maudlin is thus quite correct when he says: "It is also notable that in the electromagnetic case the relevant fields are defined on, and propagate over, space-time. The wave-function is defined on configuration space. Cramer does not seem to take account of this, writing always as if his offer and confirmation waves were simply being sent through space. Any theory which seeks to make the wave-function directly a medium of backwards causation ought to take this into account." [6, p. 203]. This weakness in Cramer's approach is corrected in RTI.

see Figure 1). It envisions a 'slow-moving OW' (assumed traveling at speed $v < c$) emitted at $t = 0$ in a superposition of rightward and leftward momentum states. On the right at some distance d is a fixed detector R, and positioned behind R (initially on the right) is a moveable detector L. If, after a suitable time has passed ($t_1 = \frac{d}{v}$), there is no detection at R, L is quickly swung around to intercept the OW on the left, where (so the proposal goes) a left-hand CW is generated and the particle must be detected at L with certainty. Thus, this is intended to be a 'contingent absorber experiment' ([4, Chapter 5]): it is assumed that the existence of a confirmation from the left-hand side is contingent on the transaction between the source and R failing.

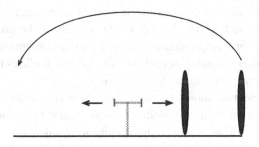

Fig. 1. Maudlin thought experiment.

Maudlin's intent was to provide a counterexample to the picture provided in [1], in which there are well-defined OW/CW matchups for all possible detection sites. The challenge presented for the original TI was twofold: (i) the probability of $1/2$ for the leftward transaction was thought to be inconsistent with the fact that whenever the left-hand CW was present that transaction would always be actualized; and (ii) the situation at $t = 0$ appeared ill-defined, since (if the CW is really contingent as imagined) it is uncertain whether or not the (backward-evolving) CW will be emitted from the left.

Both these concerns have been addressed and resolved elsewhere (cf. [7] and [4, Chapter 5]). However, these responses assumed that the Maudlin experiment could in-principle be carried out. The purpose of this paper is to observe that in fact this is not the case; no such experiment can actually be done, and therefore the challenge disappears.

2. Applicability of the 'offer wave' concept

Since it has been shown that any quantized field theory can be re-expressed as a direct action theory [8], RTI takes all such field excitations as offer waves. That is, any field for which the basic Davies model holds is a component of the transactional model, and transfers of real quanta of those fields can be understood as the result of actualized transactions. However, transactions occur in different ways depending on whether the field is a source of other fields. This issue will be explored in what follows.

In addition, this model has intrinsic restrictions on what sorts of 'particles' constitute offer waves. That is, some objects describable as quantum systems, such as atoms, do not constitute offer waves, in that they are not excitations of a specific quantum field–instead, they are bound states [9]. On the other hand, some types of offer waves can participate in actualized transactions indirectly, through confirmations of the products of their interactions, rather than by generating confirmations themselves. This work discusses both these situations, and then applies the findings to the Maudlin challenge to see why it cannot be mounted.

First, as indicated above, the 'offer wave' concept refers to the excited states of a quantum field. A specific example would be a one-photon Fock state $|k\rangle$. On the other hand, if the system at hand is not a specific field excitation of this sort, even though it may still be described by an effective quantum state, it is not an offer wave. It therefore does not generate a corresponding confirmation wave. As noted above, an example of such a system would be an atom, which is a bound state of several different quantum fields as opposed to an excitation of a single quantum field.

At this point the relevance for the Maudlin challenge is already evident: the latter proposes a 'slow-moving quantum' subject to contingent absorption. The 'slow-moving quantum' cannot be anything other than a field excitation for a quantum with nonvanishing mass if it is to constitute an offer wave, so an atom cannot instantiate the experiment. In any case, in the possibilist ontology, OW do not propagate within spacetime at subluminal speeds: they are phase waves, as opposed to group waves. It is only the actualized quantum that propagates at the subluminal group wave velocity–that in itself nullifies the Maudlin challenge. But suppose we overlook that point for now. In order to obtain an offer corresponding to a subluminal quantum, one must use a matter field, such as the Dirac field. The latter will be a source of bosonic fields, which brings us to the second important point: the asymmetry between field sources and their generated fields gives rise to a situation in which a field source participates in transactions only indirectly, by way of its emitted field.

3. Field sources are actualized without matching confirmations

In quantum electrodynamics, the (fermionic) Dirac field is the source of the (bosonic) electromagnetic field, but the following considerations apply to any quantum field and its sources. It is well known that in interactions between fields, the field source has a different physical character from the field of which it is a source. This distinction is reflected in the fact that gauge bosons are the force carriers, as opposed to the fermionic matter fields which are sources of gauge bosons. The asymmetry in question is exhibited for example in the basic QED vertex, which has only one photon line, plus an incoming and outgoing fermion line, due to the nature of the coupling between the Dirac Field and the electromagnetic field, given by $eA_\mu \times \bar{\Psi}\gamma^\mu\Psi$.

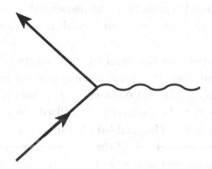

Fig. 2.   QED vertex.

Due to this asymmetry, not all offer waves generate their own confirmations when participating in transactions. Fermionic field sources participate in transactions indirectly, by way of confirmations of the fields of which they are a source.[b] For example (see Figure 2), an electron OW is liberated from a bound state by absorbing an incoming photon from another

---

[b]Even if fermionic quantized fields can be formally recast as direct-action fields, only bosonic fields (subject to a 'gauge field' description) engage in transactions by way of their own confirmations. The deep physical meaning behind this is that only the bosonic

208

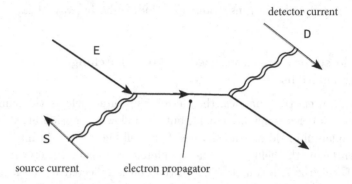

detector current

D

E

S

source current   electron propagator

Fig. 3.   Electron detection.

charged current S, and then emits a photon offer wave, which is confirmed
by another charged source field D (typically an electron) in the detector. In
the resulting actualized transaction, the associated outgoing electron OW
component is actualized as well, even though it was not confirmed by an
'electron CW.'[c]

The asymmetry between the fermionic field source (the electron E) and
its emitted/absorbed fields (the photon lines) is again evident here in that
it is an electron propagator that connects to two interaction vertices. This
allows the electron E to be indirectly actualized via its interaction with
the electromagnetic field. (The doubled photon lines indicate that a CW is
generated.) Thus, upon detection of the emitted photon by D, the electron
E is actualized without ever having generated its own confirmation.[d]

---

fields correspond to spacetime symmetries. Thus, when a Dirac field quantum such as
an electron is detected, that detection is always indirect, by way of its interaction with
the electromagnetic field.
[c]It should be kept in mind that these are all offer wave components, subject to the Born
Rule in that they will only be actualized with the corresponding probability. Also, a
given process involving particular incoming and outgoing quanta is a superposition of
all possible diagrams connecting those incoming and outgoing quanta. Here we consider
just the lowest-order diagram for simplicity, but the transaction is a sum of all such
diagrams.
[d]The electron propagator is still summed over all possible spacetime locations, as in
the usual Feynman diagram calculation. But now we have a reason for the pole in the
Feynman electron propagator: it represents the real, as opposed to virtual, electron
resulting from an actualized transaction in which real negative charge is transferred

4. OW and CW symmetry

The other new and important feature of the relativistic development of TI (RTI) is that no offer—i.e. quantum state $|\Psi\rangle$—will be emitted in the first place without the basic direct connection (i.e., the time-symmetric field correlation or direct-action propagator) between the potentially emitting system and at least one potentially absorbing system (for details, see [10]). That means that one must have a basic interaction with at least one absorber in order to have an offer at all; such an interaction is a necessary, but not sufficient, condition for an offer to be emitted. The sufficient condition for an offer to be emitted is that a confirmation also exist. Thus, there is no such thing as an isolated OW.

This may seem surprising, but it is because the offer corresponds to a real photon as opposed to a virtual photon (propagator). The only way one can have a real photon in the direct action picture is as a result of absorber response (this is discussed in [10] and implicitly in Davies [3]). The real photon corresponds to the pole in the Feynman propagator, which is only present when there is absorber response (in general, more than one absorber will respond). The pole is what corresponds to a Fock state $|k\rangle$; i.e., an offer wave (or offer wave component). Thus, to get an offer, one always has to have absorber response. This means that the picture of a sequence of stages in which there is first an OW and then one or more CW is not accurate at the fully relativistic level (which is the level at which Nature really operates).

Moreover, since there has been absorber response (usually from many absorbers), technically the correct description of the real photon is the density operator (weighted set of projection operators) corresponding to von Neumann's 'Process 1' (measurement transition). It is only when we consider the emission process by itself, without taking into account that it only occurs because of absorber response, that we label the emission by a ket $|\Psi\rangle$, thinking of it as an independent entity (as in the Maudlin experiment). But it is not. In the direct-action theory, the ket is only a partial description of the process. So it is simply not accurate in the direct-action picture to view an offer as something that is independently emitted; rather, emission is always a two-way process, with the absorber crucially participating in any emission.

---

from the emitting bound state (cathode) to the receiving bound state (e.g., a conduction band) by way of the actualized incoming and outgoing photons.)

5. Why the Maudlin challenge evaporates

Returning now specifically to the Maudlin experiment: in order for the 'slow-moving quantum' to be considered an offer, it would have to be a non-composite matter field excitation of some sort, such as an electron state $|\psi>$. The latter is a source of a bosonic field (the electromagnetic field), which is the mediator of electron detection, as described above.[e] Recall that a free electron can neither emit nor absorb a real photon offer wave (due to energy conservation). Thus, an electron subject to detection is always liberated from some bound state (by absorbing electromagnetic energy) and detected via its becoming part of a new bound state (by emitting electromagnetic energy), not through being confirmed by a matching 'electron confirmation.' For the latter would mean that it was a free electron.

To see this, refer again to Figure 2. However, it should be kept in mind that the described processes do not occur in a temporal sequence within spacetime; rather, they occur at the pre-spacetime level, as a negotiation of possibilities. On the incoming side on the left, a photon offer is emitted from another charged current (labelled S) and absorbed by the electron offer E, which is thereby liberated from its initial bound state. Meanwhile, at the outgoing side, E emits another photon offer. The latter is confirmed by a charged current D in the detector, the confirmation propagating back through the electron propagator and actualizing the absorbed photon that liberates E (which is why both photon lines are doubled). The actualization of the emissions and absorptions of both photons actualizes E as an emitter/absorber, since (recall from Section 1) all actualized transactions actualize three things: the emission event, the absorption event, and the transferred quantum. Finally, the outgoing current E becomes incorporated into another bound state (such as a conduction band in a metal) rather than prompting its own confirmation. None of these processes can occur without confirmation of the photon OWs, since (as discussed in the previous section) a necessary condition for a photon OW to exist at all is that a matching CW be generated. A rough analogy is a tug-of-war between two parties: there is no tug-of-war until both parties pick up their respective ends and begin pulling. The 'possible' processes described above are just the placing of the rope between them in preparation for the actual game.

So suppose we tried to do the Maudlin experiment with an 'offer wave'

---

[e]The weak field is a massive boson, but its range is far too short to be useful for the Maudlin experiment.

corresponding to a subluminal electron state. We could arrange for the electron offer to be in a superposition of rightward and leftward directions, but in order to it to be liberated at all, real energy would have to be supplied–i.e., a real photon would have to be absorbed by the bound electron subject to emission. The only way you get a real photon is through absorber response–otherwise you don't get the pole in the Feynman propagator corresponding to a Fock space state, i.e., the ket that you need for a photon offer wave. And you must have at least two QED interaction vertices for energy conservation. So the entire process presupposes absorber response at the detector end in order for the electron offer ever to be emitted at all. You simply don't get the electron offer in the first place without the complete photon incipient transaction (i.e., both incoming and outgoing photons must be confirmed). So the situation involving a contingent confirmation never exists; there are always photon confirmations for any offers in play, even if those offers are states of a subluminal fermionic matter field.

6. Conclusion

New developments of the Relativistic Transactional Interpretation (RTI) have been presented which nullify the Maudlin challenge for the Transactional Interpretation (TI). These new development are: (1) offer waves are excitations of quantum fields, so slow-moving composite quantum objects such as atoms are not eligible for the experiment; (2) fermionic matter fields describable as field excitations are not actualized by way of their own matching confirmations, but by confirmations of their emitted/absorbed fields; and (3) a necessary and sufficient condition for the existence of a photon offer wave (OW) is that a confirmation (CW) also be generated. (All these processes involving OW and CW are pre-spacetime processes.) These developments result in the evaporation of the Maudlin challenge, since there is no 'slow-moving offer wave' to begin with, unless it is a non-composite fermion such as an electron.[f] In the latter case, an electron offer is actualized by way of photon OW of which it is a source. And in that process, a condition for a photon OW in the first place is that a photon CW is also generated. Thus, even if we had a slow-moving electron OW, there would always be accompanying photon OW and CW, and there would therefore never be a contingent CW situation–i.e., never a situation in which a CW is only generated based on some prior non-detection. Finally,

---

[f]Moreover, according to the possibilist ontology, all OW are phase waves, and they are pre-spacetime objects, so they are not subluminal anyway.

these observations should not be mistaken as *ad hoc* maneuvers to evade the Maudlin challenge; rather, they arise directly from the Davies QED absorber theory upon which RTI is based, but had not been previously taken into account.

Acknowledgments: The author is grateful to an anonymous referee for helpful suggestions for improvement of the presentation.

References

1. Cramer J G. The Transactional Interpretation of Quantum Mechanics. *Reviews of Modern Physics 58*, 647-688, 1986.
2. Feynman, R P and Wheeler, J A. "Interaction with the Absorber as the Mechanism of Radiation", *Reviews of Modern Physics, 17* 157-161 (1945); and "Classical Electrodynamics in Terms of Direct Interparticle Action", *Reviews of Modern Physics 21*, 425-433 (1949).
3. Davies, PCW. "Extension of Wheeler-Feynman Quantum Theory to the Relativistic Domain I. Scattering Processes", *J. Phys. A: Gen. Phys. 4*, 836 (1971); and "Extension of Wheeler-Feynman Quantum Theory to the Relativistic Domain II. Emission Processes", *J. Phys. A: Gen. Phys. 5*, 1025-1036 (1972).
4. Kastner R E. *The Transactional Interpretation of Quantum Mechanics: The Reality of Possibility*. Cambridge: Cambridge University Press (2012).
5. Kastner R E. "The Emergence of Spacetime: Transactions and Causal Sets," in Licata, I. (Ed.), *Beyond Peaceful Coexistence: The Emergence of Space, Time and Quantum*. London: Imperial College Press (2016). Preprint version: https://arxiv.org/abs/1411.2072
6. Maudlin T. *Quantum Nonlocality and Relativity*, 3rd Edition. Oxford: Blackwell (2011). pp. 184-185.
7. Marchildon, L. "Causal Loops and Collapse in the Transactional Interpretation of Quantum Mechanics" *Physics Essays 19*, 422-9 (2006)
8. Narlikar, J. V. "On the general correspondence between field theories and the theories of direct particle interaction." *Proc. Cam. Phil. Soc. 64*, 1071 (1968).
9. Kastner, R.E. "Bound States as Emergent Quantum Structures." in Kastner, R.E., Dugic J., and Jaroszkiewicz, G (Eds.), *Quantum Structural Studies*. Singapore: World Scientific (2016). Preprint version: https://arxiv.org/abs/1601.07169
10. Kastner, R.E. "On Real and Virtual Photons in the Davies Theory of Time-Symmetric Quantum Electrodynamics." *Electronic Journal of Theoretical Physics 11*, 30: 75-86 (2014). http://www.ejtp.com/articles/ejtpv11i30p75.pdf

THE INEFFABLE NATURE OF BEING.
IN MEMORIAM: BERNARD D'ESPAGNAT

JAIRO ROLDÁN-CHARRIA

Departamento de Física, Facultad de Ciencias
Universidad del Valle, Cali, Colombia
E-mail: jairoroldan@gmail.com

In this article, I present a brief personal testimony about Bernard d'Espagnat, as well as an analysis of his main ideas. d'Espagnat explicitly introduced the conceptual difference between proper and improper mixtures and pointed out that such a difference is fundamental when one looks for an interpretation of the formalism. After an exhaustive analysis of the notion of non-separability, d'Espagnat concludes that the concepts of space, time, space-time, events, and even the positions of things are nothing more than mere tools for the description of phenomena. He proposes to make a distinction between the *empirical reality*, which is the set of phenomena, that is, the totality of what human experience, seconded by science, yields access to, and the independent or *ontological reality*, that exists independently of our existences. His *Axiom of Empirical Reality* illuminates the significance of the models of coherence for the understanding of the reality of macroscopic bodies. The Axiom is proposed to replace –in the identification of the pure case involved in the measurement process by a proper mixture– the "for all practical purposes" argument by a postulate, that means something theoretical and not just practical. It must be emphasized that the postulate makes sense only within the empirical reality conception. His main contribution in the more philosophical domain is his thesis of the *Veiled Reality*. In this paper, I present a critical analysis of that thesis comparing it with the ideas of Kant and Spinoza. In the end, I briefly mention some of other of his contribution in the domain of the conceptual foundation of quantum mechanics.

Keywords: D'Espagnat; proper and improper mixtures; ontological reality; empirical reality; quantum measurement; decoherence; axiom of empirical reality; veiled reality.

1. My personal experience with Bernard d'Espagnat

I met Bernard d'Espagnat in November 1984. In the previous months, we had had an exchange of letters that had begun with one in which I asked him if he would accept to be my director of Ph.D. Thesis. I had spent several years of research on various topics in theoretical physics, and my interest had finally concentrated on the conceptual foundations of quantum mechanics. d'Espagnat was one of the great world experts in the area, which is why I was interested in having the opportunity to work under his direction. To my luck, d'Espagnat agreed to advise me, and I then moved with my family to Paris in late 1984. My work began with a suggestion: to examine in depth the ideas of Bohr and Heisenberg on the interpretation of quantum mechanics, and ended in 1990 with the writing of my Thesis: *Language, Mécanique Quantique et Réalité.*[a]

When I was to return to Colombia in 1987 to continue working on my thesis, immersed again in the teaching duties of a full-time professor, he told me that he was about to retire and that he would no longer have any teaching or administrative obligations. He had been appointed Professor Emeritus which allowed him to continue using his office in Orsay and was happy to be able to devote himself entirely to research; he told me in particular that there was still much to be said about the quantum measurement theory.

During my years of direct interaction with d'Espagnat, I was able to know and enjoy his exquisite courtesy, the penetration of his mind, his immense curiosity and his high culture. I was fortunate enough to be received several times by him and his wife in his apartment on the rue d'Assaz where there were as an adornment of his library several small paintings of his father the post-impressionist painter Georges d'Espagnat. In addition to his books on the foundations of quantum mechanics, he edited one about his father's work. I remember that, when I mentioned his various interests, he jokingly replied that he had not yet undertaken the search for gold as his uncle, Pierre d'Espagnat. His uncle had been an indefatigable traveler and had written a book about his trip in Colombia between June 1897 and May of 1898 when it traveled extensively to all the places of the country. In one of my last visits, d'Espagnat had obtained a copy of his uncle's book: *Souvenirs de la Nouvelle Grenadel*[b] and seemed to be enjoying his reading.

Once I finished my thesis I saw him briefly in 1993 at a congress in Paris. In 1998, coming from Israel, I passed some few days in Paris and called him expressing my congratulations for his membership since 1996 in the Academy of Political and Moral Sciences of the Institute of France. He thanked me with his usual courtesy, and when I asked him how he was,

---

[a] Language, Quantum Mechanics and Reality.
[b] Memories of New Grenada.

he responded with a phrase I have never forgotten: *"La vie continue sans aucune surprise."*[c]

In May 2009 I sent him a letter with my felicitations for the Templeton Prize he had recently received and proposed him a visit for the fall of that year. He responded by email thanking me for my congratulations and accepting my visit. During the followings months, the idea of writing a book about his thought arose in my mind. On my visit to his apartment in Paris, which took place on November 25, I was able to see with joy that at 88 years his mind was as clear and penetrating as ever. He liked my project of writing a book about his thinking and commented that I would surely want to ask him questions about it. I gladly accepted his suggestion and expressed my desire to ask him questions by email and revisit him to discuss the project.

On October 13, 2010, I visited him. I showed him the title and the general structure of the book. Both the structure and the title were to his liking. The title would be *L'ineffable nature de l'Être. An essai sur la pensée de Bernard d'Espagnat.*[d] We had a dialogue that he kindly agreed to let me record.

On Wednesday 29 January 2011 we had our last meeting in Paris. One thing became apparent to me from that last conversation: that the way he wanted me to analyze his thinking implied a change in the audience to which the text would be directed, not the general public, as I initially thought, but the academic. That change would imply a much greater effort than I had anticipated. The time passed on, and among the various academic obligations of mine, the project was fading away. In August of 2015, I learned of the sensible death of d'Espagnat and knew that with his passing, of the project would only survive the title, which I use today to write with nostalgia this article in memoriam of one who was my doctoral thesis advisor and one of the major influences in my intellectual evolution.

Since he explicitly said that, before any published use of it, he wanted to review the recorded dialogue mentioned above, I am not allowed to transcribe it here. I can, however, mention two points supposing he wouldn't disagree with my mentioning them. In an interview that he gave to *Les Humaines Associés*, to the question if it was his research that has given him his worldview, his answer was that he had started with a vision of departure. I asked him then about the way he arrived in his early days to

---

[c]Life continues without any surprise.
[d]The Ineffable Nature of Being. An Essay on Bernard d'Espagnat's thought.

that vision. He mentions that his father and his mother loved each other a lot, but they were very different since his father was not a believer and his mother was. From the very beginning of his existence, he had seemed then that there was a problem in the conception of reality, whereas the majority of the people around him did not appear to raise this question either because they had an education as believers, or because they had one as unbelievers. The second point he mentioned was that, in high school, he did both the scientific baccalaureate and the philosophical baccalaureate at the same time, since he had not yet chosen his way. Finally, he took the scientific way, but with the idea that he wanted to understand what can one honestly deduce from science. In other words, he did not choose science by the science itself but science by the philosophy.

2. Biographical sketch

He was born on 22 August 1921 in Fourmagnat, France. He passed away on August 1, 2015, in Paris. He studied at the École Polytechnique and then did his doctoral thesis at the Henry Poincaré Institute under the direction of Louis de Broglie. Together with Jacques Prentky and Roland Omnès founded in 1954 the group of theoretical physics of CERN. He was Professor at the Sorbonne and from 1965 was Director of the Laboratoire de physique théorique et particules élémentaires of the University of Orsay. From 1965 onwards, worked intensely on the refutation of Bell's inequalities by quantum mechanics, which Alain Aspect verified between 1980 and 1982. He also worked on quantum entanglement phenomena, the various interpretations of quantum mechanics, the problem of measurement and the nature of reality. Since 1975 he was a member of the International Academy of Philosophy of Sciences in Brussels and since 1996 a member of the Academy of Moral and Political Sciences of the Institute of France. In 2009 he received the Templeton Prize and decided to found the Collège de Physique et Philosophie.

3. Distinction between proper and improper mixtures

d'Espagnat explicitly introduced the conceptual difference between proper mixtures and improper mixtures (pure cases) in 1995 [1] and 1996 [2]. In several of his writings d'Espagnat pointed out that such a difference is fundamental when one looks for an interpretation of the formalism. Let us recall the notions of proper and improper mixtures and their conceptual differences.

3.1. *Pure case*

Let E_0 be an ensemble of N_0 systems and $A_1, A_2, ..., A_j, ...$ a number of observables whose operators $\left\{ \hat{A}_1, \hat{A}_2, ..., \hat{A}_j, ... \right\}$ form a complete set with eigenvectors $\left| \psi_{a_{1k}, a_{2l}, ... a_{jm}, ...} \right\rangle$. To simplify the notation, we will write a single index n to represents the collective $a_{1k}, a_{2l}, ... a_{jm}, ...$, whereby the eigenvectors will be simply noted by $|\psi_n\rangle$. We know that if at t_0 the systems of the ensemble are subjected to a simultaneous first kind measurement of the observables in question, the result will be one of the n. Immediately after the measurement, the systems remain in some of the $|\psi_n\rangle$. The ensemble E_n of those N_n systems that correspond to the same ket $|\psi_n\rangle$ form what is called a pure case.

It is said in this case that one has all the possible information, which means that the state is "prepared" by making a measurement of a complete set of operators and selecting those systems that correspond to the same eigenvector. Or that one has all the information equivalent to the simultaneous first kind measurement of the observables that form a complete set.

For example: let's consider an experiment of dispersion of muons by hydrogen atoms. It is necessary to assign a ket to both the hydrogen atom and the muon incident on it. Now, the experimentalist does not actually measure the energy of the particular atoms on which the dispersion takes place, but considers that he knows the energy of those atoms because he knows that the temperature is small and that therefore those atoms must be in the ground state. That is: the observables are the energy, the angular momentum and the projection of the angular momentum along an axis; the systems constituting the ensemble are hydrogen atoms. The reason why it is assumed that the result of simultaneous first kind measurements of the observables in question is known with certainty is that the temperature is known. It should be added that the spectrum of the hydrogen atom is also known. From the knowledge one has, it is assumed that, if the temperature is small, the carrying out of a measurement of observables would give the result: $n = 1$, $l = 0$, $m = 0$ and that result would be the same for all elements of the ensemble. In this case, the ensemble is described by a ket: that particular eigenvector of the complete set of observables corresponding to the values $n = 1$, $l = 0$, $m = 0$.

3.2. *Mixture*

If for the same ensemble E_0 of N_0 systems, a simultaneous first kind measurement of all observables whose operators form a complete set is not made, it is not possible by the above procedure to construct a pure case. In fact, let us consider that by measuring the complete set of operators $\left\{\hat{A}_1, \hat{A}_2, ..., \hat{A}_j, ...\right\}$ the pure case is prepared where all the systems of the ensemble E_n of N_n systems remain immediately after the measurement in the ket $|\psi_n\rangle$.

Let us now consider another complete set of operators $\left\{\hat{B}_1, \hat{B}_2, ..., \hat{B}_t, ...\right\}$ such that $\left[\hat{A}_j, \hat{B}_t\right] \neq 0 \quad \forall \quad j, t$ and the measurement of a single variable B_1. If the result of its measurement is the value b_{1k} the system is immediately after the measurement in the vector:

$$|\Phi_{1k}\rangle = \sum_r d_{b_{1k},r} |\psi_{b_{1k},r}\rangle$$

where the $|\psi_{b_{1k},b_{2l},...,b_{tm},...}\rangle$ are the eigenvectors of the operators $\left\{\hat{B}_1, \hat{B}_2, ..., \hat{B}_t, ...\right\}$.

The state immediately before the measurement of B_1 is then $|\psi_n\rangle$. The probability to obtain in the measurement each possible eigenvalue b_{1k} will depend on the coefficients of $|\psi_n\rangle$ in the base of the $|\psi_{b_{1k},r}\rangle$. If

$$|\psi_n\rangle = \sum_{k,r} f_{b_{1k},r} |\psi_{b_{1k},r}\rangle$$

the probability to obtain b_{1k} after the measurement will be

$$p_{b_{1k}} = \sum_r |f_{b_{1k},r}|^2$$

One can express everything in terms of ensembles. E_n is the ensemble of systems before the measurement of B_1. The number of systems in the ensemble is N_n. Also, it is a pure case because each system of the ensemble is described by the same vector $|\psi_n\rangle$. What happens to the systems in the ensemble E_n after the measurement of B_1? The probability of obtaining b_{1k} for the systems in E_n is

$$p_{b_{1k}} = \frac{N_{1k}}{N_n}$$

where N_{1k} is the number of systems for which the same result b_{1k} is obtained after the measurement. N_{1k} systems will be left with vector $|\Phi_{1k}\rangle$ and form a sub-ensemble E_{1k}. Thus,

$$\sum_r |f_{b_{1k},r}|^2 = \frac{N_{1k}}{N_n}$$

The ensemble E_n will give rise, after the measurement, to a *mixture* composed of sub- ensembles E_{1k}. The sub ensembles E_{1k} has N_{1k} systems each of which is associated with the vector $|\Phi_{1k}\rangle$. This mixture describes then the following situation: With the information one has, which is the result of the measurement of B_1, one can only know that:

i) The possible eigenvectors consistent with the information are $|\Phi_{11}\rangle, |\Phi_{12}\rangle, ..., |\Phi_{1k}\rangle, ...$

ii) The probability that a system of the ensemble is described by $|\Phi_{1k}\rangle$ is $p_{b_{1k}}$

If a system of the mixture is chosen at random, the probability that it is described by $|\Phi_{1k}\rangle$ is clearly $p_{b_{1k}}$.

Now, for a pure case described by a ket for any observable A,

$$\langle A \rangle = \langle \psi | \hat{A} | \psi \rangle$$

How are the averages calculated in the case of a mixture?

3.2.1. Calculations of averages for a mixture

Let's consider a general mixture composed of sub ensembles E_k. The sub ensemble E_k has N_k systems each of which is associated with the vector $|\psi_k\rangle$. The total number of systems in the ensemble is:

$$N = \sum_k N_k$$

If a system of the mixture is chosen at random, the probability that it is described by $|\psi_k\rangle$ is clearly

$$p_k = \frac{N_k}{N}$$

How are calculations done in this ensemble?

The *statistical operator* $\hat{\rho}$ is defined as

$$\hat{\rho} \equiv \sum_k \frac{N_k}{N} |\psi_k\rangle \langle \psi_k| \tag{1}$$

Usually, the statistical operator associated with an ensemble is called the *density matrix* associated with the ensemble. It can be shown that

$$\langle A \rangle = \text{Tr}\left(\hat{\rho}\hat{A}\right) = \text{Tr}\left(\hat{A}\hat{\rho}\right)$$

3.2.2. *Statistical operator for a pure case*

It is a particular case in which all systems of the ensemble are described by the same vector $|\psi\rangle = \sum_n a_n |u_n\rangle$. There is no sum then over k. Then, as expected:

$$\langle A \rangle = \overline{A} = \langle \psi | \hat{A} | \psi \rangle$$

And according to (1),

$$\hat{\rho} = |\psi\rangle\langle\psi|$$

Also,

$$\text{Tr}\left(\hat{\rho}\hat{A}\right) = \langle \psi | \hat{A} | \psi \rangle$$

3.3. *Application to a composite system*

Let us extend the notion of density matrix to an ensemble E_T of N systems composed of subsystems S_U and S_V, then: $\mathcal{H}_T = \mathcal{H}_U \otimes \mathcal{H}_V$.

3.3.1. *Pure Case*

Every system of the ensemble E_T, N is described by the same vector $|\Psi\rangle = |\psi\rangle |\phi\rangle$ where $|\psi\rangle \in \mathcal{H}_U$ and $|\phi\rangle \in \mathcal{H}_V$. $|\psi\rangle |\phi\rangle \in \mathcal{H}_U \otimes \mathcal{H}_V$. Then,

$$\hat{\rho} = |\Psi\rangle\langle\Psi| = |\psi\rangle |\phi\rangle \langle\psi| \langle\phi|$$

3.3.2. *Mixture*

Ensemble E_T, N containing sub ensembles E_{ij} where each system is described by the same vector $|\Psi_{ij}\rangle = |\psi_i\rangle |\phi_j\rangle$ where $|\psi_i\rangle \in \mathcal{H}_U$ and $|\phi_j\rangle \in \mathcal{H}_V$. $|\psi_i\rangle |\phi_j\rangle \in \mathcal{H}_U \otimes \mathcal{H}_V$. Then,

$$\sum_{ij} N_{ij} = N$$

$$\hat{\rho} = \sum_{ij} \frac{N_{ij}}{N} |\Psi_{ij}\rangle\langle\Psi_{ij}| = \sum_{ij} \frac{N_{ij}}{N} |\psi_i\rangle |\phi_j\rangle \langle\psi_i| \langle\phi_j|$$

$$\left\langle \hat{O} \right\rangle = \mathrm{Tr}\left(\hat{\rho}\hat{O} \right)$$

Each E_{ij} in the mixture is a pure case with $\hat{\rho} = |\psi_i\rangle |\phi_j\rangle \langle\psi_i| \langle\phi_j|$.

3.4. Partial traces

Let E_U be an ensemble of systems S_U and E_V an ensemble of systems S_V. The Hilbert space of the ensemble E_U is $\mathcal{H}_{\mathcal{U}}$ and an orthonormal basis is $\{|u_i\rangle\}$. For the ensemble E_V there is similarly $\mathcal{H}_V$ and $\{|v_i\rangle\}$ respectively. Let's consider an ensemble E_T of total systems S composed of the subsystems S_U and S_V. The Hilbert space of S is $\mathcal{H}_{\mathcal{U}} \otimes \mathcal{H}_V$. An orthonormal basis is $\{|u_i\rangle |v_j\rangle\}$. The *partial trace* of an operator $\hat{O}$ with respect to $\mathcal{H}_{\mathcal{U}}$ is defined as

$$\mathrm{Tr}^U \hat{O} \equiv \sum_t \langle u_t|\hat{O}|u_t\rangle$$

Similarly, the *partial trace* of $\hat{O}$ with respect to $\mathcal{H}_V$ is defined as

$$\mathrm{Tr}^V \hat{O} \equiv \sum_t \langle v_t|\hat{O}|v_t\rangle$$

3.4.1. Improper Mixtures

Let's consider again an ensemble E_T of total systems S composed of the subsystems S_U and S_V, and suppose that every system is described by the vector

$$|\Psi\rangle = \sum_{ij} c_{ij} |u_i\rangle |v_j\rangle$$

E_T is then a pure case with

$$\hat{\rho} = |\Psi\rangle\langle\Psi|$$

Let's suppose an operator $\hat{A}$ belongs to S_U. It can be shown that its mean value over E_T, which of course is also its mean value over, is

$$\left\langle \hat{A} \right\rangle = \mathrm{Tr}\left(\hat{\rho}_U \hat{A} \right) \qquad (2)$$

where

$$\hat{\rho}_U = \mathrm{Tr}^V (\hat{\rho}) \qquad (3)$$

Similarly, for an operator $\hat{B}$ which belongs to S_V, it can be shown that its mean value over E_T, which of course is also its mean value over E_V, is

$$\left\langle \hat{B} \right\rangle = \mathrm{Tr}\left(\hat{\rho}_U \hat{B} \right) \qquad (4)$$

where

$$\hat{\rho}_V = \mathrm{Tr}^U(\hat{\rho}) \tag{5}$$

Because of equations (2) to (5), the ensembles and are called mixtures. Are the two ensembles genuine mixtures?

To answer the question one has to consider that there exists a mixture E_U' of systems S_U whose statistical operator is $\hat{\rho}_U = \mathrm{Tr}^V(\hat{\rho})$. Consider in effect a mixture E_U' of systems S_U with statistical operator $\hat{\rho}_U'$ equal to

$$\hat{\rho}_U' = \sum_\alpha \frac{N_\alpha^U}{N} |\psi_\alpha\rangle\langle\psi_\alpha|$$

Expressing the $|\psi_\alpha\rangle$ in terms of the base $\{|u_i\rangle\}$ it can be shown that

$$\hat{\rho}_U' = \hat{\rho}_U$$

which proves the affirmation.

There also exists a mixture E_V' of systems S_V whose statistical operator is $\hat{\rho}_V = \mathrm{Tr}^U(\hat{\rho})$. Consider in effect a mixture E_V' of systems S_V with statistical operator $\hat{\rho}_V'$ equal to

$$\hat{\rho}_V' = \sum_\beta \frac{N_\beta^V}{N} |\phi_\beta\rangle\langle\phi_\beta|$$

Expressing the $|\phi_\beta\rangle$ in terms of the base $\{|v_j\rangle\}$ it can be shown that

$$\hat{\rho}_V' = \hat{\rho}_V$$

Can one say that E_U is equivalent to a mixture E_U' and E_V is equivalent to a mixture E_V'?

Concerning the calculation of the expected values of an operator $\hat{A}$ acting on $\mathcal{H}_\mathcal{U}$ and an operator $\hat{B}$ acting on $\mathcal{H}_\mathcal{V}$ the identification mentioned above is correct. For this reason, the question may arise if such identification is not then valid in general, which would mean that we can then identify the pure case with a mixture. If, however, operators such as $\hat{A}B$ acting on $\mathcal{H}_\mathcal{U}$ and $\mathcal{H}_\mathcal{V}$ are considered, then the identification in question is generally not correct. Such operators are those that show correlations. Indeed: Let's suppose that E_U and E_V are equivalent to E_U' and E_V' respectively and consider the sub-ensemble $E_{i,j}$ of the systems of the total ensemble E_T whose sub-system S_U belongs to E_i^U and whose subsystem S_V belongs to E_j^V. Let $N_{i,j}$ be the total number of them. $E_{i,j}$ is a pure case with vector $|u_i\rangle |v_j\rangle$. E_T will then be the union of all $E_{i,j}$. It would result then that the pure case E_T with $|\Psi\rangle = \sum_{ij} c_{ij} |u_i\rangle |v_j\rangle$, would be equivalent to the mixture E_T' with

$$\hat{\rho}' = \sum_{ij} \frac{N_{ij}}{N} |u_i\rangle |v_j\rangle \langle u_i| \langle v_j|$$

Now

$$\hat{\rho} = |\Psi\rangle\langle\Psi| = \sum_{ij} |c_{ij}|^2 |u_i\rangle |v_j\rangle \langle u_i| \langle v_j| + \sum_{\substack{r,s,i,j \\ r \neq i \\ s \neq j}} c_{rs}^* c_{ij} |u_i\rangle |v_j\rangle \langle u_r| \langle v_s|$$

If one could argue that the cross-terms are not relevant, one would have that the pure case E_T would be equivalent to a mixture E_T' with statistical operator

$$\hat{\rho}' = \sum_{ij} \frac{N_{ij}}{N} |u_i\rangle |v_j\rangle \langle u_i| \langle v_j|$$

Doing $N_{ij}/N = |c_{ij}|^2$,

$$\hat{\rho}' = \sum_{ij} |c_{ij}|^2 |u_i\rangle |v_j\rangle \langle u_i| \langle v_j|$$

Every system of the total ensemble E_T would then be described by a vector: there would be N_{ij} of them described by the vector $|u_i\rangle |v_j\rangle$. It is also concluded that each sub-system S_U of the total ensemble E_T would then be described by a vector $|u_i\rangle$ and each sub-system S_V would be described by a vector $|v_j\rangle$.

It is not possible, however, to make the identification between E_T and E_T'. Let's show that in fact E_T and E_T' can be differentiated by measuring correlations. We define the cross-terms as

$$\hat{\gamma} \equiv \sum_{\substack{r,s,i,j \\ r \neq i \\ s \neq j}} c_{rs}^* c_{ij} |u_i\rangle |v_j\rangle \langle u_r| \langle v_s|$$

Let us suppose an operator $\hat{A}$ acting only on $\mathcal{H}_U$. It can be shown that

$$\mathrm{Tr}\left(\hat{\gamma}\hat{A}\right) = 0$$

It can be shown similarly that for operator $\hat{B}$ acting only on $\mathcal{H}_V$,

$$\mathrm{Tr}\left(\hat{\gamma}\hat{B}\right) = 0$$

It is concluded then that

$$\mathrm{Tr}\left(\hat{\rho}\hat{A}\right) = \mathrm{Tr}\left(\hat{\rho}'\,\hat{A}\right)$$

$$\mathrm{Tr}\left(\hat{\rho}\hat{B}\right) = \mathrm{Tr}\left(\hat{\rho}'\,\hat{B}\right)$$

With measurements of an operator such as $\hat{A}$ and $\hat{B}$ there is no way of distinguishing between E_T and E_T'. However,

$$\mathrm{Tr}(\hat{\gamma}\hat{A}\hat{B}) = \sum_{\alpha,\beta} \sum_{\substack{r,s \\ r \neq \alpha \\ s \neq \beta}} c_{rs}^* c_{\alpha\beta} \left(\hat{A}\hat{B}\right)_{rs;\alpha\beta}$$

Now, as in general, $\left(\hat{A}\hat{B}\right)_{rs;\alpha\beta} \neq 0$ for $r \neq \alpha$ and $s \neq \beta$, then in general $\mathrm{Tr}(\hat{\gamma}\hat{A}\hat{B}) \neq 0$. Hence in general $\mathrm{Tr}(\hat{\rho}\hat{A}\hat{B}) \neq \mathrm{Tr}(\hat{\rho}'\,\hat{A}\hat{B})$.

The last two traces of the above paragraph are the average values of $\hat{A}\hat{B}$ calculated with $\hat{\rho}$ and $\hat{\rho}'$ respectively. The measurement of the operator $\hat{A}\hat{B}$ allows one to distinguish between E_T and E_T'. The value of $\left\langle \hat{A}\hat{B} \right\rangle$ is used to determine if there are correlations between $\hat{A}$ and $\hat{B}$. It is concluded that in general, the predictions about correlations between operators acting on $\mathcal{H}_\mathcal{U}$ and operators acting on $\mathcal{H}_\mathcal{U}$ allow one to distinguish between E_T and E_T'.

In conclusion, the non-diagonal or crossed terms of $\hat{\rho}$ are those that allow one to differentiate the pure case in question from a mixture. They play a role in investigating correlations.

Since the mixtures E_T and E_T' are physically different d'Espagnat explicitly introduced the conceptual difference between them and proposed to call them *proper mixtures* and *improper mixtures* respectively.

When one wants only to introduce quantum statistics from a practical point of view the distinction between the two types of mixtures is often not made; for example, in Huang's book [3] quantum statistics are introduced utilizing improper mixtures whereas in Messiah's book [4] using proper mixtures. The use, in the context of quantum statistics, of one or the other of these two different concepts can create confusion in the student. In fact, I experienced that kind of problems that only the reading of d'Espagnat writings dissipated.

4. Ontological Reality and Empirical Reality

4.1. *The notion of reality is necessary*

d'Espagnat considers that the notion of reality is necessary for the coherence of thought. It recognizes that one cannot rigorously neither demonstrate realism nor even define it through discourse but asserts that from the previous statement we must not conclude that realism is false. For him, the difficulties encountered by the opposite approach constitutes a very compelling argument in favor of the realistic thesis.

In this context, he examines what he calls the purely linguistic standpoint according to which the impossibility of constructing an operational definition of the notion of independent reality implies that this notion has no meaning.

d'Espagnat use the expression purely linguistic standpoint for designating a general philosophy that incorporates the following logical positivistic thesis:

i) *The principle of verifiability*: a statement has meaning only if it can be verified or falsified by means of some definite experimental procedure.

ii) *The purely linguistic conception*: many symbols and words referring to unobservable entities are used in scientific formulas and affirmations, but it is required that all these symbols or words serve only as intermediate links in the formulation of rules that refer exclusively to what can be observed. Carnap affirms that these terms constitute a *linguistic framework*. For this author, the question of whether the set of all entities that form a given framework "really exists" cannot be formulated in terms of the ontological existence of such entities, but has only a practical meaning, namely: Shall we decide or not TO use the considered framework? According to Carnap's ideas, the decision should not be interpreted as a belief in the ontological reality of such entities because to speak of that reality has no cognitive content.

d'Espagnat analyzes the well-known technical difficulties of the purely linguistic standpoint. One of them is the justification of the principle of verifiability. Another is the problem of induction, which is acuter in the purely linguistic approach than in the realistic approach. And finally the difficulty of avoiding solipsism.

To those difficulties d'Espagnat adds two objections of a more general nature.

The first is the difficulty of distinguishing between the laws of nature and the methods –in general approximate using which scientists develop observational predictions. In a realistic position, we recognize, on the one hand, the laws and physical principles of reality and, on the other, the approximations we use for the study of these laws and principles.

In the purely linguistic approach, we have no reason to make the distinction in question, and it may happen that the computational algorithms are progressively hypostatized into elements of a description of nature without the responsible theorists even being aware of the change of meaning. d'Espagnat cites as an example the case of the notion of virtual state in quantum field theory. (See [7] Chapters 2 and 9). In the purely linguistic position there exists the danger of a proliferation of mutually incompatible models, each of which has some success and some failure, but from which it is impossible to obtain any idea of reality that is somewhat synthetic.

The second objection of d'Espagnat is the existence of the danger of depriving scientists of the motivations to do their work if they adopt seriously the purely linguistic approach. d'Espagnat presents the case of a paleontologist whose motivation would not be the belief that such and such animals existed –that is to say, independently of the researcher herself and her fellows but the possibility that her research provides her of writing scientific articles that harmonize with other scientific papers.

In addition to the critical discussion that d'Espagnat has about the purely linguistic approach, he presents an argument which is essentially the following:

One usually thinks of something and has the certitude that in any case at the moment when one thinks something at least *exists is* namely, one's very thought. As a result, one cannot dismiss the notion of being as meaningless. To say then that the independent reality does not exist or has no meaning is equivalent finally to specify that only the thinking man exists, or at least it is to specify that the affirmation of existence can be legitimately posited only secondarily to him. d'Espagnat refuses this idea of pretending to subordinate to men the existence itself as an idea as gratuitous as that of solipsism. One can say that his analysis is Descartes like.

Indeed, in the First Meditation of his book *Meditations on First Philosophy*, Descartes proposes to doubt everything about which he has the slightest reasonable doubt.

In the Second Meditation he continues along the same path described in the previous Meditation, that is to say, that he will accept as false all that on which he has the slightest doubt, even if at the end of his reflection his only sure conclusion is that there is nothing certain. He then continues to present arguments for doubting everything and finally wonders whether he does not exist. He responds, however, that if he persuades himself that nothing exists in the world, it means that he indeed exists because he was persuaded of something. And even if he admits the existence of a deceitful genius that makes him always err on purpose, then he exists because it is he whom the genius deceives.

Descartes concluded then that it is necessarily true that he exists. It remains to know what he truly is. He wonders if it is something which belongs to the nature of the body. The answer is negative because he has arguments for doubting everything related to the body. However, there is something that cannot be denied that belongs to him and that is the thought. Indeed, it is manifest certainly that he is, that he exists, and that certainly implies that there is then the thought of it and therefore the thought exists.

What Descartes does is to affirm that at the end of a radical doubt something remains as existing: thought. In other words, it concludes that something exists: the thought.

Descartes in that Second Meditation also examines the world which is supposed to be external to the mind. But not to conclude that it exists outside of the mind but to see what the essence of that world would be. By the example of the properties of a piece of wax, he concludes that that essence is the extension.

In the Third Meditation, Descartes affirms that before looking for other certainties he has to examine whether God exists and if He is not a deceiver. To prove the existence of God, Descartes argues that if there is an idea of which he cannot be the cause then there exist something other than himself that is the cause of it. Otherwise, he would have no argument to conclude that there is something else besides him. Once defined what he understands by God, Descartes concludes that the idea of God cannot come from himself and that therefore God exists. The definition of God given by Descartes implies that God is not a deceiver.

It is in the Sixth Meditation that, based on the existence of God, Descartes presents his argument to accept the existence of the *res extensa* as different from the *res cogitans*.

Now, if the concept of *being* would not have the *logical precedence* over that thing which one concludes that exists, one could not affirm of that thing that in fact exists. It is clear then that the affirmation *that something exists* has logical precedence with respect to *the knowledge of the existence* of the *res cogitans*, God, and the *res extensa*; and that *being* is the *epistemological foundation of the knowledge* of the three substances.

The brief analysis of the ideas of Descartes that I have just presented support my claim that the argument of d'Espagnat above presented is similar to that of Descartes in the Second Meditation. For the reasons set out above, d'Espagnat considers:

> that the position of rejecting the very notion of independent reality on the pretext that this notion would have no sense ultimately friction the incoherence of thought. [5, p. 224]

d'Espagnat also emphasizes that the arguments he presents retain their strength even if the independent reality to which they refer is not knowable in detail to man.

d'Espagnat's position contains, implicitly at least, the idea that this independent reality is an explanation of phenomena and states that the notion of explanation in question is not that of a cause but rather the notion of *raison d'etre*.[e] Independent reality is, therefore, the *raison d'etre* of phenomena. The development of this ideas lead him to the notion of what he calls *extended cause*, which I will discuss in section 6.2.

An ontology must have very general connotations that relate to our world view. In my opinion, it is for this reason that d'Espagnat analyzes the possible effects that a cosmovision centered on the idea of being would have on the vision that the present man possesses of the world, and about his role in the world. He notes that this vision is currently neither very clear nor very satisfactory because it deprives the contemporary man of all contact, real or supposed, with anything that can be called being. [6, p. 157]

4.2. The real is not physical

We have just seen that the reasons which induce d'Espagnat to accept the idea of the existence of an independent reality are reasons which can be qualified as philosophical since his scientific knowledge does not dictate them [5, p.157].

As for his thesis that the real is not physical, that is to say, is not knowable using physics, the situation is very different. d'Espagnat comes to his

---

[e]Reason or justification for existence

conclusion *a posteriori* after an exhaustive and detailed analysis of all the ontological attempts that have been made and are being made to construct a physics that is in agreement with the data of quantum mechanics.

In this article, I do not present a critical analysis of a process to which d'Espagnat has devoted several books. I shall, therefore, limit myself to showing the general arguments which d'Espagnat presents to exclude some of the different types of realism analyzed in his books.

4.2.1. *Theories that consider the wave function as a description of reality (Ψ-ontological theories)*

These theories face the problem of measurement. The conclusion of d'Espagnat is that to solve the problem of measurement all the present Ψ-ontological theories of measurement must pay the price, which consists in referring, in one stage or another, to the practical impossibility of a human being to carry out such and such particularly difficult measure. Consequently, we cannot consider that these theories reconcile quantum physics with physical realism.

I am not saying that d'Espagnat is making the affirmation that there is an *impossibility* to find a Ψ-ontological interpretation of quantum mechanics which would imply then a Ψ-ontological theory of measurement. What he found is that the present Ψ-ontological theories, when they come to the problem of measurement, they refer finally to what human beings can or cannot do, coming then to an epistemological solution and not an ontological one. In conclusion, they abandon in that point physical realism.

An additional difficulty encountered by a theory of measurement –which seeks to consider the notion of the reduction of the state vector as a real or ontological fact– is the impossibility to reconcile it with an ontological theory of relativity.

The foregoing facts lead d'Espagnat to conclude that the construction of a Ψ-ontological theory which is satisfactory is at least doubtful.

4.2.2. *Local micro-realism*

Micro realism is a conception in which the entities to which we give the name of particles have at each instant a well-defined position and "by symmetry" a well-determined velocity.

Local versions of micro-realism are those that obey the principle of separability. They are discarded by the data related to Bell's theorem.

4.2.3. *Non-local hidden variables theories*

These are theories of the de Broglie-Bohm type. They are based on the idea that quantum indeterminism is ultimately only an apparent indeterminism, due solely to the ignorance of fine details. These theories attempt to complete the quantum description by the hypothesis of the existence of additional parameters which, more often than not, differ from one physical system to the other even when these two systems have the same wave function. The theories in question are therefore deterministic.

The main argument of d'Espagnat to exclude these theories is that they constitute an unsuccessful immunization, that is to say, they are reduced to the level of *ad hoc* hypotheses. Immunization consists in the introduction of a supplementary hypothesis using which one attempts to preserve a theory which has hitherto been held to be satisfactory from the attacks of such or such particular objection based, for example, on a new discovery. Two conditions must be met by an immunization before it can be considered successful:

a) It must not diminish either the simplicity or the global synthetic power of the original theory.

b) It must be fertile, that is to say, it must allow the prediction of facts not yet observed and whose existence experience comes to confirm *a posteriori.*

Since hidden-variable theories have so far not fulfilled the second condition, they cannot be considered satisfactory.

d'Espagnat admits that one can dispute the application of the second criterion, namely that of fertility, to an interpretation of a theory. Consequently, he analyzes the consequences of non-separability, that is, the non-local character of the theories in question.

Non-local hidden variable theories must admit –to take into account the violation of the principle of locality the propagation of faster than light influences not allowing the transmission of signals, which poses tough conceptual problems because it is not possible to specify in an absolute manner which event-measurement is the first in time.

The difficulty of giving a satisfactory solution to the previous question is for d'Espagnat an indication of the fact that the first condition of a successful immunization –the requirement of simplicity and synthesis– is neither fulfilled by the non-local hidden variables theories. He adds that it is extremely probable that it will never be met except at the cost of a real transformation of the idea of determinism.

d'Espagnat analyzes other sorts of theories inscribed in the realism and his conclusion is that none of them is satisfactory. He argues that if physical realism is a correct theory, one should expect that as it develops, physics would be able to create more and more general theories that are never durably rivals and that these theories must be formulated as descriptions of reality, that is to say, they must be ontological.

He observes that in its usual formulation quantum mechanics satisfies the criterion of unicity, but not that of being ontological. Moreover, those ontological theories such as the non-local hidden variable theories, which one can say that in a certain sense are successful, do not satisfy the condition of unicity, since they are multiple and there is no way to scientifically choose between them.

4.3. *Independent Reality and Empirical Reality*

After an exhaustive analysis of the notion of non-separability, d'Espagnat concludes that the concepts of space, time, space-time, events, and even the positions of things are nothing more than mere tools for the description of phenomena. He proposes to make a distinction between the empirical reality, which is the set of phenomena, that is, the totality of what human experience, seconded by science, yields access to, and the independent or ontological reality, that exists independently of our existences [7, p. 4]. According to his conception the empirical reality is the only one of which the human mind can truly have knowledge in the sense that scientific research gives to this word. Independent reality is in no way in space-time which, just as locality, the events, and so on, are concepts that owe much to the structure of our mind.

In the following quotation, d'Espagnat presents an argument to convince the scientists of the necessity to make the distinction in question:

> nonseparability does not provide us with new means of operating at a distance. Consequently, it constitutes a feature of any sensible representation of mind-independent reality that, unquestionable and significant as it is, still does not fully and genuinely extend to the empirical reality domain. In other words, in sharp contrast with mind-independent reality, which, to repeat, can hardly be thought of as constituted of distinct parts, most of the phenomena that compose empirical reality exhibit no features that could be called nonseparable. Clearly, this conforms the necessity of at least distinguishing between the two notions of mind-independent and empirical reality. [7, p. 4]

A property, an argument or statement that does not refer at all to the collectivity of human beings, nor to their decisions or limitations, nor to their existence, is called by d'Espagnat *objective in a strong sense*. We can call them ontological.

A property, an argument or statement that is not ontological but is, however, intersubjective, which means it is true for everybody, is called by d'Espagnat *objective in a weak sense*.

The independent or *ontological reality* is constituted by everything that is objective in a strong sense.

The *empirical reality* is constituted by everything that is objective in a weak sense.

In section 6.4 I will analyze the relation of these notions with Kant's ideas.

5. The Axiom of Empirical Reality

5.1. *The problem of measurement in quantum mechanics*

Let us consider the following very natural assumptions.

Assumption a): quantum mechanics is a complete theory.

It is a natural assumption concerning a theory as successful as quantum mechanics and which has not yet failed empirically. However, if the assumption means that one denies the existence of hidden parameters or nonlinear terms in the Schrödinger equation (GWR theory), this means that one adopts an ontological option. It would be, as pointed out by d'Espagnat [8, p. 371], a crypto-ontological assumption. One cannot deny the logical possibility of constructing a theory of quantum phenomena in terms of hidden variables. In fact, the de Broglie-Bohm[9–11] theory is such a theory. What one can and should demand is that to agree with the experimental verdict the predictions of such theories coincide with those of quantum mechanics. That expression of the completeness of quantum mechanics is called by d'Espagnat the *weak completeness assumption* [8, p. 61].

Assumption b): quantum mechanics is universal.

The "macroscopic manifestations of quantum mechanics like superfluidity, superconductivity, and Bose-Einstein condensates provide indirect experimental evidence for the universality of quantum mechanics. It has also been shown that some quantum notions, such as quantum tunneling, are essentially indispensable to explain data relating to systems whose size and complexity can be called macroscopic [12,13].

Assumption c): the intuitive properties of the macroscopic bodies are ontological

In other words, the properties of macroscopic bodies, in particular, their position and their momentum (or in general the generalized coordinates and their conjugate moments: classical dynamic variables that define the state) have an ontological character. It is the ordinary, natural conception of macroscopic bodies, which comes from classical dynamics. It is part of our ordinary intuition. It comes from our everyday experience mediated by the concepts of classical physics. Just think of the difficulty we have to imagine a situation in which, for example, a clock is not at every instant with the hands in defined positions marking a definite hour but, at some time, or during a certain time, is in a superposition of states in which each of the hands are marking an hour. Such a hypothetical situation is so counterintuitive that we even consider it inconceivable. More precisely, it is not inconceivable but unimaginable. At least in the ordinary life, not in the strange realm of dreams. But physics is about ordinary life, so the superposition in question is almost automatically rejected by us.

The set of assumptions a) and b) is inconsistent with the assumption c.

To see why the statement is correct, consider a model of measurement which shows the so-called problem of measurement in quantum mechanics.

5.1.1. *Model of a measurement in quantum mechanics*

The problem of measurement in quantum mechanics can be formulated in terms of a model of measurement.

Let S and A be a system and an observable that belongs to the system respectively. Let $|\varphi_1\rangle, ..., |\varphi_n\rangle$ be the eigenvectors of $\hat{A}$: $\hat{A}|\varphi_m\rangle = a_m|\varphi_m\rangle$. We want to measure the value of A using an instrument M. Let G be the position of the needle in the instrument. We assume for simplicity that the eigenvectors of $\hat{G}$ belong only to the discrete spectrum which means they are $|g_n, r\rangle$: $\hat{G}|g_n, r\rangle = g_n|g_n, r\rangle$ where g_n is the position of the needle and r is a degeneracy index for other variables of M other than G.

Let us suppose a situation where we can associate a ket $|\varphi_m\rangle$ with an ensemble E_S of N particles S and a ket $|g_0, r\rangle$ with an ensemble E_M of N macroscopic bodies M. That is, the needle in each system M is in the position g_0. The model says that M is so constituted that the effect of the interaction with S is to bring it to the state $|g_m, s\rangle$ without changing the state of S. That is,

$$|\varphi_m\rangle\, |g_0, r\rangle \rightarrow |\varphi_m\rangle\, |g_m, s_{m,r}\rangle \tag{6}$$

We have sub-index m, r in S because the degeneracy index in the final state depends on m and r.

We observe:

i) The model is consistent with the concept of first class measurements: the state of the system to be measured is not altered, only the state of the apparatus is altered.

ii) There is a simple Hamiltonian model $\hat{H}^{M,S}$ of the interaction between S and M that results in evolution (1). In other words, it is possible to find a Hamiltonian $H^{M,S}$ such that the solution of $i\hbar\frac{\partial}{\partial t}|\varphi_m\rangle\,|g_0, r\rangle = \hat{H}^{M,S}|\varphi_m\rangle\,|g_0, r\rangle$ is $|\varphi_m\rangle\,|g_m, s_{m,r}\rangle$.

iii) We have the expected correlation of a measurement: one looks at the position of the needle and finding that it is g_m one concludes that the state vector of the system S is $|\varphi_m\rangle$ and that therefore the value of A is a_m.

Let us now consider the following more general initial situation: E_M is described by $|g_0, r\rangle$ but E_S is described by the superposition $|\phi\rangle = \sum_m c_m |\varphi_m\rangle$. As a result of (6) we have

$$|\phi\rangle\, |g_0, r\rangle = \sum_m c_m |\varphi_m\rangle\, |g_0, r\rangle \rightarrow \sum_m c_m |\varphi_m\rangle\, |g_m, s_{m,r}\rangle \tag{7}$$

Now, the superposition (7) is a superposition of macroscopically different states of all possible positions G of the needle. Such a situation implies that the ontological interpretation of the intuitive properties, i.e., classical, of macroscopic bodies cannot be maintained, in particular, it cannot be maintained that a macroscopic body always has, in the ontological sense, a definite position.

A possible solution would be that the ensemble E_T of systems whose state vector is:

$$|\Phi_f\rangle = \sum_m c_m \, |\varphi_m\rangle \, |g_m, s_{m,r}\rangle$$

and whose statistical operator is $\hat{\rho} = |\Phi_f\rangle\langle\Phi_f|$ (improper mixture) were equivalent to a proper mixture E_T' with statistical operator

$$\hat{\rho}' = \sum_m \frac{N_m}{N} \, |\varphi_m\rangle \, |g_m, s_{m,r}\rangle \, \langle\varphi_m| \, \langle g_m, s_{m,r}|$$

where $\frac{N_m}{N} = |c_m|^2$. This would imply, however, that the cross-terms γ,

$$\hat{\gamma} = \sum_{m,n \neq m} c_n^* c_m \, |\varphi_m\rangle \, |g_m, s_{m,r}\rangle \, \langle\varphi_n| \, \langle g_n, s_{n,r}|$$

could be ignored.

We know, however, that it is not possible *in principle* to leave out $\hat{\gamma}$ and that E_T can be distinguished from E_T' by measuring correlations between an operator $\hat{B}$ acting on S and an operator $\hat{D}$ acting on M such that $\hat{D} \neq \hat{G}$ and $\left[\hat{D}, \hat{G}\right]$. This is because the $|g_m, s_{m,r}\rangle$ are eigenvectors of G, and therefore it is true that $\mathrm{Tr}\left(\hat{\gamma}\hat{B}\hat{G}\right) = \mathrm{Tr}\left(\hat{\gamma}\hat{B}\hat{D}\right) = 0 \quad \forall \quad \hat{D}$ that commutes with $\hat{G}$.

In fact, this is same the reason why we cannot identify a pure case or improper mixture with a proper mixture (see section 3).

Let us suppose we had an argument to say that such measurements of correlations are extremely difficult to make and that *for all practical purposes* we can, therefore, leave aside $\hat{\gamma}$. The acceptance of such argument would be tantamount to maintaining that the properties of the macroscopic body acting as an instrument are then only appearances (*phenomenon*); that is, they are not ontological which goes against assumption c). As will be shown in the following section, the decoherence models provide us with the said arguments to leave aside the cross-terms.

5.2. *Decoherence and the measurement problem*

At the quantum level, the macroscopic bodies cannot be considered as isolated because the quantum energy levels are extremely close even for tiny macroscopic bodies. Consequently, extremely weak fields can induce transitions. A detailed calculation shows that even a particle of dust in interstellar space cannot be considered as isolated over a sustained amount of time. Under the universality assumption, the only basic difference between microscopic and macroscopic systems is then that the last strongly interacts with its environment. It is concluded then that the previous treatment in which the $S + M$ systems were considered isolated from their environment is not realistic. The interaction of the macroscopic instrument M with the environment must be taken into account. The different models that consider that interaction (decoherence models) show that given the human limitations, for every human being the pure case can be replaced by a proper mixture (see for example [14]).

The argument is as follows. There are some measurements that could be performed in principle because there is no law of physics that forbids them and the sequence of operations using which they would be made can be precisely stated. In practice, however, they cannot be done because they are tremendously complicated, and so in practice, they can be regarded as impossible for a human being. A superhuman Laplacian "supreme intelligence" could be invoked who could distinguish the pure case and the proper mixture. The answer coherent with the argument is again that physics is for human beings and not for a hypothetical superhuman "supreme intelligence". Since the argument refers in an essential way to human possibilities, the conclusion is that decoherence solves the problem but in terms of empirical reality: the reality of macroscopic bodies is only empirical. Decoherence shows how the classical properties emerge in the macroscopic domain.

5.3. *The Axiom of Empirical Reality*

The argument provided by the decoherence models that allow us to identify the pure case involved in the measurement process with a proper mixture is made in "the spirit of keeping close to what is operationally meaningful." d'Espagnat considers that:

> in the same spirit of keeping close to what is operationally meaningful, it is quite consistent to assume the validity of an axiom that makes sense within the empirical reality conception since the last is centered on a reference to human possibilities and which is, in fact, a necessary ingredient (although it, unfortunately, is most often kept implicit) in all the measurement theories [8, p. 372].

The Axiom in question [15] is called by d'Espagnat the *Axiom of empirical reality*:

> *Axiom of empirical reality.* A theoretical systematization of the empirical view must involve either one or both of the following

postulates: (a) replacing large times by infinite times and/or very large particle numbers by infinite number is a valid abstraction; and (b) the possibility of measuring observables exceeding a certain degree of complexity is to be considered as not existing, not even theoretically; it being specifically stated that the latter position must be taken even in cases in which, according to quantum mechanics, such a possibility, in principle, actually exists." [8, p. 372]

According to the Axiom of empirical reality of d'Espagnat, the possibility of distinguishing the pure case and the proper mixture is to be considered as not existing, not even theoretically.

The Axiom of Empirical Reality is proposed to replace –in the identification of the pure case involved in the measurement process by a *proper mixture*– the *for all practical purposes* argument by a postulate, that means something theoretical and not just practical. It must be emphasized that the postulate makes sense only within the empirical reality conception.[f]

6. The conception of Veiled Reality

For d'Espagnat, the independent reality is not knowable because the meaning of this word is that of an exhaustive knowledge of the object as it really is. On the other hand, he notices that notions such as space-time, space curvature, and so on are in no way familiar notions or *a priori* modes of our sensitivity. The fact that these highly developed notions are ultimately used by the scientific community rather than others is, in his view, an indication that they are due, at least in part, to information that we receive from outside. It is natural to think, says d'Espagnat, that they reflect something of the independent reality, and we cannot, therefore, say that reality is "unknowable". For d'Espagnat the real is veiled, and the epithet in question has an intermediate meaning between those of the terms knowable and unknowable.

For d'Espagnat the ontological reality is ineffable. On the other hand, science is not mere technology, the meaning of science is not merely practical: the empirical reality is not a mere mirage, the symmetries and regularities revealed by science correspond, albeit in a profoundly hidden way, to some form of the absolute. These reflexions lead d'Espagnat to his notion

---

[f]Analizing the relation between measurement and irreversibility, I applied the Axiom to the Poincaré recurrence time for a macroscopic system [16].

of Veiled Reality:

> definitively brush aside the view according to which the significance
> of our discipline is merely practical; that pure science is nothing
> but a technology focalized on the long term. Quite on the contrary,
> I consider it most plausible that the multifarious regularities and
> symmetries science reveals in all domains corresponds –albeit in a
> highly hidden manner to some form of the absolute. Moreover, I
> consider, as will be explained in the text, that the proper domain
> of scientific knowledge, empirical reality, is far from being a mere
> mirage [7, p. 5]

To illustrate his thesis d'Espagnat presents an analogy inspired by an idea of Bertrand Russel, which consists in comparing the real in itself –or independent reality with a concert, whereas the empirical reality –all the phenomena is compared to a recording on disc or cassette of this concert. The structure of the disc is not independent of that of the concert, but the first, which is deployed in space in the form of small hollows and bumps along the furrows, is not purely and simply identifiable with the second, which is deployed over time.

In the same way that, explains d'Espagnat, for example, an extrater-restrial disembarking on Earth and discovering the disk can –if it has the sense of hearing and imagination enough to conjecture that at the origin of the hollows and bumps that he studied there is an emission of sounds –to arrive both to grasp and to taste the essential part of the concert, so we can guess and taste on a non-illusory mode the very significant traits of the reality in itself. To say, however, adds d'Espagnat, that we know it (in the exhaustive sense of this word) would be abusive.

With his Veiled Reality hypothesis, d'Espagnat wants:

> to ponder on the question why these rules [the quantum ones] are
> there and where they come from. [7 p.236]

6.1. *Extended causality*

d'Espagnat's conjecture is that the quantum laws indirectly furnish some glimpses of Ontological Reality. He introduces what he calls an "extended causality consisting on influences exerted by the Ontological Reality on phenomena. That causality is beyond the Kantian-like causality which takes place between phenomena. [7, p. 238-239]

> these structural 'extended causes' are nothing else that the very

structures of independent reality and they constitute the ultimate explanation of the very fact that the laws –that is, physics exist. [7, p. 414]

With his notion of "extended causes", d'Espagnat wants to save what he calls the "hard core of the Principle of Sufficient Reason of Leibnitz: the very notion of some ultimate reason for the laws that govern the world. The 'extended causes' are considered as being prior to laws." [8, p. 415.]

6.2. Veiled reality and structural realism

The same d'Espagnat makes clear that his position differs from what he calls the strong structural realism according to which the great physical laws describe the structures of the Real:

> The fact that, in such conceptual context, one of these laws, the one that no influence is propagated with superluminal velocity, appears as being violated (because of nonseparability) speaks against such a structural realism. [7, p. 238]

His positions can be called then a *weak structural realism*.

6.3. Kant and d'Espagnat

Kant proceeds from *a priori* philosophical arguments and d'Espagnat from *a posteriori* analysis of the present scientific knowledge. Kant introduces the notion of *a priori* concepts and modes of sensibility to answer Hume's arguments against causality.

The concepts related to d'Espagnat empirical reality are not *a priori*. They are summited to evolution. They depend on what humans being can do. They are contextual: in a Bohrian interpretation of quantum mechanics, they are complementary.

The concepts related to Kant's phenomena are *a priori*. They are not summited to evolution. They are not contextual: they are universal in his application. They do not depend on what humans can or cannot do.

For Kant, the noumenon is entirely unknowable. For d'Espagnat the Reality is veiled and not entirely unknowable.

In the Kantian idealism the notion of Ontological Reality, called the noumenon, is meaningful. For Kant, the phenomena are entirely "internal". The noumenon is unknowable in its entirety. We are total prisoners of our sensitive and conceptual apparatus. d'Espagnat make reference to:

the wall of the prison within which, as Kant's thesis suggests, our understanding locks us up [7, p. 433].

An objection to Kantian idealism concerns the meaning consistent with it of the scientific descriptions according to which the material reality (stars, galaxies, the Earth, the universe) existed long before the emergence of beings endowed with consciousness. Kantian idealists can respond that the objection attributes to time an ontological nature and, according to idealism, time is not ontological. The objection contradicts a fundamental element of the idealist scheme and is, therefore, inconsistent with such a scheme. It can be said, consistently with idealism, that the claim that there was a time when the universe existed without the existence of consciences and all similar affirmations means only that humans can organize their collective experience describing them with such affirmations. It means that when all is said and done the scientific affirmations are merely allegories.

With his Veiled Reality theory, d'Espagnat wants to offer an answer to the question of why empirical reality can be understood by our minds. In one of his writings, he considers a sort of complementarity between the mental and the physical: he considers the set of consciousness and the set of objects as two complementary aspects of independent reality and explains that neither one exists in itself, but that they exist only one by the other. [6 p.101].

If we admit, as d'Espagnat says, that our minds and empirical reality are complementary aspects of one and the same reality, it should not seem very surprising that the general structures of this reality are reflected in the mathematics we construct and that on the other hand, manifest themselves in empirical reality.

This co-existence or cogeneration seems to suggest two entities each generating the other. It seems then to indicate that the empirical reality is not internal to the Mind and that therefore the position of d'Espagnat differs from that of Kant at that point.

I want to remark that the proposed complementarity of d'Espagnat between mind and matter looks Spinoza like. His complementarity is not bohrian: I do not see which languages are mutually exclusive but necessary; it seems rather some kind of relation between two entities and complementarity is a linguistic relationship.

However, in some other quotations, d'Espagnat speaks of the allegorical character of the cogeneration of empirical reality and consciousness [8, p. 424].

In [7 his position is more clearly expressed:

The conception I developed states that matter –what we make experiments on is but an empirical reality, that is, not in the least a basic entity but merely the set of phenomena in the Kantian sense of the world, which implies that it is, at least partly, molded by us. And it conceives of mind as emerging (but atemporally) from a Something (thereby making empirical reality also emerge from the latter). This Something, to which I gave the name (which I thought 'neutral') 'Independent Reality', alias 'the Real', is thus conceptually prior to the mind-matter scission. For me, therefore, it is not at all the object of the precise, discursive quantitative knowledge we normally refer to when we the word 'knowledge'. [7, p. 378]

His argument is that, while we can say that empirical reality is generated by the mind, the generation of the latter by the first (materialist position) is not acceptable.

He argues that the various parts of our bodies, and hence also our neurons, are constituted of or are themselves elements of empirical reality; in consequence, it's hard to imagine that such reality which is relative to consciousness might generate the latter [7, p. 418].

In constituting empirical reality the role of consciousness or mind is obviously primordial since the said reality essentially is a representation [7, p. 419]

The affirmation of a co-emergence is then only metaphorical or allegorical:

as I wrote some time ago, consciousness and empirical reality exist in virtue of one another [...] or equivalently hey generate reciprocally one another [...]
Without bluntly repudiating these assertions I would like to stress that they should be taken to be evocative pictures rather than literally true statements, for understanding they that way would create a problem. The difficulty does not bear on the view that empirical reality emerges from consciousness [...] Rather it has to do with the [...] view that (empirical) reality generates consciousness. The point is that, within the framework of ideas upheld in this book, the view in question is not defensible. How [...] could mere 'appearances to consciousness' generate consciousness? [7 p.425]

Therefore, his answer to the above-mentioned objection to Kant is similar to the Kantian one:

> If men disappeared the stars would go on in their course. This statement [...] might [...] well be used against the Veiled Reality conception since [...] in the latter, forms, positions and other contingent properties of things essentially are but projections of our modes of apprehending. [...] the idealists reply to the said objection is well known and fully consistent, even though, intuitively, we find it is difficult to accept. It consists in pointing out that the objection in question implicitly raises the concept of space, time and objects to the level of the externally given, whereas, within the idealist approach, to claim that starts exist and were there before human beings appeared merely means that we may conveniently describe our present experience by expressing ourselves in such a manner. Obviously this replay also holds good within the framework of the Veiled Reality conception [7, p. 434]

Must we conclude then that, as for Kant, for d'Espagnat the empirical reality is wholly internal to the mind and that, therefore, we are prisoners of our sensory and conceptual apparatus? About this point at least for me the answer is not entirely clear. It seems that it is negative because, for d'Espagnat, the Veiled Reality has some structures and is not completely hidden as Kant's noumenon since the physical laws, part of the empirical reality, show, in form if you want vague and distorted, some of these structures.

My conclusion is this: with the title of the article *The ineffable nature of being*, I do not pretend that for d'Espagnat the OR, which is for Kant the noumenon, is absolutely ineffable. For d'Espagnat, although the being is veiled to the discursive reason, the latter can glimpse the general structures of it. I use the adjective ineffable because, even with that glimpse, language cannot describe being. Here there is a fundamental difference with Kant, in addition to the approach since d'Espagnat does not accept the idea of concepts and forms of sensitivity considered with an *a priori* character. In what to me Kant and d'Espagnat coincide is in the conception of the *phenomena* as internal to consciousness.

6.4. *Spinoza and d'Espagnat*

d'Espagnat considers that Independent Reality plays, in a way, the role of Spinoza's God, or Substance:

Independent reality is structured in a way we cannot actually know and, via extended causality these structures give rise to the ordinary, observed cause-and-effect relationships [...] [A]s basic attributes of God Spinoza ranks both extension and thought, on equal footing. And this also is a point where the two notions of Independent Reality and the Spinoza's God have something in common since what has been allegorically described as a mutual generation of mind and empirical reality within and from Independent Reality obviously parallels the coexistence of extension and thought as God's attributes which is so essential in Spinoza's philosophy [8, p. 428]

I want to emphasize that d'Espagnat talks only in an allegorical way of the cogeneration of mind and thought within, and from Independent Reality. Neither of them has an ontological character. The attributes of thought and extension in Spinoza, on the contrary, are essential attributes of Substance and are therefore ontological. The parallel is only a comparison.

6.5. Final comments

d'Espagnat considers as not so incoherent to exclude all uses of concepts such as reality, existence and so forth that do not conform to a strictly operational code. He points out that his argument has similarities with that developed independently by certain philosophers and which is based on the idea of distinguishing between the meaning and the referent of a concept. And he criticizes the operationalist for not making this distinction. Indeed, if one does not make the distinction in question, one can argue that the notion of independent reality has no meaning.

Given the preceding, we can say that there is a linguistic consideration in d'Espagnat's thought, which consists of making the difference between the meaning and the referent of a concept.

On the other hand, if one examines the argument that d'Espagnat presents in support of his thesis that reality in itself is not accessible to science, one can see that the argumentation in question is based on the fact that non-separability implies that the concepts of space-time and causality are finally only tools for the description of phenomena. In my opinion, there is in d'Espagnat's thought, in a way maybe implicit, the idea that even if science will continue developing new and more abstract, refined and distant concepts, the fact that concepts as fundamental as those of localization and causality have proved to be inadequate to describe reality in itself, is an indication that science, and in general the discursive reason, will never be able to develop the concepts necessary to describe reality in itself.

I have already pointed out that d'Espagnat arrives at his conclusion *a posteriori* after having made an exhaustive analysis of all the attempts that have been made to construct a theory to understand the quantum phenomena on the basis of the strong objectivity. Nevertheless, the very fact that such attempts exist and that physicists such as Bohm continue to try to create new concepts which they believe can be used to describe reality in itself, is an indication that, finally, d'Espagnat's conclusion is a logical possibility, based on facts and considerations about the nature of language and the relationship between the latter and reality, which can be challenged if one does –as in the case of Bohm other considerations about language.

7. Some others of his ideas and contributions

I. **Theoretical and practical work about Bell's inequalities** He played a leadership role that triggered experimental work in the 80s and 90s, especially in France, Austria, and Switzerland. Many of the efforts in current research on quantum nonlocality are due to their early contributions.

II. **In-depth analysis of the conceptual implications of quantum entanglement** Although Schrödinger and Einstein had already examined the issue of the quantum entanglement and its implications, d'Espagnat was the one who analyzed in depth the important philosophical points that underlie this phenomenon.

III. **Serious divulgation of conceptual implication of quantum mechanics** He wrote important works in the serious divulgation for the general public of the conceptual ideas of quantum mechanics.

IV. **Consolidation of the presently respected areas of conceptual foundation of quantum mechanics and quantum information** His works helped to consolidate the now respected area of the conceptual foundations of quantum mechanics and quantum information.

All those dedicated to research in the foundation of quantum mechanics would do well to read their books carefully to avoid reinventing the wheel. In many of the articles that come to my hands, I find, for example, how some researchers devote their efforts to looking for Ψ-ontological and also for hidden variables theories without realizing the problems that such theories have and that d'Espagnat has pointed out in depth and with intellectual clarity in his books. I do not want to say that such efforts are futile, –in fact, they are quite respectable but to point out that it is necessary to carry them out in full awareness of the difficulties they entail.

References

1. d'Espagnat, B., *Conceptions de la physique contemporaine*. Hermann, Paris (1965).
2. d'Espagnat, B., *Contribution to Preludes in Theoretical Physics: In Honour of V.F. Weisskopf*, De Shalit, A., Feschbach, H., and Van Hove, L. (Eds.) North-Holland, Amsterdam, (1966) p. 185.
3. Huang, K., *Statistical Mechanics*. John Wiley & Sons, New York, 2Ed, Chap. 8 (1987).
4. Messiah, A., *Quantum Mechanics, Vol 1*, North-Holland, Amsterdam, Chap. 5 (1961).
5. d'Espagnat, B., *Une incertaine Réalité*. Gauthier-Villars, Paris, (1985).
6. d'Espagnat, B., *A la recherche du réel*. Gauthier-Villars, Paris, 2Ed, (1983).
7. d'Espagnat, B., *On Physics and Philosophy*. Princeton University Press, Princeton (2006).
8. d'Espagnat, B., *Veiled Reality: An Analysis of Present-Day QuantumMechanical Concepts*. Addison-Wesley Publishing Company, Reading (1995).
9. de Broglie, L., *La mécanique ondulatoire et la structure atomique de la matière et du rayonnement*. J. Phys. Radium 8(5), 225–241 (1927).
10. Bohm, D., "A suggested interpretation of the quantum theory in terms of hidden variables", *Physical Review*, **85**, 165-180 (1952).
11. Bohm, D., Hiley, B., *The Undivided Universe: An Ontological Interpretation of Quantum Mechanics*. Routledge and Kegan Paul, London (1993).
12. Leggett, A.J. "Schrodinger's cat and her laboratory cousins." *Contemporary Physics*, **25**, 583 (1984)
13. Leggett, A.J. "The current status of quantum mechanics at the macroscopic level", in *Foundations of Quantum Mechanics in the Light of New Technology. Advanced Series in Applied Physics, vol. 4.*, Nakajima, S., Murayama, Y., Tonomura, A. (eds.), World Scientific, Singapore (1997)
14. Joos, E., et al. *Decoherence and the Appearance of a Classical World in Quantum Theory*. Springer, Heidelberg (1996).
15. d'Espagnat, B., "Empirical reality, empirical causality, and the measurement problem." *Foundations of Physics*, 17(5), 507–529 (1987).
16. Roldán-Charria, J., "Indivisibility, Complementarity and Ontology. A Bohrian Interpretation of Quantum Mechanics." *Foundations of Physics*, **44**, 1336–1356 (2014).

QBISM, BOHR, AND THE QUANTUM OMELETTE
TOSSED BY DE RONDE

ULRICH MOHRHOFF

Sri Aurobindo International Centre of Education
Pondicherry 605002 India.
Email: ujm@auromail.net

In his recent paper "QBism, FAPP and the Quantum Omelette," de Ronde makes a variety of questionable claims concerning QBism, Bohr, and the present author's critical appraisal of QBism. These claims are examined. Subsequently an outline is presented of what one might see if one looks into the quantum domain through the window provided by the quantum-mechanical correlations between outcome-indicating events in the classical domain.

Keywords: QBism; Bohr; Kant; manifestation; causality.

1. Introduction

The formalism of quantum mechanics (QM) was characterized by Jaynes[1] as "a peculiar mixture describing in part realities of Nature, in part incomplete human information about Nature—all scrambled up by Heisenberg and Bohr into an omelette that nobody has seen how to unscramble." The (alleged) improper scrambling of ontic ("objective") and epistemic ("subjective") perspectives is the focus of a recent paper by de Ronde.[2] What mainly concerns me here is his defense of QBism, which he regards as "one of the most honest, consistent and clear approaches to QM" and as "completely safe from several (ontological) criticisms it has recently received," including one by the present author,[3] notwithstanding that it "does not solve the problems of QM, it simply dissolves them." In their response to Nauenberg,[4] another critic of QBism, the QBist triumvirate Fuchs et al.[5] "welcome criticism, but urge critics to pay some attention to what we are saying." I could say the same of de Ronde's throwaway remarks on my critique of QBism.

What follows is divided into two parts. The first part deals with Bohr, QBism, and what de Ronde has to say about Bohr, QBism, and my critical appraisal of QBism. The second part outlines what one might see if one looks into the quantum domain through the window provided by the quantum-mechanical correlations between outcome-indicating events in the classical domain.

2. QBism, Bohr, and the quantum omelette

All de Ronde quotes from my critical appraisal of QBism is the distinction I made between

(I) a transcendental reality external to the subject, undisclosed in experience, which Kant looked upon as the intrinsically unknowable cause of subjective experience, and

(II) the product of a mental synthesis—a synthesis based on the spatiotemporal structure of experience, achieved with the help of spatiotemporal concepts, and resulting in an objective reality from which the objectifying subject can abstract itself.

It is true that within the Kantian scheme, as de Ronde explains, "transcendental reality amounts to reality as it is, 'the thing in itself'." It ought to be noted, however, that I carefully avoided the vacuous expression "reality as it is" (in and by itself, out of relation to our experience and our categorial schemes). To acknowledge a transcendental reality is but to recognize that there is more to reality than what is disclosed in human experience and can be captured by mathematical models or mental constructs. An objective reality constructed by us is the one we physicists will ever be concerned with, whether we want it or not, whether or not we think of it (rightly or wrongly but in any case irrelevantly[a]) as a faithful representation of "reality as it is."

De Ronde flatly denies "that QM can be considered in terms of 'objective reality' within the Bohrian scheme." Why? Because "[t]he subject cannot abstract himself from the definition of reality provided by QM in terms of waves, particles or even definite valued properties." It is news to me that QM provides a definition of (objective) reality, let alone one in terms of possessed definite properties or classical models like waves or particles. Where Bohr is concerned, objective reality[b] is made up of

---

[a] As Xenophanes observed some twenty-five centuries ago, even if my conceptions represented the world exactly as it is, I could never know that this is the case.

[b] I am of course not speaking of transcendental reality, whose relation to QM did not concern Bohr.

two things: (i) the experimental arrangement—the system preparation, the measurement apparatus, and the indicated outcome—all of which have to be described in ordinary ("classical") language if we want to be able to communicate "what we have done and what we have learned" ([6, pp. 3, 24] and [7, pp. 39, 72, 89]), and (ii) "statistical laws governing observations obtained under conditions specified in plain language." That's all there is to it: "the physical content of quantum mechanics is exhausted by its power to formulate" such laws [6, p. 12].

De Ronde argues that because Bohr's notion of complementarity involves a subject's choice, it is inconsistent with an objective conception of reality: "Physical reality can be only represented in an objective manner if the subject plays no essential role within that representation." Was Bohr then mistaken in writing that the "description of atomic phenomena has ... a perfectly objective character, in the sense that no explicit reference is made to any individual observer," and that "all subjectivity is avoided by proper attention to the circumstances required for the well-defined use of elementary physical concepts" [6, pp. 3, 7]? By no means, for it is not the case that complementarity implies a choice.

To illustrate his point, de Ronde considers a double slit experiment, which Subject 1 performs with both slits open and Subject 2 performs with one slit shut. Subject 1 (who, like Subject 2, appears to owe his information about QM to the popular science media) concludes that the "quantum object" is a wave, while Subject 2 concludes that it is a particle. For de Ronde this means that quantum reality, giving rise as it does to subject-dependent conclusions about one and the same object, cannot be objective: "The real (objective) existence of waves and particles cannot be dependent on a (subjective) choice of an experimenter." In point of fact, what we are dealing with here is not a single situation involving a subject's choice but two distinct physical situations within a single objective reality. Nor are the objects studied in these experiments either classical waves or classical particles. They are particles only in the sense that they can produce "clicks" in counters,[c] and they are waves only in the sense that the clicks can exhibit interference fringes.

De Ronde's mention of "the Bohrian metaphysical premise according to which the description must be given in terms of classical physics by waves or particles" suggests that he actually believes that this is what Bohr had in mind when he insisted on the use of the language of classical

---

[c]On the inadequacy of *this* language see Ulfbeck and Bohr[8] and my paper.[9]

physics. Which in turn suggests that some of the most central tenets of Bohr's philosophy are lost on de Ronde, such as the necessity of defining observables in terms of the experimental arrangements by which they are measured: the "procedure of measurement has an essential influence on the conditions on which the very definition of the physical quantities in question rests".[10] In other words, to paraphrase a famous dictum by Wheeler,[d] no property (of a quantum system) or value (of a quantum observable) is a possessed property or value unless it is a measured property or value. (For extensive discussions of this point see my.[12,13])

"One of the main constituents of the present quantum omelette," de Ronde points out, "is the idea that 'measurement' is a process which has a special status within QM." While 'measurement' has a special status to be sure, it isn't a process. In fact, QM knows nothing about processes. It is about measurement outcomes—actual, possible, or counterfactual ones, performed on the same system at different times or on different systems in spacelike relation—and their correlations. What happens between a system preparation and a measurement is anybody's guess, as the proliferation of interpretations of QM proves. Given a system preparation, QM gives us the probabilities with which outcome-indicating events happen, not processes by which they come about.

Two problems, according to de Ronde, "make explicit how QM has turned into a 'quantum omelette' with no clear limit between an ontological account and an epistemological one"—"two problems in which the intrusion of a choosing subject appears explicitly in the determination of what is considered to be (classically) real—or actual." The first is the basis problem, which, so de Ronde,

> attempts to explain how is Nature capable of making a choice between different incompatible bases. Which is the objective physical process that leads to a particular basis instead of a another one? If one could explain this path through an objective physical process, then the choice of the experimenter could be regarded as well as part of an objective process.... Unfortunately, still today the problem remains with no solution within the limits of the orthodox formalism.

That this problem remains unsolved should not come as a surprise. It

---

[d] "No elementary phenomenon is a phenomenon until it is a registered (observed) phenomenon".[11]

is in the nature of pseudo-problems to lack solutions—real solutions, as against gratuitous ones. The reason this problem is a pseudo-problem is that what happens between a system preparation and a measurement is a phenomenon[e] that cannot be dissected into the unitary evolution of a quantum state and its subsequent "collapse." There is no Nature making choices, whether between bases or between possible outcomes. There is no objective physical process selecting a particular basis. What determines a particular basis is the measurement apparatus. The fact that the apparatus is usually chosen by an experimenter, however, is of no consequence as far as the interpretation of the formalism is concerned. What matters is that the apparatus is needed not only to indicate the possession of a property by a quantum system but also—and in the first place—to make a set of properties available for attribution to the system. Whether it is anyone's intention to obtain a particular kind of information, or whether anyone is around to take cognizance of it, is perfectly irrelevant.

The second problem, in the words of de Ronde, is this:

Given a specific basis (context or framework), QM describes mathematically a state in terms of a superposition (of states). Since the evolution described by QM allows us to predict that the quantum system will get entangled with the apparatus and thus its pointer positions will also become a superposition, the question is why do we observe a single outcome instead of a superposition of them? It is interesting to notice that for Bohr, the measurement problem was never considered. The reason is that through his presuppositions, Bohr begun the analysis of QM presupposing "right from the start" classical single outcomes.

In fact, what allows us to predict the probabilities of the possible outcomes of a measurement is not the evolution described by QM, for QM describes no evolution. There are at least nine different formulations of QM. The better known among them are Heisenberg's matrix formulation, Schrödinger's wave-function formulation, Feynman's path-integral formulation, the density-matrix formulation, and Wigner's phase-space

---

[e] "[A]ll unambiguous interpretation of the quantum mechanical formalism involves the fixation of the external conditions, defining the initial state of the atomic system concerned and the character of the possible predictions as regards subsequent observable properties of that system. Any measurement in quantum theory can in fact only refer either to a fixation of the initial state or to the test of such predictions, and it is first the combination of measurements of both kinds which constitutes a well-defined phenomenon."[14]

formulation.[15] Not all of them feature an evolving quantum state. Yet it stands to reason that the interpretation of QM ought to depend on what is common to all formulations of the theory (and thus has a chance of being objective) rather than on the idiosyncrasies of a particular formulation such as Schrödinger's.[f]

Another reason QM describes no evolution is that the quantum calculus of correlations is time-symmetric. It allows us to assign probabilities not only to the possible outcomes of a later measurement on the basis of an earlier measurement but also to the possible outcomes of an earlier measurement on the basis of a later one.[g] It is therefore just as possible to postulate that quantum states evolve backward in time as it is to postulate that they evolve forward in time. If the former postulate contributes nothing to our understanding of QM, then neither does the latter.

Nor does QM allow us to predict that the quantum system will get entangled with the apparatus pointer—and it had better not, for in the face of overwhelming evidence that measurements tend to have outcomes, this would be absurd. What is common to all formulations of QM is that it serves as a calculus of correlations between measurement outcomes. The reason we observe a single outcome is therefore simply that without single outcomes the quantum calculus of correlations would have no application. There would be nothing to correlate. "[P]resupposing 'right from the start' classical single outcomes" is therefore the only sound way to proceed. Thus, contrary to what was claimed by de Ronde, the two (pseudo-) problems fail to make explicit how (or even that) "QM has turned into a 'quantum omelette' with no clear limit between an ontological account and an epistemological one."

Should (or can) there be a clear limit between the two accounts? Is it even possible to give an ontological account free of any trace of epistemology, or an epistemological account free of any trace of ontology? While it is obviously beyond the scope of the present paper to enter into a discussion of philosophical issues about which countless volumes have been written,

---

[f]What is to blame here is the manner in which quantum mechanics is generally taught. While junior-level classical mechanics courses devote a considerable amount of time to different formulations of classical mechanics (such as Newtonian, Lagrangian, Hamiltonian, least action), even graduate-level quantum mechanics courses emphasize the wave-function formulation almost to the exclusion of all variants. This is not only how ψ-ontologists come to think of quantum states as evolving physical states but also how QBists come to think of them as evolving states of belief.

[g]It even allows us to assign probabilities to the possible outcomes of a measurement on the basis of both earlier and later outcomes using the ABL rule[16] rather than Born's.

off the cuff I would say that no ontological account is complete (or even meaningful) without an epistemological justification, and that no epistemological account is complete (or serves any purpose) if it does not relate to an ontological account (epistemology being about knowledge, and knowledge being about a reality of some kind).

With regard to QM there are two ways to deny this: that of the quantum-state realist, who is cavalier about epistemological concerns, and that of the QBist, who is cavalier about ontological concerns. Nothing much needs to be said about quantum-state realism, inasmuch as this is essentially self-defeating. Any interpretation of QM that needs to account for the existence of measurement outcomes—and thus for the existence of measurements, since no measurement is a measurement if it doesn't have an outcome—is thwarted by the non-objectification theorems proved by Mittelstaedt [18, Sect. 4.3(b)] and the insolubility theorem for the objectification problem due to Busch et al. [19, Sect. III.6.2].

What about QBism? QBists are right in being cavalier about ontological concerns if this means being unconcerned about the relation (if any) between QM and transcendental reality, but they are wrong in being cavalier about the relation between QM and an objective reality. Most if not all of their arguments presuppose such a reality,[h] whose existence they cannot therefore consistently deny, just as the philosophical skeptic cannot deny a version of realism whose truth she presupposes in defending her stance. To bring home this crucial point, let us assume with Searle [21, pp. 286–87]

> that there is an intelligible discourse shared publicly by different speakers / hearers. We assume that people actually communicate with each other in a public language about public objects and states of affairs in the world. We then show that a condition of the possibility of such communication is some form of direct realism.

The argument Searle is about to present is directed against the sense-datum theory of perception, according to which all we ever perceive directly—without the mediation of inferential processes—is our own subjective experiences, called "ideas" by Locke, "impressions" by Hume, and "representations" by Kant. In one form or another the sense-datum theory was held by most of the great philosophers in the history of the subject. (QBism may be seen as a throwback to these bygone days.) The argument begins by assuming that we successfully communicate with other human

---

[h]For examples the reader is invited to consult my.[3,20]

beings at least some of the time, using publicly available meanings in a public language.

But in order to succeed in communicating in a public language, we have to assume common, publicly available objects of reference. So, for example, when I use the expression "this table" I have to assume that you understand the expression in the same way that I intend it. I have to assume we are both referring to the same table, and when you understand me in my utterance of "this table" you take it as referring to the same object you refer to in this context in your utterance of "this table."

The implication is that "you and I share a perceptual access to one and the same object." However, saying that "you and I are both perceiving the same public object" does *not* mean that you and I perceive the transcendental object or "thing in itself." The "direct realism" Searle is defending is two removes from this naïve view. By the sense-datum theory we get away from it, but then we realize that

Once you claim that we do not see publicly available objects but only sense data, then it looks like solipsism is going to follow rather swiftly. If I can only talk meaningfully about objects that are in principle epistemically available to me, and the only epistemically available objects are private sense data, then there is no way that I can succeed in communicating in a public language, because there is no way that I can share the same object of reference with other speakers.

What else is this public language than the ordinary language the necessity of whose use Bohr was at such pains to stress? And what else is the general object of reference of this language than the objective reality which the QBists fail to recognize as the proper object of scientific inquiry, and which de Ronde fails to recognize as the sole reality accessible to scientific inquiry? By throwing out the baby of objective reality with the bathwater of transcendental reality, QBists have landed themselves on the horns of a dilemma: insofar as they claim to be exclusively concerned with the subjective experiences of individual "agents" or "users," they have no way of communicating their views,[i] and insofar as they succeed

---

[i]This, and not merely the obtuseness of their detractors, appears to be the reason they seem to have such a hard time making themselves understood.

in communicating, they implicitly acknowledge an objective reality. It no doubt is an interesting project to find out how far the Bayesian interpretation of probability can be carried in the context of QM, but to deny that QM refers to measurement outcomes indicated by instruments situated in an objective reality is overkill. It is an overreaction against the realism of the ψ-ontologist, grounded in a common failure to distinguish between the two kinds of reality.

A result of this failure is the frequent occurrence of the fallacy known as "false dilemma": Either we take a transcendental realist stance or we must accept that QM does not make reference to anything but beliefs of "users." Either we embrace ψ-ontology or we "remain on the surface of intersubjectivity," using an epistemic approach that restricts our discourse to the way we interact by communicating empirical findings, leaving aside "the relation of these interactions to the world and reality themselves" [2, p. 7]. The possibility which remains unconsidered is that QM makes reference to an objective reality that, while not being the reality of the ψ-ontologist, is essential to the expression of our beliefs and the communication of our empirical findings.

De Ronde's ambivalent assessment of QBism reflects the QBists' dilemma. Addressing the horn of solipsism, he writes that QBists

dissolve all important and interesting questions that physical thought has produced since the origin itself of the theory of quanta. Taking to its most extreme limit several of the main Bohrian ideas, QBism has turned physics into a solipsistic realm of personal experience in which no falsification can be produced; and even more worrying, where there are no physical problems or debates left. QBism does not solve the problems of QM, it simply dissolves them.

Addressing the other horn (i.e., accepting QBists' ability to communicate in public language), he claims that "QBists have produced a consistent scheme that might allow us to begin to unscramble—at least part of—the 'quantum omelette'," though he gives no indication how this might be done. Nor does he bother to substantiate his extravagant claim that "QBism has seen much better than Bohr himself the difficult problems involved when applying an epistemological stance to understand QM"—a claim strangely at odds with his statement that "[e]ven today [Bohr's scheme] seems to us one of the strongest approaches to QM."

If it were true that "QBism cannot be proven to be wrong," as de Ronde claims, QBism would be *not even wrong*. QBism, however, makes numerous claims, and some of the fourteen examined in my[3] *are* wrong, for example the claim that there are no external criteria—external to the individual "user's" private theater of subjective experiences—for declaring a probability judgment right or wrong. In fact, there are objective data—external to the individual "user" though not, of course, external in the transcendental sense—on the basis of which probabilities are assigned, notwithstanding that the choice of these data and hence the probability assignments depend on the "user."

De Ronde claims to "show why the epistemic QBist approach is safe from several (ontological) criticisms it has recently received," including my own. What appears to have escaped his notice is that none of my criticisms were ontological in his sense. While Marchildon, Nauenberg, and myself are collectively indicted for asking "QBists to answer ontological questions they have explicitly left aside right from the start," he offers not a shred of evidence that his indictment has merit in my case. Our attacks are said to "come either from the reintroduction of ontological problems," which is not true in my case, "or from the unwillingness to understand the radicalness of the QBist proposal." Concerning the latter, I beg forgiveness for quoting from an email I received from Chris Fuchs after posting my:[3] "Thanks for your paper tonight. I will read it very carefully in the coming days. Your Section 4 [titled "The central affirmations of QBism"] so impressed me that I know I *must* read it." In a message to his QBist colleagues, forwarded to me in the same email, he further wrote: "The 14 things he lists in Section 4 are remarkably accurate ... unless I've had too much wine tonight."

In an attempt at defending ψ-ontology, de Ronde points out that "[t]he foundational discussions that have taken place during the last decades [concerning, among other things, the EPR paper, Bell inequalities, and the Kochen-Specker theorem] are in strict relation to a realist account of the theory." If so, what is the conclusion to be drawn from these discussions if not that Bohr was right: realist accounts of QM do not work. In the quantum domain, no property or value is a possessed property or value unless its existence is implied by—indicated by, can be inferred from—an event or state of affairs in the classical domain. The distinction between a classical and a quantum domain is thus an inevitable feature of QM. It needs to be understood, not swept under the rug or explained away.

Again, according to de Ronde, "[a]ll interesting problems which we have been discussing in the philosophy of science and foundations community for more than a Century ... have been in fact the conditions of possibility for the development of a new quantum technological era"; these problems "allowed us to produce outstanding developments such as quantum teleportation, quantum cryptography and quantum computation." Here de Ronde seems to be speaking off the top of his head, considering that important contributions to these fields came from QBists and other physicists with no transcendental realist leanings.

3. Manifestation

Echoing Kant's famous dictum that "[t]houghts without content are empty, intuitions without concepts are blind" [22, p. 193], Bohr could have said that without measurements the formal apparatus of quantum mechanics is empty, while measurements without the formal apparatus of quantum mechanics are blind. What allows us to peer beyond the classical domain with its apparatuses is the combination of measurement outcomes and their quantum-mechanical correlations. And what we find if we peer into the quantum domain is that intrinsically the things we call "particles" are identical with each other in the strong sense of numerical identity.[12,13] They are one and the same intrinsically undifferentiated Being, transcendent of spatial and temporal distinctions, which by entering into reflexive spatial relations gives rise to

(1) what looks like a multiplicity of relata if the reflexive quality of the relations is ignored, and
(2) what looks like a substantial expanse if the spatial quality of the relations is reified.

In the words of Leibniz: *omnibus ex nihilo ducendis sufficit unum*—one is enough to create everything from nothing.

As said, the distinction between a classical and a quantum domain needs to be understood, and this (if possible) beyond the linguistic necessity of speaking about the quantum domain in terms of correlations between events in the classical domain. One reason it is so hard to beat sense into QM is that it answers a question we are not in the habit of asking. Instead of asking what the ultimate constituents of matter are and how they interact and combine, we need to broaden our repertoire of explanatory concepts and inquire into the manifestation of the familiar world of everyday experience. Since the kinematical properties of microphysical objects—their positions, momenta, energies, etc.—only exist if and when they are indicated by the behavior of macrophysical objects, microphysical objects cannot play the role of constituent parts. They can only play an instrumental role in the manifestation of macrophysical objects. Essentially, therefore, the distinction between the two domains is a distinction between the *manifested world* and its *manifestation*.

The manifestation of the familiar world of everyday experience consists in a transition from the undifferentiated state of Being to a state that allows itself to be described in the classical language of interacting objects and causally related events. This transition passes through several stages, across which the world's differentiation into distinguishable regions of space and distinguishable objects with definite properties is progressively realized. There is a stage at which Being presents itself as a multitude of formless particles. This stage is probed by high-energy physics and known to us through correlations between the counterfactual clicks of imagined detectors, i.e., in terms of transition probabilities between in-states and out-states. There are stages that mark the emergence of form, albeit as a type of form that cannot yet be visualized. The forms of nucleons, nuclei, and atoms can only be mathematically described, as probability distributions over abstract spaces of increasingly higher dimensions. At energies low enough for atoms to be stable, it becomes possible to conceive of objects with fixed numbers of components, and these we describe in terms of correlations between the possible outcomes of unperformed measurements. The next stage—closest to the manifested world—contains the first objects with forms that can be visualized—the atomic configurations of molecules. But it is only the final stage—the manifested world—that contains the actual detector clicks and the actual measurement outcomes which have made it possible to discover and study the correlations that govern the quantum domain.

One begins to understand why the general theoretical framework of contemporary physics is a probability calculus, and why the probabilities are assigned to measurement outcomes. If quantum mechanics concerns a transition through which the differentiation of the world into distinguishable objects and distinguishable regions of space is gradually realized, the question arises as to how the intermediate stages are to be described—the stages at which the differentiation is incomplete and the distinguishability between objects or regions of space is only partially realized. The answer is that whatever is not completely distinguishable can only be described by assigning probabilities to what is completely distinguishable, namely to the possible outcomes of a measurement. What is instrumental in the manifestation of the world can only be described in terms of (correlations between) events that happen or could happen in the manifested world.

The atemporal causality by which Being manifests the world must be distinguished from its more familiar temporal cousin. The usefulness of the latter, which links states or events across time or spacetime, is confined to the world drama; it plays no part in setting the stage for it. It helps us make sense of the manifested world as well as of the cognate world of classical physics, but it throws no light on the process of manifestation nor on the quantum correlations that are instrumental in the process. That other causality, on the other hand, throws new light on the nonlocality of QM, which the QBists so nonchalantly dismiss. The atemporal process by which Being enters into reflexive relations and matter and space come into being, is the nonlocal event *par excellence*. Depending on one's point of view, it is either coextensive with spacetime (i.e., completely delocalized) or "outside" of spacetime (i.e., not localized at all). Occurring in an anterior relation to space and time, it is the common cause of all correlations, not only of the seemingly inexplicable ones between simultaneous events in different locations but also of the seemingly explicable ones between successive events in the same location.[j]

The objection may be raised that in positing an intrinsically undifferentiated Being and an atemporal process of manifestation, I have ventured into transcendental territory. But this is not the case. While Bohr went beyond Kant only in that he opened up the Kantian world-as-we-know-it, providing a window on what lies beyond,[k] I go beyond Bohr only in that I use QM to look through this window. It is still essentially the Kantian categories that I use when speaking of the manifested world as a system of interacting and causally evolving bundles of possessed properties, and it is still the quantum-mechanical correlations between outcome-indicating events in this world that I use to draw my inferences. It is no doubt tempting to think of Being and the manifested world transcendentally, as if they existed out of relation to our experience, but of what exists out of relation to our experience we know zilch. The manifested world exists in relation to our experience—it is manifested *to us*—and so does the Being which manifests it.

---

[j]The diachronic correlations between events in timelike relation are as spooky as the synchronic correlations between events in spacelike relation. While we know how to calculate either kind, we know as little of a physical process by which an event here and now contributes to determine the probability of a *later* event *here* as we know of a physical process by which an event here and now contributes to determine the probability of a *distant* event *now*.

[k]What is responsible for the closure of the objective Kantian world is (i) Kant's apriorism, which requires (among other things) the universal validity of the law of causality, and (ii) Kant's principle of thoroughgoing determination, which asserts that "among all possible predicates of things, insofar as they are compared with their opposites, one must apply to [each thing] as to its possibility" [22, p. 553].

260

References

1. E.T. Jaynes: Probability in quantum theory, in *Complexity, Entropy and the Physics of Information*, edited by W.H. Zurek (Addison-Wesley, Redwood City, CA, 1990), pp. 381–400.
2. C. de Ronde: QBism, FAPP and the quantum omelette, preprint: arXiv:1608.00548v1 [quant-ph].
3. U. Mohrhoff: QBism: a critical appraisal, preprint: arXiv:1409.3312v1 [quant-ph].
4. M. Nauenberg: Comment on QBism and locality in quantum mechanics, *American Journal of Physics* 83, 197–198 (2015), preprint: arXiv:1502.00123v1 [quant-ph].
5. C.A. Fuchs, N.D. Mermin, and R. Schack: Reading QBism: a reply to Nauenberg, *American Journal of Physics* 83, 198 (2015).
6. N. Bohr: *Essays 1958–1962 on Atomic Physics and Human Knowledge* (John Wiley & Sons, New York, 1963).
7. N. Bohr: *Atomic Physics and Human Knowledge* (John Wiley & Sons, New York, 1958).
8. O. Ulfbeck and A. Bohr: Genuine fortuitousness: where did that click come from? *Foundations of Physics* 31, 757–774 (2001).
9. U. Mohrhoff: Making sense of a world of clicks, *Foundations of Physics* 32, 1295–1311 (2002).
10. N. Bohr: Quantum mechanics and physical reality, *Nature* 136, 65 (1935).
11. J.A. Wheeler: Law without law, in *Quantum Theory and Measurement*, edited by J.A. Wheeler and W.H. Zurek (Princeton University Press, Princeton, NJ, 1983) 182–213.
12. U. Mohrhoff: Manifesting the quantum world, *Foundations of Physics* 44, 641–677 (2014).
13. U. Mohrhoff: Quantum mechanics in a new light. *Foundations of Science* DOI 10.1007/s10699-016-9487-6 (2016); preprint: http://bit.ly/2b5OVY5
14. N. Bohr: in *New Theories in Physics: Conference Organized in Collaboration with the International Union of Physics and the Polish Intellectual Co-operation Committee*, Warsaw, 30 May–3 June, 1938 (International Institute of Intellectual Co-operation, Paris, 1939), 11–45.
15. D.F. Styer, M.S. Balkin, K.M. Becker, M.R. Burns, C.E. Dudley, S.T. Forth, J.S. Gaumer, M.A. Kramer, D.C. Oertel, L.H. Park, M.T. Rinkoski, C.T. Smith, T.D. Wotherspoon: Nine formulations of quantum mechanics, *American Journal of Physics* 70, 288–297 (2002).
16. Y. Aharonov, P.G. Bergmann, and J.L. Lebowitz: Time symmetry in the quantum process of measurement, *Physical Review B* 134, 1410–1416 (1964).
17. C.A. Fuchs, N.D. Mermin, and R. Schack: An introduction to QBism with an application to the locality of quantum mechanics, *American Journal of Physics* 82, 749–754 (2014).
18. P. Mittelstaedt: *The Interpretation of Quantum Mechanics and the Measurement Process* (Cambridge University Press, Cambridge, MA, 1998).
19. P. Busch, P.J. Lahti, and P. Mittelstaedt: *The Quantum Theory of Measurement*, 2nd Revised Edition (Springer, Berlin, 1996).

20. U. Mohrhoff: Quantum mechanics and experience, preprint: arXiv:1410. 5916v2 [quant-ph].
21. J.R. Searle: *Mind: A Brief Introduction* (Oxford University Press, Oxford, UK, 2004).
22. I. Kant: *Critique of Pure Reason*, translated and edited by P. Guyer and A.W. Wood (Cambridge University Press, New York, 1999).

ONTIC STRUCTRAL REALISM AND
QUANTUM MECHANICS

JOAO L. CORDOVIL

Center for Philosophy of Science,
University of Lisbon, Portugal.
* E-mail: jlcordovil2@hotmail.com
http://cfcul.fc.ul.pt/equipa/jcordovil.php

Radical-Ontic Structural Realism and Moderate-Ontic Structural Realism are usually distinguished by the specific way of how they address the question of the primacy between objects and relations. However, I will argue that the difference between Radical-Ontic Structural Realism and Moderate-Ontic Structural Realism runs deeper; it rests on the different programmatic assessments of the relationship between science and metaphysics.

In this sense, at least at its roots, the distinction between Radical-Ontic Structural Realism and Moderate-Ontic Structural Realism lies in the divergence between Radical Naturalistic Metaphysics and Esfeld's account of Natural Philosophy (based on the Primitive Ontology approach).

From the above distinctions, I will argue: i) even that Quantum Mechanics offers us good arguments in favor of Ontic Structural Realism, both main Ontic Structural Realism's proposals seem to struggle with QM's challenges; ii) Ontic Structural Realism's failure is not due to itself but to this metametaphysics proposals. That is, maybe some assumptions made by both NM and PO, like micro-physicalism, monism or metaphysical fundamentalism, should be reviewed or even dismissed from Ontic Structural Realism proposals.

Keywords: Ontic structural realism; naturalized metaphysics; primitive ontology; quantum mechanics' metaphysics; metaphysical fundamentalism.

1. Motivations for Ontic Structural Realism

Ontic Structural Realism has been one of the central topics in contemporary philosophy of science. According to Steven French, Ontic Structural Realism is motivated by the following idea:

two sets of problems that "standard" realism is seen to face. The first has to do with apparent ontological shifts associated with theory change that can be observed throughout the history of science.

The second is associated with the implications again ontological — of modern physics (French 2010)

This means that standard Scientific Realism faces two challenges: a) general objections against Scientific Realism; and b) metaphysical considerations raised by Modern Physics. However, these ideas do not stand for themselves; they are deeply interrelated at the ontological level. Now, Ontic Structural Realism addresses these issues not by replying to them but by embracing them, i.e. Ontic Structural Realism accepts the arguments against Scientific Realism and the implications for traditional Metaphysics of Physics stemming from Fundamental Physics.

1.1. *Contra Traditional Metaphysics of Objects*

Ontic Structural Realism literature emphasizes that modern physics implies the downfall – or is at least incompatible with – the traditional metaphysics of objects (TMO). That is, the metaphysical view that claims that a) there is a fundamental level of reality; b) it is composed of individuals; c) there are ontological independent entities – objects; and d) those objects move inside a spatiotemporal framework.

According to this metaphysical position, entities are individuals because they possess qualitative properties which distinguish them from other entities (e.g. a spatial-temporal location or other putative properties). This means individuality conditions are fully exhausted by the qualitative properties those entities instantiate. Qualitative properties "are all and only those properties whose instantiation does not depend on the existence of any particular individual; properties such as being that individual are hence excluded" (Esfeld 2003:5).

Further, this view holds that entities are independent. This is the case because they have intrinsic and monadic properties.

Being intrinsic properties are all those qualitative properties that an object possesses independently of being accompanied or alone (see Langton and Lewis (1998), Lewis (2001)). Since they are intrinsic, those properties are unchangeable and independent of any relational context. And, since they are monadic, those properties are instantiated in each individual object. In this context, standard examples are mass or charge.

The atomist stance of TMO is clear. Consider Lewis's Humean supervenience (Lewis 1986), where everything that exists in the world supervenes on the distribution of those basic intrinsic properties found all over the space. Consequently, all relations between objects also supervene on the intrinsic properties. This Metaphysical position (TMO) has been undermined – or at least severely challenged – by some crucial features of Modern Physics. In particular, both individuality and independence have been targeted (see b) and c) above). Quantum statistics or permutation symmetry question if quantum objects are discernible and, therefore, if they possess individuality - the Received View on quantum non-individuality (See e.g. French and Krause 2006; French 2014; Arenhart and Krause 2014, Arenhart 2015. See also Saunders 2006 (weak discernibility), Dorato and Morganti 2013 (identity taken as primitive) for criticism to the Received View). Furthermore, entanglement and space-time metrics of General Relativity (the so-called Hole Argument – see e.g. Dorato 2000) both indicate that physical objects cannot be considered independent entities. Consequently, it is claimed that Modern Physics is irreconcilable with TMO.[a]

1.2. *Contra Scientific Realism based on TMO*

Paralleling the previous debate are discussions within Scientific Realism. Anti-realistic arguments like "Pessimist Meta-Induction" (Laudan 1981) or "Underdetermination of Theory by Evidence", where two or more theories make the same observational predictions and, therefore, it is not possible to conclusively decide from observational evidence in favour of one or the other (Papineau 1996:7, da Costa and French 2003: 189), make a strong case against "standard realism" or "Object-oriented Scientific Realism" (French 2006: 168, Psillos 2001: S23). These same arguments run also against Epistemic Structural Realism (French 2014: 22). The worries about Object-oriented Scientific Realism are, according to French (French, 2011 and 2014), strengthened by other kinds of Underdeterminations. The "Jones Underdetermination" claims that there are different empirically equivalent formulations of the same theory (Jones 1991, Pooley 2006), while the "Metaphysical Underdetermination" states that there are different metaphysical interpretations of the same theory (see van Frassen 1991:491; French 2011 and 2014).

1.3. *Broad Ontic Structural Realism Characterization*

In general, following Ontic Structural Realism literature, both Modern Physics' metaphysical implications and the Scientific Realism Debate point to the conclusion that we should not be ontologically committed to objects in the first place. Consequently, we should reject TMO and

---

[a] I will briefly discuss assumption a) below.

Objects-Oriented Scientific Realism. Instead, we should be ontologically committed to relations and relational structures. Why? Firstly, because in cases of ontological changes related with theory shifts structures (at least some structures) are preserved (Worrall, 1989). Thus, Structural Realism is immune to Pessimist Meta-Induction. Secondly, the above-mentioned Underdeterminations become unproblematic if we only contemplate "common" structure between the theories involved (French 2011). Thirdly, given that entanglement, space-time metrics and Permutation Invariants are relations instantiated by the entities of fundamental physics, they evidence the relational nature of such entities. That is, entities of fundamental physics are elements of a relational structure.

So, Ontic Structural Realism's starting point is that modern physics give us strong arguments to dismiss the traditional metaphysics assumption that there is a fundamental level of physical reality composed by entities with intrinsic properties. Ontic Structural Realism aims both to rescue Scientific Realism from its main objections and to develop a Metaphysics driven by or compatible with Contemporary Physics. Ontic Structural Realism's main tenet consists in the idea that we should be ontologically committed primarily with relations – even if we admit objects, we should dismiss intrinsic properties!

Proponents of Ontic Structural Realism should be ontologically committed to the relational structures of our best scientific theories and, therefore, only be realistic about those very structures. At this point, however, contemporary literature splits into two dominant Ontic Structural Realism's versions.

2. Different kinds of Ontic Structural Realism.

The standard way of presenting those versions is to classify their different ways of conceiving the ontological relationship of *relations and objects.* Hence, in opposition to the traditional metaphysics of objects, Ontic Structural Realism can be broadly presented as the ontological view according to which there are either structures of relations – Radical Ontic Structural Realism – or structures of relations and *relata* – objects in Moderate Ontic Structural Realism.

2.1. *Radical-Ontic Structural Realism*

On the one hand, driven by Modern Physics features such as the QFT's group-theoretical characterization of elementary particles – where elementary particles are the irreducible representations of fundamental symmetry groups (a result first achieved by Wigner 1939) – Radical-Ontic Structural Realism can be eliminativist. That is there is a fundamental structure and its features (laws and symmetries (French, 2014: 275)) that consequently asserts that there are no objects or intrinsic properties at all (French, 2014). There are only relations – like symmetry - and no *relata*. On the other hand, R-Ontic Structural Realism can be asymmetrical. This means that the existence of both relations and *relata* (objects) is acknowledged, but ontological priority is given to relations since objects are ontologically dependent (or somehow derived) on the relational structure. Objects are constituted by the relations in which they stand as mere nodes of relations within the structure (see e.g. Ladyman and Ross 2007, French and Ladyman 2011, Ladyman 2016). This implies that the distinction between relations and *relata* is only conceptual. *Relata* are just posited as a conceptual convenience. A relatum has no other feature than its conceptual identity, i.e. "to be in relation". The crucial point for Radical-Ontic Structural Realism is that objects are, at best, constituted by pre-given relations.

2.2. Moderate Ontic Structural Realism

Now, Moderate-Ontic Structural Realism can be spelled out, at least, in two different versions: a symmetric and a thin-objects version. Symmetric M-Ontic Structural Realism describes objects and relations as two ontologically distinct but interdependent entities, (see e.g. Esfeld 2004 and 2008). In this context, objects and relations share the same following ontological weight:

> "Moderate structural realism proposes that there are objects, but instead of being characterized by intrinsic properties, all there is to the basic physical objects are the relations in which they stand. Admitting objects provides for an empirical anchorage of the relations. [. . .] According to this position, neither objects nor relations (structure) have an ontological priority with respect to the physical world: they are both on the same footing, belonging both to the ontological ground floor." (Esfeld 2008: 31)

Thin-Objects or Thin-Atomism Moderate-Ontic Structural Realism claim the conjugation of Moderate-Ontic Structural Realism and "Primitive Ontology" (see, for instance, Allori 2013 and 2015). Nowadays, this is the dominant version of Moderate-Ontic Structural Realism. According to Thin-Objects Moderate-Ontic Structural Realism: "there are fundamental

physical objects, namely matter points; but all there is to these objects
are the spatial relations among them. Thus, they do not have an intrinsic
nature, but a relational one" (Esfeld et al. forthcoming). Hence – just as
TMO claims – there is a fundamental level of the physical reality composed
of individual objects, namely, matter points (as in primitive ontology). The
difference with TMO is that those objects do not have intrinsic properties,
they are fully characterized by being entities-in-distance-relation. These so-
called distance relations are the only mode of existence for those objects.
Since all that exists at the putative fundamental level are matter-points-in-
distance-relation and distance-relation-between-matter-points, all objects
are in internal relations however still consisting of a plurality of objects (like
in TMO). On the other hand, the possibility of empty space (unoccupied
space-points) must be rejected. According to Thin-Objects Moderate-Ontic
Structural Realism, and contra TMO, there is no background space.

In sum, views of Ontic Structural Realism reject, revise or doubt all but
one characteristic of traditional metaphysics of objects – even the assumption that there is an independent spatio-temporal background. The only
assumption these views maintain is metaphysical fundamentalism.

3. Further Distinctions

Notwithstanding the widespread distinction between Radical-Ontic Structural Realism and Moderate-Ontic Structural Realism already recalled, we
can argue that this distinction runs deeper. More specifically, it rests in the
different programmatic or Metametaphysical assessments of the relationship between science and metaphysics.

3.1. *Radical-Ontic Structural Realism and Naturalized Metaphysics*

Radical-Ontic Structural Realism is a "radically naturalistic metaphysics"
(Ladyman and Ross 2007: 1) approach, where the ontology must be "read
off" or is almost directly derived from the formal content of our current
best scientific theories (Ladyman and Ross 2007). Against the tendency of
Analytical Philosophy, it is said that we must base our ontology *on our
best scientific theories*, and not on "rational intuitions." Against speculative metaphysics, we must adopt a science-based metaphysics and, more
radically, base it on the formal content of our best scientific theories. As
Ladyman and Ross state:

[. . .]rather than metaphysicians using rational intuition to work out exactly how the absolute comes to self-consciousness, they ought instead to turn *to science* and concentrate on explicating the deep structural claims about the nature of reality implicit in our *best theories*. (Ladyman and Ross 2007: 9)

By "our current best scientific theories" they mean our best theories of fundamental physics that today's scientific community considers *bona fide* (Ross and Ladyman, 2007). In fact, fundamental physics enjoys priority over other theories established by the "Principle of Physics Constraint"[b] (cf. Ladyman and Ross 2007: 39). As a consequence of "Principle of Naturalistic Closure", metaphysics continues (or should continue) fundamental physics (cf. Ladyman and Ross 2007: 37). According to this approach, ontology is "read off" by our current best fundamental physical theories. Thus, a strong form of scientific realism is mandatory for Naturalistic Metaphysics. It is only possible to "read off" the ontology from the content of your best scientific theories if we are realistic about that content. Since Structural Realism seem to be able to tackle the objections that perils object-based scientific realism, then it seems that the best option for a Naturalist Metaphysics (NM) is just to give up TMO endorsing Ontic Structural Realism in its radical form, leading to the dictum: "relations are all there is" (Ladyman and Ross 2007; French 2014).

However, or Nevertheless NM as elaborated by Ladyman, Ross and French does not merely defend that metaphysics must be developed from the formal content of our best scientific theories. Despite using the expression "our best scientific theories" in fact, Ladyman and Ross do not call for a metaphysical attention on the natural sciences, but only with one particular science called *Physics* (and Mathematics, we should add). And even within Physics they contemplate only the fundamental theories considerer *bona fide*. Thus, even non-fundamental physics shares the same hierarchy status of Chemistry, or Biology: is a Special Science. That is why Ladyman and Ross just identify "science" and "fundamental physics" (since even non-fundamental physics is a Special Science) and "naturalistic" with "physicalist", using these concepts interchangeably, as if they were synonymous. Nevertheless, this synonymy between science and fundamental physics is an *extra* thesis added to a Naturalistic Metaphysics. Indeed, are there

---

[b]Special science hypotheses that conflict with fundamental physics, or such consensus as there is in fundamental physics, should be rejected for that reason alone. Fundamental physical hypotheses are not symmetrically hostage to the conclusions of the special sciences. (Ladyman, Ross, et al., 2007, p. 44)

naturalistic or *scientific* arguments that justify Physicalism within a Naturalist Metaphysical view? We mean 'naturalistic' or 'scientific' arguments... not speculative philosophical arguments such as the ones typically elaborated by Analytical Philosophers! The answer seems to be: no, there is not.

3.2. Moderate-Ontic Structural Realism and Primitive Ontology

Now, Moderate-Ontic Structural Realism is a kind of "philosophy of nature" directed towards and engaging with Primitive Ontology. Here metaphysics and physics come in one package. Therefore, Moderate-Ontic Structural Realism is less ontologically tied to the formalism of our best current physical theories. However, this interpretation of Ontic Structural Realism is sensible to the fact that we cannot assume scientific realism without e.g. giving a reasonable and direct answer to the measurement problem from QM. Thus, M-Ontic Structural Realism combined with Primitive Ontology moves away from standard Quantum Theories, promoting a revival of Bohmian Mechanics (Durr et al 2012).

According to Primitive Ontology, fundamental Physical Theories should be able to explain the physical world as it appears to our senses, namely a world composed of tri-dimensional objects with well-defined properties, i.e. the Manifest Image. This explanation derives from the lowest-level descriptions of how the world is like according to this fundamental physical theory, i.e. the Scientific Image.

Two questions, among others, immediately arise: how do we get to the theory's Scientific Image? And, what is the role of formalisms? According to the Primitive Ontology approach, the Scientific Image is not gained by "reading off" a formalism, but by stating what the formalism is about, namely the Primitive Ontology. In this sense, Primitive is not derived nor inferred from a formalism, it is rather the referent of that formalism (Esfeld 2014: 99; Egg and Esfeld 2015: 3230). The Primitive Ontology and its evolution in time (its history) provide the theory's Scientific Image. In turn, the formalism of the theory contains primitive variables to describe the Primitive Ontology (and non-primitive variables to describe the dynamics of those primitive variables (Allori 2013: 60)). This means that according to this approach every fundamental physical theory is supposed to explain the Manifest Image regarding its primitive ontology.

According to Valia Allori (2015), the main proponent of PO, this approach is necessarily fundamentalist and micro-physicalist, since it includes the following features: 1) "all fundamental physical theories have a common structure"; and 2) "any satisfactory fundamental physical theory contains a metaphysical hypothesis about what constitutes physical objects, the PO, which lives in three-dimensional space or space-time and constitutes the building blocks of everything else" (Allori 2015: 107).

The elements of primitive ontology (or primitive stuff) are point particles (or matter points). Matter points are primitive not only in the broad sense of primitive ontology but, according to Deckert and Esfeld (Deckert and Esfeld *forthcoming*), also in the following manners: 1) they are not composed of anything; 2) they do not have intrinsic properties, and they are not bare substrata either; and 3) they are simply there, at the fundamental physical level.

This characterization of Primitive Stuff is clearly expressed by Moderate-Ontic Structural Realism's main thesis: "at least some central ways in which the fundamental physical objects exist are relations so that these objects do not have any existence – and in particular not any identity independently of the structure they are part of" (Esfeld and Lam 2011: 143).

According to Esfeld, our ontology is not derivable from the formal content of our best scientific theories. We must admit that ontology is primitive. This means that, in a sense, ontology is not derived, nor inferred from the formalism. By the opposite, the ontology must be put as the referent of that formalism (Esfeld 2014: 99; Egg and Esfeld 2015: 3230), i.e. it is an ontology behind the formalism of our theories that makes them possible and understandable.

Consequently, it could be claimed that Radical-Ontic Structural Realism relations are prior to objects in the same way that Radical-Ontic Structural Realism Physics' formalism is prior to Metaphysics (objects are derived from relations as metaphysics is read off from the formalism of our best scientific theories). However, Moderate-Ontic Structural Realism relations come with objects like Moderate-Ontic Structural Realism Physics comes with Metaphysics in terms of the Primitive Ontology (objects and relations come in one package just as Physics and Metaphysics do). At least at its roots, the distinction between Radical-Ontic Structural Realism and Moderate-Ontic Structural Realism is, therefore, more programmatic or Metametaphysical than Metaphysical: this distinction lies in the divergence between Radical Naturalistic Metaphysics and Esfeld's account of Natural Philosophy (based on the Primitive Ontology approach).

4. Ontic Structural Realism and Quantum Mechanics

How does Ontic Structural Realism tackle QM's challenges?

If Radical-Ontic Structural Realism stands both as a form of Scientific Realism and Scientific-committed form of Metaphysics, then it is supposed to give some account of the main challenge of Quantum Mechanics (QM), namely the infamous measurement problem. Or, at least, it should give us a clear picture of how the world is like if standard QM is right. Being committed only with the "physical structure" of our best scientific theories, Radical-Ontic Structural Realism dismisses entities like "particles" or "waves" from the metaphysical phaeton right from the start. In fact, Radical-Ontic Structural Realism only commits itself with some QM's relational features, like entanglement and symmetries groups. But, as Esfeld (2012) notes, no explanation is provided to those relations. Esfeld shows that Radical-Ontic Structural Realism is compatible with more than one interpretation and, at the same time, no interpretation is necessarily endorsed by Radical-Ontic Structural Realism. So, in regards with QM, Radical-Ontic Structural Realism partially fails both as a form of ontology and as a form of realism since it does not give us a clear picture of what is like for QM to be right and therefore of what we should be realists about.

In opposition to Radical-Ontic Structural Realism, Moderate-Ontic Structural Realism is motivated right from the beginning by the necessity of giving a clear picture of how is the world if QM to be right. Moderate-Ontic Structural Realism motivation's is to set up an ontology that accounts for the existence of measurement outcomes and, in general, the Manifest Image. This ontology avoids the infamous measurement problem by denying the so-called collapse (or projection) postulate and by stating that quantum systems have a well-defined state before being measured. This view claims that all measurements outcomes are particle-like (following the old Quantum Mechanics' *dictum*: "prepare waves; detect particles"). Since all particle measurements are, at its basis, position measurement (Falkenburg 2007), it is argued that quantum systems have a well-defined position independently of measurement. This is basically the main thesis of Bohmian Mechanics. However, the long-standing issue with Bohmian Mechanics is the problem of what is represented by the wave-function, or better what is the role of wave-function within the configuration space. At this point the Primitive Ontology approach is decisive. It detaches the theory's Ontology from its formalism and allows therefore that the wave-function may be considered as a non-primitive element due to the dynamics of primitive stuff. The wave function in question therefore does not have a primordial ontological status. On Esfeld's account of Bohmian Mechanics, the primitive ontology of this physical theory "consists in one actual distribution of matter in space at any time (no superpositions), and the elements of the primitive ontology are localized in space-time, being "local beables" in the sense of Bell (2004, chap. 7), that is, something that has a precise localization in space at a given time" (Esfeld 2014; Esfeld *et al.* forthcoming). The elements of primitive ontology (or primitive stuff) are point particles (or matter points). Matter points are primitive, not only in the broad sense of primitive ontology, but according to Deckert and Esfeld (Deckert and Esfeld forthcoming), also in the following senses: 1) they are not composed by anything; 2) they neither have intrinsic properties and nor are they bare substrata; and 3) they are simply there, at the fundamental physical level.

This approach, however, raises the following difficulties. In the first place, Moderate-Ontic Structural Realism clearly engages in a compositional form of micro-physicalism. This implies that chemical or biological properties, and even all physical properties, should be derivable from the fundamental structure. That is, everything is composed by and supervene on point-particles-in-distance-relation. However, this strong reductionism is far from obvious, to say the least. In the second place, this proposal implies that all physics must be reformulated or explained in terms of a minimal ontology. However, despite all efforts already done (for instance Egg, M. and Esfeld, M. (2015), Esfeld, M.; Deckert, D. and Oldofredi, A. (forthcoming)), there is no clear way how this can be done. The problem seems to derive not from the alleged necessity of admitting a Primitive Ontology – understated as the call for a compatibility between the Scientific and the Manifest Images trough an ontology behind the formalism – but from the particular primitive ontology endorsed by Esfeld: that is, a network of ultimate matter-points and just one type of relation – namely, a spatial relation of distance.

5. Conclusion

Ontic Structural Realism seems to be facing an impasse. Ontic Structural Realism is strongly motivated by Quantum Mechanics. Criticisms both on individuality and independence arise from Quantum Mechanics. But, on one hand, if Quantum Mechanics is right then Radical Ontic Structural Realism would be failing to provide a clear picture of the world. Moreover, Radical-Ontic Structural Realism fails to give us a solution to the measurement problem of Quantum Mechanics. On the other hand, Moderate-Ontic Structural Realism does provide an ontology, but it's also committed to a specific form of Bohmian Mechanics that still is, in some way, in its speculative

274

stage of theoretical development. That is, Moderate-Ontic Structural Realism fails to give us a broadly accepted primitive ontology to QM. So, QM offers us good arguments in favor of Ontic Structural Realism, but both main Ontic Structural Realism's proposals seem to struggle with QM's challenges. Since Radical-Ontic Structural Realism is grounded on NM and Moderate-Ontic Structural Realism is grounded on PO, then maybe Ontic Structural Realism's failure is not due to itself but to this metametaphysics proposals. That is, maybe some assumptions made by both NM and PO, like micro-physicalism, monism or metaphysical fundamentalism, should be reviewed or even dismissed from Ontic Structural Realism proposals.

Acknowledgements

This work has been done with the support of FCT Fellowship Grant: SFRH/BPD/92254/2013

I like to thank to the editor and to the anonymous reviewers for their constructive comments, which helped me to improve the manuscript

References

1. Allori, V., 2013, "Primitive Ontology and the Structure of Fundamental Physical Theories"in Alyssa Ney and David Z Albert (eds.) *The Wave Function: Essays on the Metaphysics of Quantum Mechanics,*Oxford: Oxford University Press: 58-75
2. Allori, V., 2015, "Primitive Ontology in a Nutshell" in *International Journal of Quantum Foundations,*1 (3): 107-122 .
3. Arenhart, JRB, Krause, D., 2014, "Why non-individuality? A discussion on individuality, identity, and cardinality in the quantum context" in *Erkenntnis*, 79 (1): 1-18
4. Arenhart, JRB, 2015, "The received view on quantum non-individuality: formal and metaphysical analysis" in *Synthese*: 1-25
5. Bell, J. S., 2004, *Speakable and Unspeakable in Quantum Mechanics (2nd ed)*, Cambridge University Press
6. Da Costa, N. and French, S., 2003, *Science and Partial Truth: A Unitary Approach to Models and Scientific Reasoning*, Oxford: Oxford University Press.
7. Dorato, M. and Morganti, M., 2013, "Grades of individuality. A pluralistic view of identity in quantum mechanics and in the sciences" in *Philosophical Studies* , 163: 591-610.
8. Durr, D.; Goldstein, S. and Zanghì, N., 2012, *Quantum Physics Without Quantum Philosophy* Dordrecht: Springer.
9. Esfeld, M., 2003, "Do Relations Require Underlying Intrinsic Properties? A Physical Argument for a Metaphysics of Relations" in *Metaphysica. International Journal for Ontology and Metaphysics* , 4 (1): 5-25.

10. Esfeld, M. and Lam, V., 2011, "Ontic Structural Realism as a Metaphysics of Objects"" in A. Bokulich and P. Bokulich (eds.) , *Scientific Structuralism*, Dordrecht: Springer;

11. Esfeld, M., 2014, "The primitive ontology of quantum physics: Guidelines for an assessment of the proposals" in *Studies in History and Philosophy of Modern Physics*, 47: 99-106

12. Esfeld, M.; Deckert, D. and Oldofredi, A. (n.d) "What is matter? The fundamental ontology of atomism and structural realism" forthcoming in Anna Ijjas and Barry Loewer (eds.) *A guide to the philosophy of cosmology*, Oxford University Press.

13. van Fraassen, B., 1991, *Quantum Mechanics: An Empiricist View*, Oxford: Oxford University Press

14. Egg, M. and Esfeld, M., 2015, "Primitive ontology and quantum state in the GRW matter density theory" in *Synthese*,192 (2015):3229–3245

15. Falkenburg, B., 2007, *Particles Metaphysics: A Critical Account of Subatomic Reality* , Dordrecht: Springer

16. French, S. and Krause, D., 2006, *Identity in Physics: A Historical, Philosophical and Formal Analysis*, Oxford: Oxford University Press

17. French, S., 2006, "Structure as a Weapon of the Realist" in *Proceedings of the Aristotelian Society*, 106 (2): 167-185

18. French, S., 2010, "The interdependence of structure, objects and dependence" in *Synthese*, 175:90

19. French, S., 2011, "Metaphysical Underdetermination: Why Worry?" in *Synthese*, 180: 205-221.

20. French, S. and Ladyman, J., 2011, "In defence of Ontic Structural Realism"" in Alisa Bokulich and Peter Bokulich (eds.) *Scientific Structuralism*, Dordrecht: Springer: 25-42

21. French, S., 2014, *The Structure of the World: Metaphysics and Representation*, Oxford: Oxford University Press.

22. Jones, R., 1991, "Realism about what?" in *Philosophy of Science* 58: 185-202

23. Ladyman, J. and Ross, D., 2007, *Everything must go: Metaphysics naturalized*, Oxford: Oxford University Press.

24. Ladyman, J., 2016, "The Foundations of Structuralism and the Metaphysics of Relations" in Anna Marmodoro and David Yates (eds) *The Metaphysics of Relations*, Oxford: Oxford University Press: 177-197

25. Langton, R. and Lewis, D., 1998, "Defining Intrinsic" in *Philosophy and Phenomenological Research*, 58 (2): 333-345.

26. Laudan, L., 1981, "A Confutation of Convergent Realism" in *Philosophy of Science*,48: 19–48.

27. Lewis, D., 1986, *On the Plurality of Worlds*, Oxford: Blackwell.

28. Lewis, D., 2001, "Redefining 'Intrinsic'" in *Philosophy and Phenomenological Research* 63: 381–398

29. Papineau, D. (ed.), 1996, *Philosophy of Science*, Oxford: Oxford University Press

30. Pooley, O., 2006, "Points, Particles and Structural Realism" in D. Rickles, S. French and J. Saatsi (eds.) *Structural foundations of quantum gravity*, Oxford: Oxford University press: 83-120

31. Psillos, S., 2001, "Is Structural Realism Possible?"" in *Philosophy of Science*, 68: S13–S24.

32. Saunders, S., 2006, "Are quantum particles objects?" in *Analysis*, 66: 52–63
33. Wigner, E. P., 1939, "On unitary representations of the inhomogeneous Lorentz group" in *Annals of Mathematics*, 40 (1): 149–204
34. Worrall, 1989, "Structural Realism: the Best of Both Worlds?" in *dialectica*, 43: 99-124 (reprinted in D. Papineau (ed.), *The Philosophy of Science* Oxford: Oxford University Press, 1996: 139-165)

EPISTEMOLOGICAL VS. ONTOLOGICAL RELATIONALISM IN QUANTUM MECHANICS: RELATIVISM OR REALISM?

CHRISTIAN DE RONDE*

CONICET, Institute of Philosophy "Dr. A. Korn"
Buenos Aires University - Argentina
Center Leo Apostel and Foundations of the Exact Sciences
Brussels Free University - Belgium
E-mail: cderonde@vub.ac.be

RAIMUNDO FERNÁNDEZ MOUJÁN

Institute of Philosophy "Dr. A. Korn"
Buenos Aires University - Argentina
Center Leo Apostel, Brussels Free University - Belgium
E-mail: raifer86@gmail.com

In this paper we investigate the history of relationalism and its present use in some interpretations of quantum mechanics. In the first part of this article we will provide a conceptual analysis of the relation between substantivalism, relationalism and relativism in the history of both physics and philosophy. In the second part, we will address some relational interpretations of quantum mechanics, namely, Bohr's relational approach, the modal interpretation by Kochen, the perspectival modal version by Bene and Dieks and the relational interpretation by Rovelli. We will argue that all these interpretations ground their understanding of relations in epistemological terms. By taking into account the analysis on the first part of our work, we intend to highlight the fact that there is a different possibility for understanding quantum mechanics in relational terms which has not been yet considered within the foundational literature. This possibility is to consider relations in (non-relativist) ontological terms. We will argue that such an understanding might be capable of providing a novel approach to the problem of representing what quantum mechanics is really talking about.

Keywords: Relationalism; relativism; epistemic view; ontic view; quantum mechanics.

Introduction

In this article we attempt to discuss the possibility of providing a relational account of quantum mechanics. For such purpose we intend to clarify which are the main distinctions between substantivalism, relationalism and relativism. We will argue that, apart from the numerous interpretations which consider relationalism from an epistemological perspective, there is also the possibility to understand relations from a (non-relativist) ontological viewpoint. The paper is organized as follows. In the first section we provide a short account of the relation between physics, philosophy and sophistry as related to realism and relativism. In section 2 we address relational and substantivalist approaches within Greek philosophy. We consider the relational theory of Plato as presented in *Sophist* and the substantivalist theory of atoms as presented by Democritus and Leucippus. Section 3 reconsiders the relational-substantivalist debate in Modernity, more in particular, the relational scheme proposed by Spinoza and the triumph of the atomist metaphysics through its implementation in Newtonian physics. In section 4 we will introduce quantum theory as deeply related, in the 20th century, to both atomism and positivism. In section 5 we will discuss several relational accounts of quantum mechanics such as those of Bohr, Bene-Dieks and Rovelli. Finally, in section 6, taking into account our previous analysis, we will present an ontological account of relations with which we attempt to provide a new approach for representing the theory of quanta.

1. Philosophy, Physics and Sophistry: Realism or Relativism?

Let us remember once again the Greek moment, the origin of both physics and philosophy. And let's remember, to emphasize this common origin, the name that Aristotle uses to refer to the first philosophers: the "physicists". This denomination comes from the object that, according to multiple sources, they all intended to describe: *phúsis*. A term that is unanimously translated as "nature" and whose meaning covers what we refer to when we talk about "the nature of reality" (its essence), as well as what we commonly, broadly and in an extensive way refer to as nature: the reality in which we take part. *Phúsis* is, for physicists, something dynamic which —at the same time— responds to some sort of internal order or formula. Some of these first philosophers proposed an "element" (or a series of them) from which —and according to which— all reality develops and can be explained. But among them there are also others who didn't follow this strategy. In particular, there are two of them who are particularly important for our analysis: Heraclitus of Ephesus and Parmenides of Elea.

Heraclitus redirected the search for the fundament of *phúsis* no longer to an "element" but to the description of a formula, an internal order that rules *phúsis*. He described this formula and called it *lógos*. This denomination is very significant for the development of philosophy. Until Heraclitus' use of the word, logos had a meaning exclusively related to language: discourse, argumentation, account, even tale. In all of those translations we can see already something that will be essential to all meanings and nuances of logos, even when it doesn't refer to language: a significant combination, a reunion with criterion, a collection with purpose. *Lógos* never means an isolated word, or a meaningless sentence, or dispersed and ineffective ensembles of words. It always refers to a combination that is able to produce an effect or a meaning. We now begin to understand why Heraclitus chooses this specific word to name the internal order of phúsis. He sees in *phúsis* exactly that: a combination that responds to a formula, a criterion. This double meaning of *lógos* —formula of *phúsis* and human discourse— implies an affinity between language and reality that allows for philosophical knowledge: it is in a linguistic manner that we are capable of exposing the internal order of reality. Thus, there is an affinity between the *lógos* of men and the lógos of *phúsis*. However, it is a difficult task to expose the true *lógos* since, as remarked by Heraclitus, "*phúsis* loves to hide." [f. 123 DK]. Doing so requires hard work and sensibility, but —following Heraclitus— the latter can be revealed in the former. In a particular *lógos* one can "listen" something that exceeds it, that is not only that personal discourse but the lógos of *phúsis*: "Listening not to me but to the *lógos* it is wise to agree that all things are one" [f. 50 DK]. We are thus able to represent *phúsis*, to exhibit its *lógos*.

In another part of the ancient Greek world, Parmenides makes a discovery that, as all great discoveries in the history of philosophy, is beautifully simple: *there is being (and not being is impossible)*. No matter the "element" or the formula that you may choose as fundamental for exposing *phúsis*, the truth is that anything, any element, order, etc., must necessarily and "previously" *be*. The "fact of being" —according to Néstor Cordero's formula [12— comes before any determination that we can predicate of whatever. Parmenides' philosophy begins with the evidence of this all encompassing and irreducible fact of being, and searches what we can say starting from it. As Heraclitus, and maybe in a more explicit manner, he will affirm a natural relation between being, thought and language; a relation that will become, at least for some years, a sort of dogma.

But time passes, things change, and philosophy, already in its youth, will encounter its first opponent, its first battle. We are in the V century before Christ and Athens is the most powerful (politically, culturally and intellectually) *polis* of the Greek world. Thinkers from all over the region come to Athens. Some of these foreigners begin to make a living by teaching Athenian citizens. But, unlike physicists and philosophers, they don't teach how to know the true nature of reality; instead they teach techniques of argumentation and persuasion, techniques of great utility in the *ágora*. In private lessons and public conferences, the sophists prove the persuasive power of *lógos*, understood now exclusively as human discourse —and independent of the *lógos of phúsis*. It is quite a blow for philosophy, which in a somewhat naïve way had affirmed a privileged, unbreakable relation between truth and *lógos*. But, contrary to what their fame indicates, it is not only the taste for controversy what lies behind sophistry, there are some originally sophistic positions —which evidently arose from an opposition to philosophy as it was known— that justify their *praxis*. Some strongly sceptical postures. These are positions that undermined the basis of the philosophical attempt to represent *phúsis*. In order to develop the fundamental aspects of these positions, it is useful to focus on two of the most famous and prodigious sophists: Protagoras and Gorgias. From the former an original phrase remains —that will be used many centuries later by the Vienna Circle in their manifesto— which has become famous: "Man is the measure of all things, of the things that are, that they are, of the things that are not, that they are not" [DK 80B1] . We don't possess much more of Protagoras' text but we do have some comments about his philosophy that date back to antiquity, and they all seem to coincide: he proposes a relativistic view. According to this stance, there is no such thing as 'a reality of things' —or at least, we are not able to grasp it. We can only refer to our own perception. Things do not have a reality independent of subjects; and even in the case such a reality would exist we simply cannot have access to it. We, individuals, have only a relative knowledge dependent of our perception: "Italy is beautiful —for me". This is a stance that has undoubtedly an intuitive appeal and that we will encounter again in some of the — more contemporary— positions we will analyze in this article. Gorgias, on the other hand, a quite talented orator, does a systematic critique of the founding principles of philosophy —in particular of the Parmenidean philosophy. In his famous discourse, *On Not Being*, he tries to dismantle the relations that Parmenides establishes between being, thought and *lógos*. A brief summary of the discourse would be: there is not a being of things (or even: "nothing exists"); even if there is a being of things, it would not be accessible to our thought; and finally if our thought would be able to grasp the being of things, we still wouldn't be able to communicate it.

Evidently, philosophy could not continue without resolving, or at least responding, the deep arguments of the sophists. The attempt to save and precise the nature and method of philosophical knowledge will give birth, among other things, to two of the most important works of western thought. We refer —of course— to the philosophies of Plato and Aristotle. Even if the positions of sophistry and philosophy are in great measure irreconcilable, and if this opposition then continues —more or less implicitly, as we will see— throughout the history of thought, the truth is that sophistry has accomplished an important role in the history of philosophy for it constitutes the first critical moment that has proven to be fundamental for the development of philosophy itself.

2. Relationalism vs. Substantivalism in Greek Thought

We will come back to the reappearance, within the history of Western thought, of the strong critical arguments that gave sophistry its importance, but first we want to establish another opposition that appears inside philosophy (and physics), this is, an opposition between two ways of representing *phúsis*, two fundamentally different views regarding reality. Both confident in the human capability of expressing *phúsis* through a discourse, but each one arriving at a very different representation of the world. We are talking about substantivalism and relationalism. The former is a view of the world as populated by multiple individual substances, the latter refuses to introduce ontological separation within reality and defines its elements as relations —not as separated existents. We want to clearly establish which are the fundamental differences between these two worldviews so that we can produce a solid basis for the discussion and analysis of what should be considered the substantivalism-relationalism debate.

2.1. Plato's Relationalism in the Sophist

The interest in relationalism is relatively recent. During the last century, different developments in various disciplines seem to tend towards the need of a relational understanding of reality. But the truth is that, if we look back, we find that in the history of both physics and philosophy there have always been some doctrines that we might catalog as "relationalist"

—even if they are not explicitly described in that manner. We can even say that what we will call here "substantivalism" —the view opposed to relationalism—, that is, the conception according to which the world is made of individual and independent separated substances, has been seen in philosophy with suspicious eyes. The foundation of this view was recognized as weak or even unsustainable, even if some of its characteristics might seem to coincide today with a "common sense" understanding of reality.

From a chronological point of view, the first philosopher who defined being as relation (more specifically: as capacity or power of relation) was Plato. Returned from his second trip to Sicily, already an old man but more lucid than ever, Plato transforms his own philosophy. This is especially visible in a series of three dialogues which conclude with the difficult yet prodigious *Sophist*. First, in the *Parmenides*, Plato uses the figure of the Eleatic philosopher to carry out a critique of an orthodox version of the 'Theory of Ideas' —the one we can find in his previous dialogues. Through a fictitious Parmenides, Plato does some autocriticism, leaving the character of Socrates —in a very young version— mourning the loss of his beloved, and now refuted, Theory of Ideas. But Parmenides tries to fight Socrates' anguish with some words of encouragement: he praises the Socratic attempt to direct his search for knowledge towards the intelligible and he gives him confirmation that, despite his previous failure, it is necessary —if we do not want to discard knowledge as impossible— to sustain the existence of Ideas. This means we still need a Theory of Ideas, just not the same of the previous dialogues. In the next dialogue, the *Theaetetus*, Socrates searches the possibility of defining true scientific knowledge (*epistéme*) without postulating Ideas. All attempts fail, and the characters of the dialogue decide to meet again next morning for the continuation of the discussion. What happens the next morning is what the *Sophist* describes. Socrates, Theaetetus and Theodore meet up, and one of them has invited a new participant: the mysterious Eleatic Stranger. This Stranger will be Plato's spokesman for the majority of the dialogue, and he will expose the new version of the Platonic Theory of Ideas. Meanwhile, Socrates listens in silence. It is in this context that Plato will define being, for the first time in his philosophy. He will define it as *dúnamis* of affecting and being affected, *dúnamis* of communicating, *dúnamis* of relation. Being means to posses this *dúnamis* (see [36,37). This Greek term is usually translated as potency (especially when it is Aristotle who uses it), possibility, capacity, power. In each one of these translations we emphasize either the passive or the active aspect which coexist within the Greek term. But before exploring the meaning of this definition, let us begin by pointing out something evident: the transformation of the platonic philosophy that we witness in these dialogues has a strong Eleatic background. Let us remember that it is Parmenides who, speaking in the name of Plato, refutes the previous Theory of Ideas, and afterwards it is also an Eleatic who accomplishes the task of changing the theory. Undoubtedly Parmenides is, next to Socrates, the philosopher that most influenced Plato. The pejorative characterizations, more or less ironic, that Plato uses to describe the thought of the previous and contemporary philosophers, are left asides when he talks about Parmenides. He shows him respect, sometimes veneration (even if he is forced, in the *Sophist*, to contradict him in some aspects). It is useful then to briefly remember Parmenides' philosophy. His main thesis can be simply expressed (but not so simply interpreted): *there is being (or 'there is what is', or 'there is the fact of being') and not being is impossible.* If the previous philosophers, or those contemporary to Parmenides, gave privilege to one or several 'elements' as origin and foundation of nature, or dedicated their thought to decipher the hidden order that governs reality (the case of Heraclitus), Parmenides starts by reflecting on a previous truth: any 'element', any 'order', anything of any nature, must be, first, something that is. The simple yet universal, all encompassing fact of being is the origin of the Parmenidean wonder. And as evident as the fact of being is the impossibility of its contrary: not-being. One of the ways in which Parmenides phrases this impossibility is the one that identifies not-being with separation: "you will not sever what is from holding to what is" [f. 4]; "it is wholly continuous; for what is, is in contact with what is" [f. 8.25]; "Nor is it divisible, since it is all alike" [f. 8.22]. In this sense, "There is being and not-being is impossible" means that there is no cut, no strip, no ditch, inside being —through which not-being would pass. Being has no cracks within, no interstices. But how should we then continue? What else to say besides "there is being"? Which content can we predicate of the fact of being? It seems —at first sight— like an empty discovery. Plato, with a truly parmenidean spirit (despite the parricidal declaration in the *Sophist*), tries to go a little further, or to be a bit more specific: he proposes a definition of being as *dúnamis*. "And I hold that the definition of being is simply *dúnamis*" [*Sophist*, 247e]. Plato says that being is the *dúnamis* to act and be acted upon, the capacity to affect and be affected. The fundamental reality of everything is, for Plato, this inherent tendency to relation, this potency of relation. Néstor Cordero, in a commentary to the *Sophist* [13, p. 155], describes this *dúnamis* as "the capacity of an entity, any entity, for relating with another (either affecting or being affected) and for that reason a few pages later Plato replaces 'acting' and 'suffering' by a single verb, 'communicating', and he talks about *dúnamis koinonas*, 'the possibility or potentiality of communication' [*Op. cit.*, 251e]. And since being is communicating, something that doesn't communicate doesn't exist. (...) Plato assimilates this potentiality to the fact of being and he gives precisions about it: it is the possibility of communicating, that is, (...) to produce reciprocal bonds". This being in everything that is, it is *dúnamis* of affecting and being affected, of interacting, it is an inherent tendency towards relation. It is the participation of a universal relationability. If all things, from any kind, are, and —according to Parmenides' lesson— there can't be not-being between them, if always "what is, is in contact with what is", it is then —Plato adds— this same universal communicability, this unbreakable basic tendency towards relation, the nature on the fact of being. Nothing can exist, for Plato, which does not possess this *dúnamis*. Anything, any "something" is, first, potency of relation, impossibility of being something isolated, it is part of a universal communicability. Where there is potency of relation there is being, and vice versa. For the first time we encounter a relational ontology. Parmenides, with good reasons, denied the possibility of identifying being with a given qualification, because this would relativize or limit the fact of being. But Plato finds a qualification that can be applied to being without limiting or relativizing it, that leaves nothing outside being, that doesn't identify being with a determined kind of entity, that every existent, of every kind, shares: the capacity of relation. This capacity, like Parmenides' being, has no possible contrary. There are no possible ontological separations. It is then a universal qualification, with no opposite, without limits, and which also allows for a more specific knowledge of reality, of which is given among being. It restores its dynamism and variety without introducing not-being, without postulating separated substances. Because not every relation is the same relation; relations have variable intensities, different qualities; sometimes there is affecting, sometimes there is suffering. Relation allows us to articulate the general —and immutable— truth of being (and the absence of not-being) with the specific —and mutable— reality of experience.

Although this deduction from the Parmenidean fact of being to Plato's being as *dúnamis koinonas* —following the guide of the impossibility of separation— is reasonable when following the mentioned texts, the truth is that Plato does not describe his path towards the definition of being in this manner. Instead, he does so by putting together a brief revision of the previous philosophies; what might be considered to be the first "history

of philosophy" within the history of philosophy. By this revision he tries
to discover what "being" could mean. He wonders: which characterizations
can be extracted from previous philosophies? What he finds is that the
past philosophers mistake being with some particular determinations, either
quantitative or qualitative ones, that limited it. Determinations that cannot
bear the universal applicability (and absence of contrary) that corresponds
to being. Some identified what is to a determined opposition, a duality,
as for example the hot and the cold. Others identified being with the One
(denying multiplicity). Then he verifies that some of them referred being
to material things, and others (Plato himself, before the *Sophist*) identified
it with intelligible entities. All of them qualified or quantified being, they
identified being as some kind of limited determination. And that created
problems for them. They always seemed to leave some things outside being.
What Plato is proposing to them is that they should expand, broaden, their
conception of being, that they should not limit being to a particular kind of
entity, excluding from existence —for that reason— things that also exist.
In the context of the discussion with the materialist (who denied incorporeal
realities) he proposes a solution, a settlement: let's just say that all that
has the possibility of interacting exists. "Anything which possesses any
sort of *dúnamis* to affect another, or to be affected by another, if only for a
single moment, however trifling the cause and however slight the effect, has
real existence; and I hold that the definition of being is simply *dúnamis*"
[*Op. cit.*, 247e]. The lesson that we can extract from this brief history
of philosophy is the necessity of not limiting being to a specific type of
entity, or to certain number of entities. Instead, we must conceive being
in a way inclusive enough so that everything that proves itself existent, is
covered by such proposed metaphysical account. And what we undoubtedly
know about everything that is, is that, in a greater or smaller degree, it is
communicated with the rest, it has relations, it is included in a universal
communicability. Relation is thus irreducible and primary.

However, there is still a point in which the Parmenidean heritage
remains uncomfortable. We need to be able to say that everything is, that
the fact of being is universally applicable, yes, but at the same time we also
know that everything is not the same. Experience tells us that being is full
with differences. One of the main problems for a truly relational conception
of being, one that denies ontological separation and takes relation as fundamental, is to justify the differences inside being without producing separated substantial individuals. Here enters the main concept of the Sophist:
difference. There is not an absolute not-being, the contrary of being does

not exist, there are no separations, but there are differences. These differences don't amount to separations (in fact difference is a relation), but they do account for variety and multiplicity among being. Plato introduces in this manner —and for the first time in the history of philosophy— a relative notion of not-being. There is no absolute not-being, no void, the nothing does not exist, but everything that is, at the same time is not an infinity of other things (it is different from them). This table, for instance is not (is different from) a chair. Difference, as well as Sameness, and some other determinations, articulates the identity and specificity of things that exist, but this only on the irreducible base of being, that is to say, on the base of the potency of relation that subtends, comprehend and renders them dynamic.

2.2. Greek Atomism: A Substantivalism

Among the first philosophers a particular school formed to which we now turn our attention. It began with Leucippus (of whom we have no original texts) and his pupil Democritus (of whom we have several original fragments from his apparently numerous books). They proposed, as many of the first philosophers, some fundamental elements out of which *phúsis* was made of. These elements where being and not-being, which in term they interpreted as 'the full' and 'the void'. Contrary to Parmenides, they conceived the existence of a non-being which is the contrary of being. In their minds, there is void within nature. Being, or the full, consisted to them of indivisible bodies, indivisible fragments of mass with a minimum size. Simple bodies. They used an adjective to describe these bodies: *átomos*, which means, literally, "not divided". That adjective became an -ism and this school was called "atomism".

Atoms couldn't be infinitely small, there's a limit to how small they are. If not, atomism wouldn't work, we wouldn't be able to say that "atoms have mass", and we would be stuck with a difficult conclusion for them (a conclusion that Zeno of Elea already pointed out astutely): bodies would be made of zeros, mass would be made of not-mass. A conclusion that for atomism would be catastrophic. The admission that atoms are indivisible bodies with a small amount of mass is for atomists an axiom. They don't justify it, they only postulate it. They sacrificed the justification of this admission in order to produce an explanation of reality that seemed coherent to them. It is easy to demonstrate that this axiom is problematic, for it entails the existence of a mass that can't be divided. If we want to be fair, we can say that Parmenides also started from a postulate he wouldn't justify: there's being and not-being is impossible. If we compare the amount of presuppositions made by one and the others: Parmenides presupposes being (and the inexistence of not-being), while the atomists presuppose being, not being, and that being is made of indivisible small bodies with mass that travel in void. But let's leave the comparison aside, for it is not the economy of presuppositions that interests us. For atomists, then, everything is made of these simple bodies that move around in void. Atoms have shapes, and according to these shapes they can unite to form more complex bodies. In conclusion, what it's important to us is that atomism is without a doubt a substantivalism: there are small individual substances, and these substances are separated from each other by not-being. The world for atomists is made of these separated substances.

3. Relationalism vs. Substantivalism in Modernity

The tension between relationalism and substantivalism remained through the history of western thought, sometimes in more central arenas, sometimes in more marginal ones. However, there is in modernity a triumph of substantivalism which would change the balance between these two opposite accounts of reality. Even though relationalism still remained an important viewpoint within modernity, the power of Newtonian mechanics interpreted on the lines of atomism would determine the fate of what would be later on recognized as "classical common sense". But before considering the triumph of the newborn "mechanical atomism" imposed by Newton, let us begin by recalling what we consider to be a particular interesting development on the lines of relationalism. A development which will prove particularly interesting for the discussion regarding the meaning and understanding of relations.

3.1. Spinoza's Relationalism

If we would pay attention only to terminology, it wouldn't seem logical to pick Spinoza's philosophy as our next example of relationalism, since he intends explicitly an analysis of substance. But one always has to pay attention to what each philosopher does with the terms he uses within his own philosophy, what does he takes from the traditional meaning, what does he change. This is particularly useful when faced with Spinoza, since it is evident —even for those who, in his own time, didn't understand him— that he took the terminology of Cartesian and scholastic traditions, but in order to say —with those same terms— something completely different. The most emblematic case is, undoubtedly, the term "substance". The meaning is unprecedented: for Spinoza there are no multiple substances, there is only one single substance, with infinite attributes. And there's nothing else besides it. It is, in general terms, an equivalent of the Parmenidean being. All the variety that we encounter is the infinite variety of modifications of only one substance, according to its different attributes. But before analyzing the nature of these modes of the substance (where relation appears), let us see, even in a brief manner, how unity and continuity of substance —and impossibility of separation— represent the basic aspects of his ontology.

Spinoza's account of nature pictures a similar landscape to the one we drew from Parmenides and Plato, but now developed with a modern terminology: attribute, mode, real distinction, modal distinction, etc. For Spinoza, nor the difference between attributes (extension, thought), nor the difference between modes (things, bodies, ideas, souls, etc.), entail substantial divisions. There aren't any differences, in modes or in attributes, which imply separations. For there to be separations there would have to be distinct substances, and Spinoza proves, especially in the first book of the *Ethics*, that —first— there are no multiple substances of the same attribute, and —second— that there is only one substance for all attributes. For Spinoza there's no separation, no not-being, no void in reality. It is true that the differences between attributes are real differences, but these don't amount to substantial distinctions, only to qualitative differences. Attributes are the qualitative natures according to which the same substance expresses itself. Also modal differences don't amount to separations, only to the different modifications inside the one and only substance: "As regards the parts in Nature, we maintain that division, as has also been said before, never takes place in substance, but always and only in the mode of substance" [*Short Treatise I*, chap. II, 19-22].

But physics is not the science of *being qua being*, nor of God as an absolutely infinite substance, but —to put it in Spinozian terms— the science of the modes of the substance. As we have seen, modes are not substances but modifications of a unique substance, according to its different attributes. One of the main issues of a non-substantivalist ontology is —as we said before— to give an account of the singularity of each individual (the many) without producing separated substances (the one). This means arriving to the multiple individuals without losing the unity of being. What defines an individual, that which essentially characterizes its singularity and distinguishes one from the others, is —according to Spinoza— a relation. A "part" of the divine potency, or in other words, a degree of potency that expresses itself as a relation. Thus, following the physics of his time, each individual is characterized as a specific relation of movement and rest. Spinoza tells us that these relations can be larger or smaller, more or less "perfect". There are differences among them, but these differences can't be thought of as an extensive quantity (e.g. the quantity of mass). There is still a quantitative difference that distinguishes relations, but it is no longer an extensive quantity, it is an intensive one, a degree of potency. In the order of relations that don't depend on their terms —in the context of a relational ontology— what distinguishes and quantifies different relations are their intensities of potency. Individuals, in Spinoza's philosophy, are essentially potency and relation. They are quantities of potency because they are modifications, of the same univocal being. They are relations because those potencies are expressed in relations, different relations, specific relations. And, according to Spinoza, from simpler to more complex relations, made of the composition of those simpler relations, we arrive at more complex individuals. And we also arrive to the idea that the totality of the relations between these relations would give us the totality of nature. An individual composed of all individuals. The total relation of relations. Since separation is not possible, within a relational ontology we do not have, we can't have, individual substances. What we have instead are modifications of being, elements that can be different from each other but that can't truly be separated from each other. But still, every one of those individuals can be described, can be quantified, can be experienced as a particular individual, we are not lost in an indefinite totality.

3.2. Newtonian Physics and the Triumph of Atomism

It was Isaac Newton who was able to translate into a closed mathematical formalism both the ontological presuppositions present in Aristotelian logic together with the materialistic reduction of reality to res extensa — taking in this way actuality as the unique mode of existence of things. He did so with the aid of atomistic metaphysics. In the V and IV centuries B.C., Leucippus and Democritus had imagined existence as consisting of small simple bodies with mass. According to their metaphysical theory, atoms were conceived as small individual substances, indivisible and separated by void. The building blocks of our material world. Many centuries later, Newton had been able not only to mathematize atoms as points in phase space, he had also constructed an equation of motion which allowed to determine the evolution of such "elementary particles". The picture of the world described by Newtonian mechanics was that of small completely determined particles bouncing between each other in space in an absolutely deterministic manner. The obvious and most frightening conclusion implied by the conjunction of Greek atomism and Newton's use of the effective cause was derived by the mathematician Pierre Simon Laplace:

> We may regard the present state of the universe as the effect of its past and the cause of its future. An intellect which at a certain moment would know all forces that set nature in motion, and all positions of all items of which nature is composed, if this intellect were also vast enough to submit these data to analysis, it would embrace in a single formula the movements of the greatest bodies of the universe and those of the tiniest atom; for such an intellect nothing would be uncertain and the future just like the past would be present before its eyes. [45, p. 4]

In the XVII Century, in the newly proposed mechanical description of the world, the very possibility of the indetermination supposed by the potential realm of being had been erased from physical reality. In classical mechanics, every physical system may be described exclusively by means of its actual, coexistent (in a non-contradictory way) and determined properties. A point in phase space is related to the set of values of properties that characterize the system. In fact, an actual property can be made to correspond to the set of states (points in phase space) for which this property is actual. Thus, the change of the system may be described by the change of its actual —meaning, preexistent or independent of observation— properties. Potential or possible properties are then considered as the points to which the system might (or might not) arrive in a future instant of time. Such properties are thought in terms of irrational potentiality; as properties which might possibly become actual in the future. As also noted by Dieks:

> In classical physics the most fundamental description of a physical system (a point in phase space) reflects only the actual, and nothing that is merely possible. It is true that sometimes states involving probabilities occur in classical physics: think of the probability distributions in statistical mechanics. But the occurrence of possibilities in such cases merely reflects our ignorance about what is actual. The statistical states do not correspond to features of the actual system (unlike the case of the quantum mechanical superpositions), but quantify our lack of knowledge of those actual features. [31, p. 124]

Classical mechanics tells us via the equation of motion how the state of the system moves in phase space along the curve determined by the initial conditions and thus, any mechanical property may be expressed in terms of phase space variables. Needless to say, in the classical realm the measurement process plays no role within the description of the state of affairs and actual properties fit the definition of elements of physical reality in the sense of the EPR paper [41]. Moreover, the structure in which actual properties may be organized is the (Boolean) algebra of classical logic. With Newtonian physics, modernity embraced —at least for the physical realm— a substantivalist and materialistic representation of the world commanded by the efficient cause. A view that was nourished by atomism, by Aristotle's both logical and ontological principles of existence, non-contradiction and identity, and by the reduction of existence only in the restrictive terms of the actual mode of being. A physics of pure actuality. The Newtonian metaphysical representation of the world as an "actual state of affairs" remained a dictum that still traverses not only classical physics, but also relativity theory. It was only the appearance of the theory of quanta that disrupted the classical —actualist and atomist— representation of the world, producing a revolution that —as Constantin Piron [49] has remarked— has not yet fully taken place. If classical physics has sustained for quite some time the limitation of what exists to actual substantial entities, quantum physics came to break that limitation, and forces us now —following the example of the Eleatic Stranger— to consider the broadening of our understanding of reality maybe even beyond substantivalism and the actual mode of existence. But to apprehend the development of the physics of quanta we must understand how Machian positivism, by way of a deconstruction of the Newtonian *a priori* notions of absolute space and time, was able to place physics within a new critical moment.

4. Quantum Theory: Between Atomism and Positivism

Positivism was born in the XIX century, taking elements from both English empiricism and French Enlightenment. On the one hand, in a reaction against metaphysics, it stood on the idea of founding knowledge on sensible data; on the other hand, it maintained a generalized trust on the progress of reason and science. Positivism derived from thinkers like Laplace and many others, but was first systematically theorized by August Compte, who saw in the "scientific method" the possibility of replacing metaphysics in the history of thought. Just a century before, Kant had also fought what he saw as the "dogmatic" metaphysics of his time, developing a new system capable not only of resolving the dispute between rationalists and empiricists of the XVII and XVIII centuries, but also capable of justifying Newtonian physics as "objective knowledge". However, a century later, the categories and forms of intuition had become —according to many— exactly what Kant had striven to attack: dogmatic and unquestioned ideal elements of thought. Against metaphysics, positivism stated that the only authentic knowledge is knowledge that is based on actual sense experience. Such knowledge can only come from the affirmation of theories conceived in terms of what was believed to be a "strict scientific method". Metaphysical speculation —understood now as a discourse attempting to go beyond the observed phenomena— should be always avoided and even erased from scientific inquiry and research.

Ernst Mach is maybe one of the most influential positivist thinkers of the XIX century. His criticisms might be regarded as the conditions of possibility for the development of physics that took place at the beginning of the XX century. He developed a meticulous deconstruction of the fundamental concepts of Newtonian physics; a critique that produced a crisis in the fundament of scientific thought itself. This crisis was certainly a standpoint not only for the birth of relativity —as recognized by Einstein himself— but also played an essential role in the development of the theory of quanta. Mach, a physicist himself, was primarily interested in the nature of physical knowledge. His investigations led him to the conclusion that science is nothing but the systematic and synoptical recording of data of experience. In his *Analysis of Sensations*, Mach concluded that primary sensations constitute the ultimate building blocks of science, inferring at the same time that scientific concepts are only admissible if they can be defined in terms of sensations.

> Nature consists of the elements given by the senses. Primitive man first takes out of them certain complexes of these elements that present themselves with a certain stability and are most important to him. The first and oldest words are names for 'things'. [...] The sensations are no 'symbols of things'. On the contrary the 'thing' is a mental symbol for a sensation-complex of relative stability. Not the things, the bodies, but colors, sounds, pressures, times (what we usually call sensations) are the true elements of the world. [48]

In Machian positivism there is thus no room for *a priori* concepts, nor for unobservable entities —like atoms. Talking about entities that can't be observed is to fall in the trap of metaphysics, to go beyond phenomena producing a discourse with no meaning nor reference, to detach our discourse from the only possible true reference: sensations. In what was one of the main scientific controversies of his time, Mach firmly opposed to accept the existence of atoms. However, as a result of the experimental and theoretical work developed by physicists like Dalton, Maxwell and Boltzman, towards the end of the XIX century, atomism —not without resentment of a newborn community which went back to a wave type description of reality— had won the battle and occupied once again the dominant position in scientific communities. It is in this same context that the theory of quanta would make its appearance, producing very soon a paradoxical entanglement between two mutually incompatible positions, namely, atomist substantivalism —that maintained, in metaphysical terms, the existence of unobservable atoms— and Machian positivism —which grounding itself in observed phenomena affirmed the need to eradicate all metaphysical notions from physics, including of course that of "atom".

Quantum physics was born together with the XX century, after the introduction by Max Planck in 1900 of the "quantum postulate" —in order to solve a problem related to the emission of radiation by hot bodies. Quantum theory begun its history as a theory about atoms. Its development continued through the first three decades of the XX Century, when it finally became what we know today as "Quantum Mechanics". But once the formalism of quantum mechanics had become a closed mathematical scheme, it also became very soon evident to the founding fathers of the theory that there were too many problems to conceive the theory as describing physical reality in terms of atoms —as "tiny elementary particles living in space-time". According to Heisenberg [39, p. 3], "the change in the concept of reality manifesting itself in quantum theory is not simply a continuation of the past; it seems to be a real break in the structure of modern science". Quantum contextuality, the existence of strange superpositions, the measurement problem, the problem of quantum individuality and the problem of non-locality, among many others, showed the limits of attempting to understand quantum physics in terms of an atomist ontology.

Concomitant to quantum mechanics, in the first decades of the XX century logical positivism was also developed attempting to fight metaphysical thought through the development of Mach's ideas and his empiricist standpoint. Congregated in what was called the Vienna Circle, in their famous manifesto [9] they argued that: "Everything is accessible to man; and man is the measure of all things. Here is an affinity with the Sophists, not with the Platonists; with the Epicureans, not with the Pythagoreans; with all those who stand for earthly being and the here and now." Their main attack to metaphysics was based in the idea that one should focus in "statements as they are made by empirical science; their meaning can be determined by logical analysis or, more precisely, through reduction to the simplest statements about the empirically given." Their architectonic stood on the distinction between "empirical terms", the empirically "given" in physical theories, and "theoretical terms", their translation into simple statements. This separation and correspondence between theoretical statements and empirical observation left aside metaphysical considerations, regarded now merely as a discourse about un-observable entities, pure blabla. One of the major consequences of this empiricist perspective towards observation is that physical concepts become only supplementary elements in the analysis of physical theories. At most, an economy to account for physical phenomena. When a physical phenomenon is understood as a self-evident given (independent of physical concepts and metaphysical presuppositions), empirical terms configure an objective set of data which can be directly related —without any metaphysical constraint— to a formal scheme. Actual empirical observations become then the very fundament of physical theories which, following Mach, should be understood as providing an "economical" account of such observational data. As a consequence, metaphysics, understood as a conceptual and systematic representation of *phúsis*, was completely excluded of the main positivist picture attempting to describe scientific theories.

Empirical Observable Data ————— *Theoretical Terms*

(Supplementary Interpretation)

According to this scheme, physical concepts are not essentially needed, since the analysis of a theory can be done by addressing only the logical structure which accounts for the empirical data. The role of concepts becomes then accessory: "adding" metaphysics might help us to picture what is going on according to a theory. It might be interesting to know what the world is like according to an interpretation of a formalism but, as remarked by van Fraassen [53, p. 242]: "However we may answer these questions, believing in the theory being true or false is something of a different level." The important point we would like to remark here is that according to this empiricist viewpoint, since the world is unproblematically "described" in terms of our "common sense" understanding of phenomena an adequate empirical theory can perfectly account for experiments without the need of an interpretation.[a] However, the project of articulating the empirical-formal relation through the distinction between *theoretical terms* and *observational terms* never accomplished the promise of justifying the independence of those realms —specially with respect to categorical or metaphysical definition of concepts. The fundamental reason had already been discussed by Kant in the *Critique of Pure Reason*: the "actual observations" (or empirical terms) can't be considered as "givens", the observation cannot be understood nor considered without previously taking into account a categorical structure that allows to account for phenomena. The description of phenomena always presupposes —implicitly or explicitly— metaphysical elements. Identity or non-contradiction are not "things" we see in the world but rather the very conditions of possibility of classical experience; we presuppose them in order to make sense of the world. This categorical systematization, allowing for a theoretical-conceptual representation, is in itself metaphysical. As the philosopher from Königsberg would have said, it is the representational framework of the transcendental subject, articulating categories and forms of intuition, that which allows for an objective empirical experience.

After the second World War, and establishing a continuity with positivism, the Anglo-Saxon thought consolidated in what was called "analytical philosophy". This new philosophy was originated in opposition to another supposed philosophical "school", called "continental" —meaning the European continent. Even if the branching of analytical philosophy advanced in a vertiginous manner in the academic world, and even with the internal critiques in the 60's and 70's by figures as Lakatos, Feyerabend and Kuhn (among others), we can safely say that the fundamental presuppositions remained those of classical logical positivism. Contemporary philosophy of science (not only of physics) continues to rely on two fundamental distinctions: one between "theoretical terms" and "empirical terms", and the other between "observables" and "unobservables". About the first of those distinctions, Curd and Cover [14, p. 1228] affirm: "Logical positivism is dead and logical empiricism is no longer an avowed school of philosophical thought. But despite our historical and philosophical distance from logical positivism and empiricism, their influence can be felt. An

---

[a]It is important to remark that the problem of interpretation in the context of philosophy of physics has been a deep problem since its origin. The relation between empirical observation has been a difficult subject of analysis since Carnap, Neurath, Popper, Hempel and many others tried to escape the metaphysical characterization of physical concepts. See [10].

important part of their legacy is observational-theoretical distinction itself, which continues to play a central role in debates about scientific realism." And about the distinction between observables and unobservables, Musgrave [14, p. 1221] explains: "In traditional discussions of scientific realism, common sense realism regarding tables and chairs (or the moon) is accepted as unproblematic by both sides. Attention is focused on the difficulties of scientific realism regarding 'unobservables' like electrons." This perspective has very deep consequences for research not only in philosophy but also in physics. In particular, it closes the door to the development of radically new physical representations, since it assumes that we already know what reality is in terms of the (naive) "common sense" observation of tables and chairs —also known, following Sellars, as the "manifest image of the world". It is in this frame that the problem of realism has been reconfigured —inside the limits established by the newborn philosophy of science— around the question of the scientific justification of a "given" reality, exhibited always through the "common sense" language that we use to give an account of what we observe, and not around the question of the means to produce a systematic theoretical representation or expression of reality. Realism is then situated inside the limits imposed by a perspective according to which it is the "self-evidence" of what is observed by individuals —and not the representation of *phúsis*, of reality— the true fundament of knowledge.

It is by these multiple paths that we arrive to the current situation in which the philosophy of quantum mechanics is at the center of a perfect storm created around the questions of its meaning and reference. These questions are articulated in a paradoxical manner, by sustaining two mutually incompatible perspectives, in what we could call a curious "sophistic substantivalism". On one hand, the philosophy of physics tries to produce a bridge between, first, a language assumed by physicists in terms of unobservable elementary particles —namely, the atoms, protons, electrons, quarks, etc.—, and, secondly, a "common sense" language where "tables and chairs" are taken a-critically as "self-evident" unproblematic existents. On the other hand, the referentiality of theories is considered under a double standard, where we still consider science as an economy of the experience of subjects (of experiments and measurement outcomes), and at the same time we ask —with little conceptual support— about the reality of the world beyond measurement results. So it seems, the very foundations of the project rest on the paradoxical entanglement of a substantivalist metaphysics that refers to unobservable particles, with an observational empiricism that, while aiming at leaving aside the metaphysical-conceptual debate, tries to justify at the same time the existence of our "common sense" (but still metaphysical) classical representation of the world.

5. Epistemic Relationalism in Quantum Mechanics

What might be called in a broad sense "the epistemic view of quantum mechanics" has become one of the main viewpoints accepted not only within philosophy of quantum mechanics, but also within physics itself. According to this perspective, in line with empiricism, observation is not considered as something problematic. Observation is considered not only as the ground but also as the condition of possibility that allows us to gain knowledge about the world that surround us. "Common sense" plays here an important role securing the parameters of a "common language" which produces the illusion of an unproblematic (intersubjective) discourse about "common observations". From this viewpoint, the nature of observation should not be questioned. Accordingly, it is argued that if we begin by raw empirical data alone we are then starting by something "pure", "uncontaminated" and thus "objective".[b] From this perspective, the orthodox philosophy of physics project focuses in trying to "bridge the gap" between our "best (mathematical) theories" and our common sense "manifest image of the world" [33] —an image derivative of the representation produced by classical physics in the XVII century. Following some of the main elements present within this very general line of thought, we might characterize then the epistemic view in terms of three main points. As the reader might recognize, each one of these points is quite commonly —implicitly— presupposed within many philosophical and foundational debates about quantum theory.

I. **Prediction of Measurement Outcomes:** *Physical theories provide predictions about ("self evident") observable measurement outcomes and they do not necessarily provide a representation of physical reality.*

II. **Mathematical Formalism and Empirical Adequacy:** *A physical theory is a mathematical formalism which can be considered —via a set of minimal interpretational rules— as empirically adequate (or not).*

III. **Interpretations are Superfluous:** *The interpretation of an empirically adequate theory is superfluous. It is a metaphysical exercise which*

---

[b]Here the notion of objectivity is confused with that of intersubjectivity. A common mistake within philosophy of quantum mechanics since Bohr's account of physics.

cannot change the formalism of the theory nor the nature of the observations predicted by it.

Quantum mechanics has been characterized many times as one of "our best physical theories". It is empirically adequate and possesses a closed mathematical formalism. But after more than a century we do not possess —up to the present— any coherent representation of what the theory is really talking about. From the epistemic viewpoint this question might be regarded as an unimportant metaphysical enterprise which attempts to talk about something beyond observability. However, as we discussed above, the question of interpretation is not completely vanished even from epistemic and empiricist viewpoints.[c] An empiricist such as van Fraassen might be interested for some reason or another in trying to find out what is the particular representation of physical reality provided by a particular interpretation (e.g., see van Fraassen's analysis of Rovelli's interpretation [54]) even though he might not believe the theory to be true.

Following our definitions, we might provide a general characterization of epistemic relationalism as a view that understands relations as derivative of observations, as a way to relate in a more coherent way such data. Once the data are observed, only then relations are introduced as a means of "better resolving" the strange problems which appear within the theory of quanta; this being done without actually questioning those fundamental —conceptual— presuppositions. Relations are not then taken as the basic elements out of which the world is made of, as elements of a systematic ontology capable of representing reality, but as an hypothesis added after the observations have been performed, in order "to save" in a —maybe— more precise way what we have already essentially supposed (i.e. classical phenomena). As we shall see, such observations might be characterized in different ways: in terms of 'experimental arrangements', in terms of 'facts', 'perspectives' or even other 'systems'. As we will now show, there are several interpretations of quantum mechanics which have developed in different ways this particular understanding of (epistemic) relationalism.

5.1. *Bohr's Instrumental Relationalism*

Many elements present within the epistemic view we have just characterized might be associated to Bohr's pragmatic view of physics [6] according to which: "Physics is to be regarded not so much as the study of something a

---

[c]The question of interpretation reappears in QM in different levels. See [10].

priori given, but rather as the development of methods of ordering and surveying human experience." Bohr —in perfect line with the epistemic view— considered quantum theory as an abstract symbolic formalism which had to be reduced through a limit (i.e., the correspondence principle) to classical physics and phenomena [8]. In this respect, Bohr also shared with the epistemic viewpoint —even though in terms of a neo-Kantian perspective— the idea that observable quantum phenomena are essentially "classical phenomena". According to the Danish physicist [59, p. 7]: "[...] the unambiguous interpretation of any measurement must be essentially framed in terms of classical physical theories, and we may say that in this sense the language of Newton and Maxwell will remain the language of physicists for all time." In this respect [Op. cit., p. 7], "it would be a misconception to believe that the difficulties of the atomic theory may be evaded by eventually replacing the concepts of classical physics by new conceptual forms." Thus, taking distance from ontological and metaphysical problems, Bohr was maybe the first to develop an epistemic type of relationalism grounded on classical experimental situations. In his book, The Philosophy of Quantum Mechanics, Max Jammer discussed the attempt of Bohr to understand quantum mechanics is analogous fashion to relativity theory.

In 1929 Berliner decided to dedicate an issue of his journal to Max Planck in commemoration of the golden anniversary of his doctorate; he asked Sommerfeld, Rutherford, Schrödinger, Heisenberg, Jordan, Compton, London and Bohr to contribute papers and his request was answered in all cases. Bohr used this opportunity to expound in greater detail the epistemological background of his new interpretation of quantum mechanics. In his article he compared in three different aspects his approach with Einstein's theory of relativity. [...] Concerning the first two points of comparison Bohr was certainly right. But as to the third point of comparison, based on the assertion that relativity theory reveals 'the subjective character of all concepts of classical physics' or, as Bohr declared again in the fall of 1929 in an address in Copenhagen, that 'the theory of relativity remind us of the subjective character of all physical phenomena, a character which depends essentially upon the motion of the observer,' [...] Bohr overlooked that the theory of relativity is also a theory of invariants and that, above all, its notion of 'events,' such as the collision of two particles, denotes something absolute, entirely independent of the reference frame of the observer and hence logically prior to the assignment of metrical attributes. [42, p. 132]

Jammer continues to say:

> [...] in Bohr's relational theory, the question 'What is the position (or momentum) of a certain particle' presupposes, to be meaningful, the reference to a specified physical arrangement [...] one may formulate a theory of 'perspectives', the term perspective denoting a coordinated collection of measuring instruments either in the sense of reference systems as applied in relativity or in the sense of experimental arrangements as conceived by Bohr. The important point now is to understand that although a perspective may be occupied by an observer, it also exists without such an occupancy [...] A 'relativistic frame of reference' may be regarded as a geometrical or rather kinematical perspective; Bohr's 'experimental arrangement' is an instrumental perspective. [42, p. 201]

Indeed, in relativity theory (like in classical mechanics) all events can be conceived as perfectly well defined events, meaning they can be always placed in a structure which allows us to think consistently of the actual existence of all present events. However, as expressed by the Kochen-Specker theorem [44], this possibility is precluded in the orthodox quantum formalism. As we know, the multiple projection operators of a quantum state cannot be mapped to a global valuation of the Boolean elements $\{0, 1\}$ (see for discussion [20]).

As one of us argued in [16] Bohr might be regarded as responsible for introducing the linguistic turn into physics, confronting in this way the very naive conceptions of the praxis and original meaning of physics itself. Physics was then understood as being fundamentally grounded in language. Accordingly, *phúsis* and reality had to be considered only as words —created by humans. A direct consequence of this development was that the ontological questions to which quantum mechanics was confronted in the first decades of the XX century had to be "suspended". Bohr's philosophy of physics played in this respect an essential role: "We are suspended in language in such a way that we cannot say what is up and what is down. The word 'reality' is also a word, a word which we must learn to use correctly." There is no quantum world but only a classical language in which we are trapped. As Wittgenstein had claimed: "The limits of my language mean the limits of my world." Or, rephrasing it in Bohr's own terms: "We must be clear that when it comes to atoms, language can be used only as in poetry. The poet, too, is not nearly so concerned with describing facts as with creating images and establishing mental connections." The Bohrian linguistic turn in physics was able to deconstruct reality through language. As a consequence, objectivity —which presupposed a moment of unity related to an object— became mere intersubjective agreement. Ontological questions in quantum mechanics were —even more— blurred, the relation between the experimental arrangement as described classically and that of which the mathematical formalism of quantum mechanics was talking about became then "unspeakable".

According to Bohr's interpretation of quantum mechanics —in analogous terms to positivism— the (supposedly) objective character of the theory was secured by our classical language, a language which allowed us to refer to (classical) experimental apparatuses and phenomena:

> On the lines of objective description, [I advocate using] the word *phenomenon* to refer only to observations obtained under circumstances whose description includes an account of the whole experimental arrangement.[...] The experimental conditions can be varied in many ways, but the point is that in each case we must be able to communicate to others what we have done and what we have learned, and that therefore the functioning of the measuring instruments must be described within the framework of classical physical ideas. [59, p. 3]

Bohr wanted to bring together the multiple incompatible contexts through his own concept of complementarity. However, he was never able to answer the ontological questions which Einstein had posed to him once and again. He escaped the issue by always translating Einstein's ontological concerns into his own epistemological scheme of thought. But, as it is said, a translator is also a traitor. When Bohr's translation was finished ontology had been completely erased from the main discussion; once the job was done, he could then explain everything exclusively in terms of (classically described) experimental and measurement situations.[d]

5.2. Modal and Perspectival Relationalism

During one of the famous conferences in Johensu organized by Kalervo Laurikainen in the eighties, Simon Kochen presented a relational type

---

[d]A particularly good example of Bohr's methodology can be found in his famous reply to Einstein Podolsky and Rosen [4] where he also applied his complementarity principle and the idea that measurement situations define the representation of the state of affairs.

modal interpretation [43]. The ideas presented there were regarded by Carl Friedrich von Weizsäcker and Thomas Gornitz [38] as "an illuminating clarification of the mathematical structure of the theory, especially apt to describe the measuring process. We would however feel that it means not an alternative but a continuation to the Copenhagen interpretation (Bohr and, to some extent, Heisenberg)." Kochen had proposed an ascription of properties based on the so called Schmidt theorem, inaugurating —together with van Fraassen and Dieks— what would become to be known, some years later, as "modal interpretations" (see [56] for a detailed anlaysis). Within this framework, one is able to ascribe properties to the subsystems of a composite system in a pure state. The biorthogonal decomposition theorem (also called Schmidt theorem) is able to account for correlations between the quantum system and the apparatus considering the measurement and the actual observation as a special case of this representation.

Theorem 5.1. *Given a state $|\Psi_{\alpha\beta}\rangle$ in $\mathcal{H} = \mathcal{H}_\alpha \otimes \mathcal{H}_\beta$. The Schmidt theorem assures there always exist orthonormal bases for $\mathcal{H}_\alpha$ and $\mathcal{H}_\beta$, $\{|a_i\rangle\}$ and $\{|b_j\rangle\}$ such that $|\Psi_{\alpha\beta}\rangle$ can be written as*

$$|\Psi_{\alpha\beta}\rangle = \sum c_j |a_j\rangle \otimes |b_j\rangle.$$

The different values in $\{|c_j|^2\}$ represent the spectrum of the state. Every λ_j represents a projection in $\mathcal{H}_\alpha$ and a projection in $\mathcal{H}_\beta$ defined as $P_\alpha(\lambda_j) = \sum |a_j\rangle\langle a_j|$ and $P_\beta(\lambda_j) = \sum |b_j\rangle\langle b_j|$, respectively. Furthermore, if the $\{|c_j|^2\}$ are non degenerate, there is a one-to-one correlation between the projections $P_\alpha = \sum |a_j\rangle\langle a_j|$ and $P_\beta = \sum |b_j\rangle\langle b_j|$ pertaining to subsystems $\mathcal{H}_\alpha$ and $\mathcal{H}_\beta$ given by each value of the spectrum.

Through the Schmidt decomposition one can thus calculate the states of the subsystems (which are one-to-one correlated) obtaining:

$$\rho^\alpha = tr_\beta(|\Psi^{\alpha\beta}\rangle\langle\Psi^{\alpha\beta}|) = \sum_i |c_i|^2 |\alpha_i\rangle\langle\alpha_i| \tag{1}$$

$$\rho^\beta = tr_\alpha(|\Psi^{\alpha\beta}\rangle\langle\Psi^{\alpha\beta}|) = \sum_i |c_i|^2 |\beta_i\rangle\langle\beta_i| \tag{2}$$

These two states can be interpreted in a later stage as representing the apparatus and the quantum system, respectively. The different values $|c_i|^2$ represent the spectrum of the Schmidt decomposition given by λ_j. Every λ_j represents a projection in $\mathcal{H}^\alpha$ and a projection in $\mathcal{H}^\beta$ defined as $P^\alpha(\lambda_j) = \sum |a_j^\alpha\rangle\langle a_j^\alpha|$ and $P^\beta(\lambda_j) = \sum |b_j^\alpha\rangle\langle b_j^\alpha|$, respectively. Furthermore, if the $|c_i|^2$ are non degenerate,[e] there is a one-to-one correlation between the projections $P^\alpha(\lambda_j)$ and $P^\beta(\lambda_j)$ pertaining to subsystems α and β given by each value of the spectrum λ_j. In other words, the state of a two-particle system picks out (in most cases, uniquely) a basis (and therefore an observable) for each of the component systems. In this way, the projections of a two composite system $\alpha\beta$ defined from the Schmidt decomposition define the joint probability as:

$$p(P^\alpha(\lambda_a), P^\beta(\lambda_b)) = \delta_{ab} \tag{3}$$

So if $[P^\alpha(\lambda_j)] = 1$, which means that the projection $P^\alpha(\lambda_j)$ is certain, then $[P^\beta(\lambda_j)] = 1$ with probability 1, and *vice versa*. Kochen considers in particular the state $|\Psi^{\alpha\beta}\rangle = \sum_i c_i |a_i^\alpha\rangle \otimes |b_i^\beta\rangle$ that one obtains after a von Neumann measurement, and interprets it in terms of modalities: α *possibly possesses one of the properties* $|a_i^\alpha\rangle\langle a_i^\alpha|$, *and the actual possessed property* $|a_k^\alpha\rangle\langle a_k^\alpha|$ *is determined by the observation that the device* β *possesses the reading* $|b_k^\beta\rangle\langle b_k^\beta|$. In the biorthogonal decomposition there is thus, a bi-univocal relation between the properties of the object and the measuring device. So every pure state of a composite of two disjoint systems should receive this interpretation. In this way the *dynamical state*, with the use of the biorthogonal decomposition, generates a probability measure over the set of possible *value states*[f], namely the standard quantum mechanical measure. As noted by Kochen [43, p. 152]: "Every interaction gives rise to a unique correlation between certain canonically defined properties of the two interacting systems. These properties form a Boolean algebra and so obey the laws of classical logic." Kochen's relational interpretation defines systems as "being witnessed" by one another:

> In place of an official human observer, we assume that each system acts as witness to the state of the other... The world from this view becomes one of perspectives from different systems, with no privileged role for any one, and of properties which acquire a relational character by being realized only upon being witnessed by other systems. [43, pp. 160-164]

The relation imposed by Kochen between systems that observe other

---

[e]In the case of degeneracy it is also possible to define new multi-dimensional projections and recompose this one-to-one correlation between subsystems [26].

[f]The distinction between *dynamical state* and *value state* was introduced by Bas Van Fraassen in order to solve the inconsistencies into which one is driven by the eigenstate-eigenvalue link. See [22,53].

different systems is of a type which resembles some of the views already discussed. It is interesting to notice that, as Vermaas [*Op. cit.*, p. 49] argues: "If one accepts such relationalism; i.e. that properties are meaningful only with respect to a relative system; one can deny the need for correlations between the properties of all possible subsystems of a composite because, for Kochen, properties have a truly relational character." However, even though we agree with Vermaas' conclusion regarding the fundament of correlations within Kochen's approach, for our own purposes, at this point of our analysis it becomes of outmost importance to remark that: *relationalism is by no means equivalent nor implies necessarily relativism.* We might clarify this important point recalling the analysis we provided in the first part of this article. When we talked about sophistry, we analyzed Protagora's views in order to describe what the common ground of all relativisms is: *the reality of something is always relative to the individual perceiving subject who observes it.* "Man is the measure of all things." This statement, as recognized by the Vienna Circle in their manifesto, is also close to empiricism which, in Locke's own wording, rephrased the Protagorian statement in terms of perception: "To be, is to be perceived". Both Bohr's and Kochen's proposals fall under this category. While for Bohr, quantum systems are defined in terms of (as being *relative* to) measurement situations, for Kochen the properties of systems are defined in terms of (as being *relative* to) other systems. As we shall see, the perspectival version of Bene and Dieks [2], and maybe, to some extent, even that of Rovelli, fall also under this general sophistic umbrella. We could say that they all propose —in different ways— extended Protagorisms: relativisms with broader definitions of what can be considered as an "observer" or "perceiving subject".

Contrary to sophistry, in the sections above, we analyzed philosophical and physical theories that affirmed the possibility of representing *phúsis* beyond the dependence to actual *hic et nunc* observations. Some of them presented a world primarily made of independent substances; others, like Plato's *Sophist* and Spinoza's *Ethics*, denied ontological separation and proposed relation as fundamental within their own metaphysical representation of reality. For both Plato and Spinoza the world should be conceived as made of relations or potencies of relation. These ontological relational schemes expose the fact that relationalism does not entail relativism. Thus, at this point of our analysis, it is useful to distinguish between two different types of relations:

Epistemic relations: *Relations as modeled from the empirical subject-*

object model. What is observed or perceived (the object or system) is not only related but also is —more importantly— relative to a subject (an agent, another system or an apparatus). Epistemic relations entail relativism, since we only know how things (observed or perceived) seem to be as relative to a given perspective, and never how they really are independently of a perceiving actor (an agent, another system or an apparatus).

Ontic relations: *Relations are the metaphysical building-blocks of reality, they are essential to the representation of phúsis. Relations exist within reality right from the start and their existence is absolutely independent of observations or a perceiving subject (an agent, another system or experimental arrangement).*

Going now back to modal interpretations, we might remark that it is only in the case of *ontological*[g] *relations* that it would make sense to look for joint probability distributions. This was, in fact, the main program in which Dieks, Bacciagaluppi, Clifton, Bub and other modal researchers engaged in at the beginning of the '90 (see [32]). However, in the case of *epistemic relations* it becomes meaningless to seek for such joint probability distribution. In such case, relations are intrinsically determined —following Bohr's notion of contextuality [19]— by the choice performed by a subject of the particular experimental set-up with which the object is studied. The problem is that due to the structure of the quantum formalism different choices of contexts determine incompatible local valuations which —according to the Kochen-Specker theorem— cannot be embedded into a whole global valuation [19]. For example, if we consider a composite system $\omega = \alpha\beta\gamma$ in a pure state and we take the two subsystems $\alpha\beta$ and $\alpha\gamma$, these two subsystems have properties which are realized with respect to different relative systems. In general the projections resulting from these subsystems will not be commensurable and it will be not possible to obtain a classical distribution for joint probabilities.[h] As remarked by Vermaas the impossibility to define joint property ascriptions does diminish —at least from a realist viewpoint— the attractiveness of such interpretations.

"the spectral [K-D] modal interpretation is condemned to perspectivalism. [...] If one accepts perspectivalism as discussed [...] and as possibly

---

[g]Through this paper we use 'ontic' and 'ontological' as synonyms, in analogous manner to the way in which 'epistemic' and 'epistemological' are used in philosophy of physics.
[h]This result was explicitly addressed by Vermaas in his no-go theorem of 1997 [55].

embraced by Kochen (1985), the joint probabilities give the correlations between all the properties one can consider simultaneously. For, according to perspectivalism, one can only simultaneously consider the properties of subsystems if these systems can be considered from one and the same perspective. [By] adopting perspectivalism most of the interesting questions in quantum mechanics are simply evaded." [56, p. 255]

Indeed, one considers in somewhat sophistic terms that "the subject is the measure of all clicks", the choice of the experimental arrangement appears then as a necessary precondition to account for what will be observed. Obviously, no individual subject or agent can adopt *hic et nunc* many different perspectives. An individual cannot be present, at one and the same time, in different viewpoints (contexts or frames of reference) which allow to observe a system. We, human subjects, perform observations always from our own particular perspective. In line with Bohr's proposal, the choice of the measurement arrangement must be then regarded as a precondition to define the system and its properties (see for a detailed discussion [19]). From this relativist stance, one can deny that Kochen's modal interpretation ever needs to say something sensible about the joint occurrence of the properties of α and β. The problem, as in sophistry, is that the consequence of this move erases any reference of the theory to a representation of physical reality (or *phúsis*) beyond observations or measurement outcomes. The connection between the theory and reality is then lost.

In 2002, a new perspectival version of the modal interpretation was developed by Gyula Bene and Dennis Dieks [2] in which they continued the line of research proposed by Kochen.[i] The central point of their interpretation can be summarized in the following passage: "[...] instead of the usual treatment in which properties are supposed to correspond to monadic predicates, we will propose an analysis according to which properties have a relational character." Once again, relativism and relationalism are used as synonyms, assuming implicitly that relations must be necessarily understood as epistemic relations. As in the Kochen interpretation, in the Bene-Dieks (B-D for short) interpretation the state of a physical system S requires the specification of a "reference system" R with respect to which the state is defined. "In this 'perspectival' version of the modal interpretation properties of physical systems have a relational character and are defined with respect to another physical system that serves as a reference

---

[i] For a detailed analysis of the Bene-Dieks perspectival interpretation see [15].

[or witnessing] system."

$$\rho_S(R) = Tr_{R/S}(\rho_R(R)) \tag{4}$$

were $\rho_S(R)$ means that the system S is witnessed from the perspective R; and $Tr_{R/S}$ means the trace with respect to the degrees of freedom of *system R 'minus' S*. In the special case in which $R=S$ the state is in general a one dimensional projector; i.e. the *state of S with respect to itself*:

$$\rho_S(S) = |\Psi_S\rangle\langle\Psi_S| \tag{5}$$

The state of R *w.r.t. itself*, $\rho_R(R)$, is postulated to be one of the projectors contained in the spectral resolution of $\rho_R(U/R)$, i.e. the state ρ_R from the perspective of the *Universe 'minus' system R*; and represents the monadic properties of the system as in the K-D modal interpretation. If there is no degeneracy among the eigenvalues of $\rho_R(U/R)$ these projectors are one dimensional and the state can be represented by a vector $|\Psi_S\rangle$. The *state of R w.r.t. itself* is given by one of the eigenvectors $|R_j\rangle$ of $\rho_R(U/R)$. Kochen's witnessed state will turn out to correspond to the state of the object with respect to itself. The dynamical principle of the interpretation is that $\rho_U(U)$ evolves unitarily in time. There is "no collapse" of the $|\Psi_U\rangle$ in this approach just as in ordinary modal interpretations. The theory specifies only the probabilities of the various possibilities and in this sense is an indeterministic interpretation. Furthermore, it is assumed that the state assigned to a closed system S undergoes a unitary time evolution given by the Liouville equation:

$$i\hbar\frac{\partial}{\partial t}\rho_S(S) = [H_S, \rho_S(S)] \tag{6}$$

If the systems $S_1, S_2...S_n$ are pair-wise disjoint and U is the whole universe, then the joint probability that $|\Psi_{S_1}\rangle$ coincides with $|\varphi_{j_i}^{S_1}\rangle$, $|\Psi_{S_1}\rangle$ coincides with $|\varphi_{j_i}^{S_1}\rangle$, ..., $|\Psi_{S_1}\rangle$ coincides with $|\varphi_{j_i}^{S_1}\rangle$ is given by

$$P(j_1, j_2, ..., j_n) = Tr(\rho_U(U)\prod_{i=1}^{n}|\varphi_{j_i}^{S_i}\rangle\langle\varphi_{j_i}^{S_i}|) \tag{7}$$

To summarize, because the existence of properties is always relative to a particular (perceiving) system, according to Bene and Dieks it makes no sense to compare properties from different perspectives: "we do not define joint probabilities if the systems are not pair wise disjoint; in this way we block the no go probability theorem by Vermaas." Joint probabilities are only definable from a single definite *perspective R* and the restricted set of subsystems considered are thus commensurable ones.

5.3. Rovelli's Informational Relationalism

A similar analogy to that of Bohr between quantum mechanics and relativity was proposed by Carlo Rovelli in his now famous "Relational Quantum Mechanics" [50]. The notion explicitly rejected —also, implicitly rejected in the interpretations of Bohr, Kochen and bene-diks— is that of: absolute state or observer independent state of a system (observer-independent values of physical quantities). This notion is replaced in favor of: state relative to something. Rovelli argues that this necessity derives from the observation that the experimental evidence at the basis of quantum mechanics forces us to accept that distinct observers give different descriptions of the same events. There are two main ideas surrounding the interpretation of Rovelli, firstly, that the unease in quantum mechanics may derive from the use of a concept, which is inappropriate to describe the physical world at the quantum level; i.e. the notion of absolute state of a system. Secondly, that QM will cease to look puzzling only when we will be able to derive the formalism from a set of simple physical assertions (postulates, principles) about the world.

According to Rovelli we should derive the formalism from a set of experimentally motivated postulates just in the same way Einstein did for special relativity:

[...] Einstein's 1905 paper suddenly clarified the matter by pointing out the reason for the unease in taking Lorentz transformations 'seriously': the implicit use of a concept (observer-independent time) inappropriate to describe reality when velocities are high. Equivalently: a common deep assumption about reality (simultaneity is observer-independent) which is physically untenable. The unease with the Lorentz transformations derived from a conceptual scheme in which an incorrect notion absolute simultaneity was assumed, yielding any sort of paradoxical consequences. Once this notion was removed the physical interpretation of the Lorentz transformations stood clear, and special relativity is now considered rather uncontroversial. Here I consider the hypothesis that all 'paradoxical' situations associated with quantum mechanics as the famous and unfortunate half-dead Schrödinger cat [Schrödinger 1935] may derive from some analogous incorrect notion that we use in thinking about quantum mechanics. (Not in using quantum mechanics, since we seem to have learned to use it in a remarkably effective way.) The aim of this paper is to hunt for this incorrect notion, with the hope that by exposing it clearly to public contempt, we could free ourselves from the present unease with our best present theory of motion, and fully understand what does the theory assert about the world. [*Op. cit.*, p. 1639]

Rovelli's interpretation takes distance from Bohr's distinction between macroscopic and microscopic systems. "The disturbing aspect of Bohr's view is the inapplicability of quantum theory to macrophysics. This disturbing aspect vanishes, I believe, at the light of the discussion in this paper." Instead of the privileging certain observers (classical systems) Rovelli centers his interpretation in the concept of information.

Information indicates the usual ascription of values to quantities that founds physics, but emphasizes their relational aspect. This ascription can be described within the theory itself, as information theoretical information, namely correlation. But such a description, in turn, is quantum mechanics and observer dependent, because a universal observer-independent description of the states of affairs of the world does not exist. [*Op. cit.*]

Rovelli recognizes the impossibility of presenting an objective description in terms of systems and replaces this notion by "net of relations". According to him [*Op. cit.*]: "[...] at the present level of experimental knowledge (hypothesis 2), we are forced to accept the result that there is no objective, or more precisely observer-independent meaning to the ascription of a property to a system. Thus, the properties of the systems are to be described by an interrelated net of observations and information collected from observations." The question becomes then: what can we say about this net of relations. Rovelli, talks about the notion of information: "The notion of observer independent state of a system is replaced by the notion of information about a system that a physical system may possess." Still, as in the case of Bohr, Kochen, Bene and Dieks, the ontological question that any realist would want to answer is still present even though in a different form: information about what? Although it is possible to maintain a relational view of quantum states in terms of information, the ontological status of such information seems to remain a problematic issue —at least, from a realist perspective.

5.4. *Revisiting Epistemic Relationalism: Relativism Reloaded?*

Our analysis has attempted to understand the common epistemological ground on which the just mentioned relational interpretations of quantum mechanics have been developed. As we have shown, all such interpretations take both a sophistic and empiricist standpoint. Sophistic in the sense that there is always the presupposition of someone or something (an agent, another system or even an apparatus) playing the role of a perceiving subject. Empiricist in the sense that observations are always considered as the "self evident" givens which allow us to produce knowledge. Even in the case of quantum measurements, the observed 'clicks' in detectors are considered as being unproblematic —as providing objective data. Such an epistemological (relativist) viewpoint conceives that knowledge is always of a perspectival nature, that it is fundamentally limited by observers, perceptions, other systems or even particular measurement situations. It also takes as a fundamental standpoint the idea that measurement outcomes or data can be —in principle— discussed without the need of applying a rigorous conceptual architecture.

For our purposes, it becomes then important to remark that there is a common line of thought, a common agenda and set of presuppositions within the relational interpretations of quantum mechanics discussed above. While for Bohr the experimental arrangement is the measure of all (classical) phenomena, for Kochen it is always a system which acts as the measure (or witness) of another system. While Bene and Dieks argue — following Bohr— in favor of considering an instrumental perspective as a necessary condition of possibility to measure a quantum system, Rovelli seems to claim that all we have is the information that a system has about another system. Relationalism is then understood as implying relativism: the definition of a physical system and its properties (the object of study) is always relative to a witnessing subject, system or measurement apparatus (the subject which perceives). It is true that relativism implies always a relation between a subject and an object; but the opposite —as we have already shown explicitly in previous sections— is not true: relationalism is not necessarily committed to relativism. In fact, it is possible to understand relations in a completely non-relativist manner, namely, as ontological relations. To provide a clear understanding and definition of what is to be considered an ontic relation and how such ontological relationalism might help us to better understand the theory of quanta has been the main goal of this article.

6. Ontological Relationalism in Quantum Mechanics: A New Proposal

The ontic viewpoint —as we understand it— differs radically with respect to the epistemic account of physics. We can characterize the ontic view in terms of two main elements, firstly, the possibility of the theoretical representation of reality, and secondly, the denial of "self evident" or "common sense" observability. Returning to our introductory discussion, we might begin by stressing the positive characterization of the meaning of physics as a discipline which attempts to represent phúsis (or reality) in formal-conceptual terms. According to this viewpoint, physical theories provide, through the tight inter-relation of mathematical formalisms and networks of physical concepts, the possibility of representing both physical reality and experience.

The possibility to imagine and picture reality beyond *hic et nunc* observation is provided not only by mathematical formalisms but also by physical concepts. Mathematics does not contain physical concepts, it does not represent anything beyond its own structure. One cannot derive as a theorem physical concepts from a mathematical system. Mathematicians can obviously work without learning about physical theories or the way in which physicists are able to relate formalisms with particular representations of physical reality. In fact, most mathematicians know nothing about physics and their work can be done without ever doing any type of experiment in a lab. A laboratory is completely useless for a mathematician. They neither require meta-physical concepts for their practice. The theory of calculus does not include the physical notions of Newtonian space and time, it does not talk about 'particles', 'mass' or 'force'. In the same way, Maxwell's formalism cannot derive through a theorem the physical notion of 'field'. Within physical theories, while mathematical formalisms are capable of providing a quantitative understanding, only conceptual schemes —produced through the interrelation of many different concepts— are capable of giving a qualitative understanding of physical reality and experience.

Gedankenexperiments are a good example of the power of conceptual and formal representations within physics. In fact, thought-experiments in physics have many times escaped the technical capabilities of their time and ventured themselves into debates about possible —but unperformed— experiences. Not only that, even impossible experiences —such as those imagined by Leibinz and Newton regarding the existence of a single body in the Universe— have been of great importance for the development of physics. Such physical —possible or impossible— counterfactual experiences can be only considered and imagined through an adequate conceptual scheme. Indeed, as remarked by Heisenberg [41, p. 264]: "The history of physics is not only a sequence of experimental discoveries and observations, followed by their mathematical description; it is also a history of concepts. For an understanding of the phenomena the first condition is the introduction of adequate concepts. Only with the help of correct concepts can we really know what has been observed." According to the ontic viewpoint, reality is not something "self-evidently" exposed through observations —as positivists, empiricists and even Bohr has claimed—; on the contrary, its representation and understanding is only provided —following the first physicists and philosophers— through physical theories themselves. To avoid any misunderstanding, let us stress that the ontic viewpoint we are discussing here is not consistent with scientific realism, phenomenological realism or realism about observables all of which are in fact variants of empiricism grounded on common sense observability. As Einstein [23, p. 175] made the point: "[...] it is the purpose of theoretical physics to achieve understanding of physical reality which exists independently of the observer, and for which the distinction between 'direct observable' and 'not directly observable' has no ontological significance". Observability is secondary even though "the only decisive factor for the question whether or not to accept a particular physical theory is its empirical success." Empirical adequacy is part of a verification procedure, not that which "needs to be saved" — as van Fraassen might argue[j]. Observability is something developed within each physical theory, it is a result of a theory rather than an obvious presupposition. At the opposite corner from the epistemic standpoint, Einstein [40, p. 63] explained to Heisenberg that in fact: "It is only the theory which decides what we can observe." Following these set of general considerations we might characterize what we have called elsewhere [18] representational realism:

I. **Physical Theory:** *A physical theory is a mathematical formalism related to a set of physical concepts which only together are capable of providing a quantitative and qualitative understanding of a specific field of phenomena.*

---

[j] According to van Fraassen [52, p. 197]: "the only believe involved in accepting a scientific theory is belief that it is empirically adequate: all that is *both* actual *and* observable finds a place in some model of the theory. So far as empirical adequacy is concerned, the theory would be just as good if there existed nothing at all that was either unobservable or not actual. Acceptance of the theory does not commit us to belief in the reality of either sort of thing."

II. **Formal-Conceptual Representation of Reality:** *Physics attempts to provide theoretical, both formal and conceptual, representations of physical reality.*

III. **Observability is Created by the Theory:** *The conditions of what is meant by 'observability' are dependent on each specific theory. The understanding of observation is only possible through the development of adequate physical concepts.*

To summarize, while the epistemic view considers that the world is accessible through "common sense" observation —understood as a given—, which is also the key to develop scientific knowledge itself; the ontic viewpoint takes the opposite standpoint and argues that it is only through the creation of theories that we are capable of providing understanding of our experience in the world. According to the latter view, the physical explanation of our experience goes very much against "common sense" observability. The history of physics can be also regarded as the continuous change in our "common sense" understanding of the world. It was not obvious for the contemporaries of Newton that the same force commands the movement of the moon, the planets and a falling apple. It was not inescapable in the 18th Century that the strange phenomena of magnetism and electricity could be unified through the strange notion of electromagnetic field. And it was far from evident —before Einstein— that space and time are entangled, that objects shrink and time dilate with speed. To sum up, we maintain that what is needed is a conceptual representation of what the quantum formalism is expressing, and not merely a salvaging of the relation between our "common sense" understanding of reality and measurement outcomes —sweeping under the classical carpet the most interesting, effective and productive aspects of the formalism. What QM talks about —we argue— seems difficult to be grasped through a substantivalist (atomistic) understanding of reality that supposes individual separated substances. After more than a century trying to fit the quantum formalism into such a presupposed metaphysical representation of reality it might be time to try something new.

Our proposal is to develop, taking inspiration from some of the elements found in the revisions of both Plato's and Spinoza's philosophies, a truly relational ontology (this is, one that considers relation as being fundamental) which is capable of providing a new (representational) realist way of understanding the theory of quanta. Both philosophers' understanding of 'potency' or 'possibility' in ontological terms, as well as the connection

314

between that understanding and their relational views —which, as we saw, are capable of articulating a specific knowledge of the world without producing substantial separations—, might allow us to throw new light on some key features of the quantum formalism such as: contextuality, superposition, non-individuality, non-separability, etc. The specific consideration of these features in ontic relational terms will be addressed in future works.

Acknowledgements

This work was partially supported by the following grants: FWO project G.0405.08 and FWO-research community W0.030.06. CONICET RES. 4541-12 and the Project PIO-CONICET-UNAJ (15520150100008CO) "Quantum Superpositions in Quantum Information Processing".

References

1. Bacciagaluppi, G., 1995, "A Kochen Specker theorem in the Modal Interpretation of Quantum Mechanics", *Internal Journal of Theoretical Physics*, **34**, 1205-1216.
2. Bene, G. and Dieks, D., 2002, "A Perspectival Version of the Modal Interpretation of Quantum Mechanics and the Origin of Macroscopic Behavior", *Foundations of Physics*, *32*, 645-671, (arXiv: quant-ph/0112134).
3. Bohr, N., 1929, "The quantum of action and the description of nature" in Bohr, N., Collected works, Rudinger, E. (general ed.), *Vol. 6: Foundations of Quantum Physics I*, pp. 208-217, Kolckar, J. (ed.), Amsterdam, North-Holland, 1985.
4. Bohr, N., 1935, "Can Quantum Mechanical Description of Physical Reality be Considered Complete?", *Physical Review*, **48**, 696-702.
5. Bohr, N. 1949. "Discussions with Einstein on Epistemological Problems in Atomic Physics", in *Albert Einstein: Philosopher-Scientist*, edited by P. A. Schilpp, 201-41. La Salle: Open Court. Reprinted in (Bohr 1985, 9-49).
6. Bohr, N., 1960, *The Unity of Human Knowledge*, in *Philosophical writings of Neils Bohr*, vol. 3., Ox Bow Press, Woodbridge.
7. Bohr, N. 1985, *Collected works, Vol. 6: Foundations of Quantum Mechanics I (1926-1932)*, Edited by J. Kalckar. North-Holland, Amsterdam.
8. Bokulich, P. and Bokulich, A., 2005, "Niels Bohr's Generalization of Classical Mechanics", *Foundations of Physics*, **35**, 347-371.
9. Carnap, H., Hahn, H. and Neurath, O., 1929, "The Scientific Conception of the World: The Vienna Circle", *Wissendchaftliche Weltausffassung*.
10. Cassini, A., 2016, "El problema interpretativo de la mecánica cuántica. Interpretación minimal e interpretaciones totales", *Revista de Humanidades de Valparaíso*, **8**, 9-42.
11. Cooper, J.M. (ed.), 1997, Plato: Complete Works, Indianapolis: Hackett.
12. Cordero N.L., 2005, *Siendo, se es. La tesis de Parmenides*, Buenos Aires, Biblos.

13. Cordero, N.L., 2014, *Cuando la realidad palpitaba*, Buenos Aires, Biblos.
14. Curd, M. and Cover, J. A., 1998, *Philosophy of Science. The central issues*, Norton and Company (Eds.), Cambridge University Press, Cambridge.
15. de Ronde, C., 2003, *Perspectival Interpretation: a story about correlations and holism*, URL = http://www.vub.ac.be/CLEA/people/deronde/.
16. de Ronde, C., 2015, "Epistemological and Ontological Paraconsistency in Quantum Mechanics: For and Against Bohrian Philosophy", *The Road to Universal Logic (Volume II)*, pp. 589-604, Arnold Koslow and Arthur Buchsbaum (Eds.), Springer, Berlin.
17. de Ronde, C., 2016, "*Probabilistic Knowledge* as *Objective Knowledge* in Quantum Mechanics: *Potential Powers* Instead of *Actual Properties*", in *Probing the Meaning and Structure of Quantum Mechanics: Superpositions, Semantics, Dynamics and Identity*, pp. 141-178, D. Aerts, C. de Ronde, H. Freytes and R. Giuntini (Eds.), World Scientific, Singapore.
18. de Ronde, C., 2016, "Representational Realism, Closed Theories and the Quantum to Classical Limit", in *Quantum Structural Studies*, pp. 105-136, R. E. Kastner, J. Jeknic-Dugic and G. Jaroszkiewicz (Eds.), World Scientific, Singapore.
19. de Ronde, C., 2016, "Unscrambling the Omelette of Quantum Contextuality (PART I): Preexistent Properties or Measurement Outcomes?", preprint (arXiv:1606.03967)
20. de Ronde, C., Freytes, H. and Domenech, G., 2014, "Interpreting the Modal Kochen-Specker Theorem: Possibility and Many Worlds in Quantum Mechanics", *Studies in History and Philosophy of Modern Physics*, **45**, 11-18.
21. D'Espagnat, B., 1976, *Conceptual Foundations of Quantum Mechanics*, Benjamin, Reading MA.
22. Dickson, W. M., 1998, *Quantum Chance and Nonlocality: Probability and Nonlocality in the Interpretations of Quantum Mechanics*, Cambridge University Press, Cambridge.
23. Dieks, D., 1988, "The Formalism of Quantum Theory: An Objective description of reality", *Annalen der Physik*, **7**, 174-190.
24. Dieks, D., 1988, "Quantum Mechanics and Realism", *Conceptus XXII*, **57**, 31-47.
25. Dieks, D., 1989, "Quantum Mechanics Without the Projection Postulate and Its Realistic Interpretation", *Foundations of Physics*, **19**, 1397-1423.
26. Dieks, D., 1993, "The Modal Interpretation of Quantum Mechanics and Some of its Relativistic Aspects", *Internal Journal of Theoretical Physics*, **32**, 2363-2375.
27. Dieks, D., 1995, "Physical motivation of the modal interpretation of quantum mechanics", Physics Letters A, **197**, 367-371.
28. Dieks, D., 2005, "Quantum mechanics: an intelligible description of objective reality?", *Foundations of Physiscs*, **35**, 399-415.
29. Dieks, D., 2007, "Probability in the modal interpretation of quantum mechanics", *Studies in History and Philosophy of Modern Physics*, **38**, 292-310.
30. Dieks, D., 2009, "Quantum mechanics: an intelligible description of objective reality?", *Foundations of Physiscs*, **39**, 760-775. *****
31. Dieks, D., 2010, "Quantum Mechanics, Chance and Modality", *Philosophica*, **83**, 117-137.

316

32. Dieks, D. and Vermaas, P.E. (Eds.) 1998, *The Modal Interpretation of Quantum Mechanics*, Vol. 60 of the Western Ontario Series in the Philosophy of Science, Kluwer Academic Publishers, Dordrecht.
33. Dorato, M., 2015, "Events and the Ontology of Quantum Mechanics", *Topoi*, **34**, 369-378.
34. Einstein, A., 1916, "Ernst Mach", *Physikalische Zeitschrift*, **17**, 101-104.
35. Einstein, A., Podolsky, B. and Rosen, N., 1935, "Can Quantum-Mechanical Description be Considered Complete?", *Physical Review*, **47**, 777-780.
36. Fronterotta, F., 1995, "L'Etre et la participation de l'autre. Une nouvelle ontologie dans le Sophiste", *Les Etudes Philosophiques*, 311-353.
37. González, F.J., 2011, "Being as power in Plato's *Sophist* and beyond", in *Plato's Sophist. Proceedings of the seventh symposium Platonicum praguense*, Prague.
38. Görnitz, T. and von Weizsacker, C.F., 1987 "Remarks on S. Kochen's Interpretation of Qunatum Mechanics", in *Symposium on the foundations of Modern Physics 1987*, 357-368, P.Lathi and P. Mittelslaedt (Eds.), Singapore: World Scientific.
39. Heisenberg, W., 1958, *Physics and Philosophy*, World perspectives, George Allen and Unwin Ltd., London.
40. Heisenberg, W., 1971, *Physics and Beyond*, Harper & Row, New York.
41. Heisenberg, W., 1973, "Development of Concepts in the History of Quantum Theory", in *The Physicist's Conception of Nature*, pp. 264-275, J. Mehra (Ed.), Reidel, Dordrecht.
42. Jammer, M., 1974, *The Philosophy of Quantum Mechanics*, Wiley, New York.
43. Kochen, S., 1985, "A New Interpretation of Quantum Mechanics", in P.Lathi and P. Mittelslaedt (eds.), *Symposium on the foundations of Modern Physics 1985*, 151-169, World Scientific, Johensuu.
44. Kochen, S. and Specker, E., 1967, "On the problem of Hidden Variables in Quantum Mechanics", *Journal of Mathematics and Mechanics*, **17**, 59-87.
45. Laplace, P.S., 1951, *A Philosophical Essay on Probabilities*, translated into English from the original French 6th ed. by Truscott, F.W. and Emory, F.L., Dover Publications, New York.
46. Laudisa, F., 2017, "Relational Quantum Mechanics", preprint. (arXiv:1710.07556) ****
47. Laudisa, F. and Rovelli, C., 2005, "Relational Quantum Mechanics", *The Stanford Encyclopedia of Philosophy* (Fall 2005 Edition), Edward N. Zalta (ed.), forthcoming URL = http://plato.stanford.edu/archives/fall2005/entries/qm-relational/.
48. Mach, E., 1959, *The Analysis of Sensations*, Dover Edition, New York.
49. Piron, C., 1999, "Quanta and Relativity: Two Failed Revolutions", In *The White Book of Einstein Meets Magritte*, 107-112, D. Aerts J. Broekaert and E. Mathijs (Eds.), Kluwer Academic Publishers.
50. Rovelli, C., 1996, "Relational Quantum Mechanics", *International Journal of Theoretical Physics*, **35**, 1637-1678.
51. Spinoza, Benedictus, *The Collected Writings of Spinoza*, 2 vols., Edwin Curley, translator (Princeton: Princeton University Press, vol. 1: 1985; vol. 2: 2016). The Ethics is in vol. 1; the Theological Political Treatise is in vol. 2.
52. Van Fraassen, B.C., 1980, *The Scientific Image*, Clarendon, Oxford.

53. Van Fraassen, B.C., 1991, *Quantum Mechanics: An Empiricist View*, Oxford: Clarendon.
54. Van Fraassen, B.C., 2010, "Rovelli's World", *Foundations of Physics*, **40**, 390-417.
55. Vermaas, P.E., 1997, "A No-Go Theorem for Joint Property Ascriptions in Modal Interpretation", *Physics Review Letters*, **78**, pp.2033-2037.
56. Vermaas, P.E., 1999, *A Philosophers Understanding of Quantum Mechanics*, Cambridge University Press, Cambridge.
57. Vermaas, P.E., 1999, "Two No-Go Theorems for Modal Interpretations of Quantum Mechanics", *Studies in History and Philosophy of Modern Physics*, Vol 30, No. 3, 403-431.
58. Vermaas, P.E. and Dieks, D., (1995). The Modal Interpretation of Quantum Mechanics and Its Generalization to Density Operators. *Foundations of Physics, 25*, 145-158.
59. Wheeler, J.A. and Zurek, W., 1983, *Theory and Measurement*, (W.H. Eds.), Princeton University Press. Princeton.

QUANTUM COGNITIVE MODELING OF CONCEPTS: AN INTRODUCTION

TOMAS VELOZ[1,2,3] AND PABLO RAZETO[2,4]

[1] Universidad Andres Bello, Departamento Ciencias Biológicas, Facultad Ciencias de la vida, 8370146 Santiago, Chile.
[2] Instituto de Filosofía y Ciencias de la Complejidad, Los Alerces 3024, Ñuñoa, Santiago, Chile.
[3] Free University of Brussels, Centre Leo Apostel, Krijgskundestraat 33, B-1160 Brussels, Belgium.
[4] Universidad Diego Portales, Vicerrectoría Académica, Manuel Rodríguez Sur 415, 8370179 Santiago, Chile.
* E-mail: tveloz@gmail.com

The aim of this article is to give an introductory survey to the quantum cognitive approach to concepts. We first review the fundamental problems in the modeling of concepts. Next we show how the quantum cognition program for modeling concepts and their combinations is able to cope with these problems. Finally, we elaborate on some of the most recent developments that deepen the structural relations between quantum entities and concepts. As a conducting line, we have followed the contributions of the Brussels group of quantum cognition directed by Diederik Aerts.

Keywords: Concepts modelling; quantum cognition; non-rationality; vagueness; non-compositionality; contextuality.

1. Introduction

1.1. *From the Mind-Body Problem to Cognitive Models*

The capacity of human beings to observe elements of physical reality, identify and represent relations among these elements, and specially to hypothesize and test unobserved relations to modify such reality to convenience, is a capacity unique among all the other species. This capacity has lead to the development of a profound understanding of the physical realm, and to the exploration of its properties much further than what is expected from our biological sensorial capacities. The existence of humans however, requires a second '*Non-physical*' realm to be completed. In this realm, human manifestations such as ideas, emotions, and self-awareness reside. This realm is known as '*The Mind*'.[1] Whether or not the physical realm and the mind exist independent of each other is known as the 'the mind-body problem', and is one of the most fundamental questions in western philosophy. In modern sciences, this question has raised an interdisciplinary effort known as cognitive science.[2]

Cognitive science is defined as the scientific study of the mind and its processes. It examines what cognition is, what it does and how it works, and as any other science, aims at the development of technologies and tools that allow to study the realm of mind further. The investigations in cognitive science include multiple aspects of intelligence and behavior, especially focusing on how information is represented, processed, and transformed. The majority of cognitive scientists start from a cognitive architecture, usually the brain or a Turing machine, and try to build models of cognitive phenomena as results of processes in such cognitive architecture.

An alternative view, which is mainly dominated by a mix of applied mathematicians, physicists, and cognitive psychologists,[3] does not take a position with respect to the architecture at which the cognitive phenomena occurs. Instead, this view focuses on understanding the structural aspects of the cognitive phenomena from an abstract, and usually mathematical, perspective. This alternative, known as cognitive modeling, is the perspective we follow here.

1.2. *Cognitive Modeling and the Notion of Concept*

A cognitive model is an approximation to a cognitive phenomena for the purposes of comprehension and prediction. Cognitive models normally focus on a single cognitive phenomenon, or on how two or more phenomena interact. We recommend[4] to the reader interested in knowing how different mathematical tools such as set theory, probability, linear algebra, and others, apply to the different cognitive phenomena.

Despite the variety of mathematical tools in use, cognitive modeling approaches can be sharply divided in two main classes.

The first consists of 'ad-hoc' cognitive modeling. The ad-hoc approach produces a cognitive model of a particular phenomena in a specific domain of application. For example in,[5] a model of visual categorization of geometrical shapes based on ontologies is presented. They developed a list of features that play an important role in visual categorization and their relations, and an algorithmic procedure, based on Bayesian statistics, to categorize them. Examples of such categorization elements in the ontology are *sphere-like, round, uniform texture*, etc., and an example of a relation is (*rounded,uniform texture*)→(*sphere-like*). The algorithms in this model assume three incremental stages: i) knowledge acquisition, ii) learning, and iii) categorization. The model is useful for the task for which it was developed, especially in the case of smooth shapes. However, its design is not meant to represent anything else but visual categorization, nor is it compatible with other cognitive phenomena.

The second type of effort seeks for a general representation framework for cognitive phenomena, so called '*Concept Theory.*' Concepts are envisaged as the units that underlie cognitive phenomena.[6,7] However, a concept is a very general notion, as it may represent a concrete or abstract entity. A concept will be denoted throughout this article by single quotes in italic style with the first letter capitalized on each word, and by capital calligraphic letters when denoted in abstract form. For example, let $\mathcal{A}$ denote the concept '*Animal.*' Conceptual instances, also called exemplars, will be denoted between quotes without italics, and by letters $p, q, ...$ when denoted in abstract form. For example, we say $p =$'dog' is an exemplar of concept $\mathcal{A}$. Traditional models of concepts have focused mainly on categories that possess concrete or imaginary instances, such as 'horse', or 'dragon'.[8,9] Modern approaches have extended such focus to more abstract instances such as topics of discussion,[10] images, videos, etc. In a first approximation, a concept is a collection of instances. These instances are the members of the concept. Instances belong to the same concept whenever they share some attributes or properties that make them similar. Properties or attributes will be denoted in italics without quotes, e.g. *has four legs* is a property of the exemplar 'dog' of concept '*Animal*'.

In cognitive science, there are three main proposals for a mathematically sound theory of concepts: The first is known as classical theory, and follows the tradition of classical logic. It assumes that concepts are determined by a fixed set of attributes. Hence, any instance that holds these attributes is a member of the concept. Classical logic or some of its extensions are applied for inferential tasks, and concept combination. This theory of concepts thus assigns for example membership truth values: an instance 'is' or 'is not' associated with a particular concept. However, it is well known that membership values are not binary but graded.[11] The second theory, so called prototype theory,[6] tries to conciliate this problem proposing that concepts are not defined by a fixed set of attributes, but instead by one or multiple prototypes that incorporate the most relevant properties. Each exemplar has a degree of membership and, if the membership is positive, a degree of typicality. The prototype has the maximum degree of typicality. Prototype theory, formulated in the language of fuzzy sets,[12] is more general than the classical theory of concepts because it complements the notion of membership with other notions such as typicality, similarity, representativeness, etc [a]. The problem of this theory is that prototypes not only vary from person to person or culture to culture, but also they can depend on the situation in which a concept is being elicited. The third theory, so called exemplar theory, assumes that a concept is defined by a list of stored entities that represent the current understanding of a certain agent concerning the concept. Depending on the situation, the structure of the concept will vary. For example, an instance such as 'dog' can be more or less a member of the concept '*Pet*,' depending on the situation in which the concept is elicited. Moreover, one can assess similarity estimations among the instances and apply logical techniques to infer the similarity to new instances, as well as to combine concepts. Hence, the notion of prototype is recovered in this theory, for one can refer to some instances as more typical. However, the mathematical framework of this theory requires a number of parameters that grows with the number of exemplars, and these parameters do not have a clear interpretation.[13]

Note that these theories are built upon different philosophical, psychological, and mathematical assumptions. Unfortunately, none of them has become broadly accepted. In fact, the only consensus is that none of them is satisfactory. In particular, researchers in cognitive science agree that, in view of the complexity of the cognitive phenomena, completely novel mathematical formulations are required.[14]

1.3. From Quantum Physics to Cognition

Quantum Physics emerged at the beginning of the 20th century as an explanation for some microscopic physical phenomena that could not be explained by the current classical theories. These include the radiation profile of black bodies at different temperature levels and the measurement of electric currents in materials exposed to light. These two phenomena are known as the black body problem, and the photoelectric effect respectively. Quantum physics was able to explain and incorporate these challenging phenomena into a unified representation of the microscopic realm that proposed an entirely different way of thinking.

---

[a]When we want to speak generally about the measurements applied to concepts we will refer to these measurements as semantic estimations

From a structural standpoint, the differences between classical and quantum physics are the following: In classical physics the outcomes we observe when performing experiments i) do exist as concrete states in the system prior to measurement, and ii) are deterministically obtained from measurement, while in quantum physics the outcomes we observe when performing experiments i) do not exist as actual but potential states prior to measurement, and ii) the measurement acts as a context which co-determines the observed outcomes in a non-deterministic manner. These differences are best illustrated by taking a closer look at the notions used to represent physical systems: While in classical physics systems are described by particles, and measurements do not influence the systems, in quantum physics systems are described by superpositions of complex waves, and measurements are means to observe these waves, but cannot avoid to influence the waves. This is known as the collapse of the wave function.

Interestingly, the structural aspects that characterize the differences between classical and quantum physics seem to resemble some of the most important challenges faced by current approaches to concept modeling. Namely, there is a body of phenomena in concept research outlining the problems of current concept theories. These problematic phenomena not only challenge the accuracy of traditional models, but also point to the possibly misleading philosophical principles of traditional models, and where their mathematical realizations seem to fail. The Geneva-Brussels group on foundations of quantum mechanics noticed that some of these problematic phenomena resemble the structural problems that classical physics was facing before the development of quantum physics.[15] This profound observation motivated the development of quantum-inspired models for concepts and their combinations. These models provided successful accounts of various conceptual phenomena, and induced the development of a quantum-inspired perspective about the nature of cognitive entities and their interactions.[16,17] The area of study that applies quantum notions and quantum mathematical tools to describe the cognitive realm has been named quantum cognition.[18,19]

In this article we elaborate an introduction to quantum cognitive modeling of concepts from the perspective of one of the foundational groups, leaded by Diederik Aerts, at the Centre Leo Apostel in Brussels. In section 2 we outline the conceptual phenomena that are on the one hand incompatible with classical representational tools in cognition, and that on the other hand motivate the application of quantum structures. In section 3 we elaborate on the history and development of concept modeling in the quantum cognition research program, emphasizing on the advances developed by the Brussels group, and finally in section 4 we elaborate on the most recent developments and on novel perspectives.

2. Structural Problems of Cognitive Modeling

In their attempt to formalize the theory of concepts, scholars have encountered a number of obstacles due to the flexible structure of concepts.[20] In this section, we describe these obstacles because, since they are generally used to either support or criticize theories of concepts that have been proposed, they highlight the fundamental requirements for a mathematical theory oriented to cognitive modeling. It is important to mention that these obstacles do not appear in isolation, but form an interrelated mixture common to most cognitive phenomena. Hence, these obstacles present a difficult landscape for the development of a formal theory of concepts.

2.1. Gradeness, Subjectivity and Vagueness

Concepts we reason with in our daily life cognitive activities are not sharply defined, neither in their boundaries nor in their implications.[21] For example consider a situation where we need to reason about the concept '*Pet*'. Suppose a person thinks about whether or not a 'dog' is a member of the concept '*Pet*.' We can ensure with almost complete certainty that the answer is going to be 'yes.' However, if the same question is thought for a 'snake' or a 'robot,' we cannot be that certain. In the same way, we usually assume that anything we think is a '*Bird*' is able to *fly*. However, a 'penguin' is a '*Bird*,' but penguins do not *fly*. Cognitive psychologists, mostly during the seventies and eighties, investigated this imprecise use of concepts in reasoning. They carried out a large number of experiments to reveal how people understand the meaning of concepts we use in daily life, and concluded the way people estimate the meaning of concepts cannot be modeled using binary systems ('yes'/'no'), but requires instead graded relations that reflect their structural vagueness.[6,8]

Therefore, a first requirement for a cognitive modeling framework is to handle the gradedness of the concepts we reason with. One important notion related to this issue is that of 'exemplar': given a concept, for example '*Light Bulb*', a specific instance of it such as 'halogen lamp' and 'led light bulb' are exemplars of '*Light Bulb*'. From here, we can use the notion of exemplar to explain the vagueness of this concept by assuming that concepts are represented by their possible exemplars an that exemplars have different degrees of membership to the concepts.

Note that exemplars of a concept can also be understood as concepts on its own. For example, 'cheap light-bulb' is an exemplar of the concept *'Light Bulb'*, but is also a concept (which we denote *'Cheap Light Bulb'*). Clearly, exemplars of *'Cheap Light Bulb'* are also exemplars of *'Light Bulb'*. The latter example illustrates that concepts have a recursive structure where each level in the recursion is identified with a concept and the next level is identified with its exemplars, which in turn can be thought of as concepts on their own as well. This recursive structure introduces the notion of level of abstraction for concepts.[8]

In addition, note that concepts can be alternatively represented in terms of some salient *'Properties'* instead of a set of exemplars. In principle, the number of properties of a concept is infinite.[22] However, we tend to identify a smaller set of relevant properties which are proposed to tease out the most useful information.[23] A supplementary notion in this vein is *'Prototype'*. A Prototype is an exemplar that embodies the most salient features of a concept. In the same way than exemplars have a graded structure with respect to the membership relation, properties can be more or less relevant or applicable to a concept, and thus a prototype is an exemplar whose most salient properties are representative of the concept in question.

Summarizing, the notion of concept is vaguely defined. First, there is no clarity on whether exemplars or properties (or both) best capture the notion of concept, and even in the case we assume a certain set of relations to define what a concept is, it is not clear how to grade such relations. This raises important philosophical and interpretational issues about the nature of concepts.[14]

2.2. *Semantic Estimations and their Relations*

The notion of degree of membership has been introduced to allow for intermediate evaluations of membership when we are unable to sharply estimate the whether an exemplar is or is not a member of a concept.[21]

However, note that membership estimation is not the only type of estimation that one can make about exemplars and concepts. Another common estimation is whether or not an exemplar is typical for a concept. Clearly this relation, know as *typicality*, is also graded, and although typicality and membership are known to be correlated, there is not a clear relation between them.[11]

Another estimation is known as *semantic similarity*.[24] This estimation is generally defined as a measure of how overlapped are the property structures associated to the concept and exemplar in question[b].

Several other notions have been defined to compare concepts, exemplars and properties. Among such notions we find *semantic relatedness, representativeness, relevance, elicitedness,* etc.[14] In this work we will refer to all these estimations by a general category called *semantic estimation*.

The problem of defining semantic estimations for the modeling of concepts is two-folded. On the one hand it is not clear what are the correct semantic estimations that should be considered for a theory of concepts,[25] and on the other hand the relations among semantic estimations is not clear.

2.3. *Context dependence of semantic estimations*

Context is roughly understood as 'the circumstances in which something occurs'.[26] In our case, 'something occurs' refers to a concepts that is being elicited by a human mind. Paradoxically, the notion of context is possibly more vague than the notion of concept itself. Namely, a total of more than 150 definitions have been proposed in different areas such as linguistics, cognitive science, psychology, and philosophy. Depending on the area of application, different aspects of what constitutes a context become the focus of the definition. For concept theories, context entail all the priors at the moment of eliciting a concept. Cognitive psychologists have elaborated multiple experiments to observe how these factors affect the semantic estimations of concepts. These experiments include exemplar membership,[6] typicality,[27,28] property relevance,[29] among others. The conclusion is that context radically affects the semantic estimations of concepts.

As an example of how this conclusion is obtained we review the results of an experiment reported in:[28] In the experiment, subjects rated different exemplars of the concept '*Hat*' assuming different contexts using a 7-point Likert scale. Subjects had a short training session to be sure that they understand the meaning of the word 'context'. The results of the experiment show that the typicality of exemplars 'baseball cap' and 'pylon' are 6.32 and 0.56 under the neutral context '*is a hat*', while for the context '*is not worn by a person*' their typicalities shift to 0.64 and 3.95 respectively. In addition, the correlation coefficient for the typicality estimations across all

---

[b]This estimation has been extensively used for other information structures such as documents, images, videos, ADN, etc.

exemplars in the experiment considering these two contexts is $p = -0.93$. This strong anticorrelation shows that the two contexts imply an opposite typicality structure for the exemplars considered in the experiments. Hence, we conclude that a change of context is even able to flip the structure of a semantic estimation for a given concept.[28]

2.4. Concept Combination is Non-compositional

Early investigations in concept theory are concerned with semantic estimations for one concept only. In a general setting, a cognitive situation might include multiple concepts forming aggregated structures.[30] For example, consider the concepts '*Fruit*' and '*Vegetable*.' They can be combined to form the concept '*Fruit Or Vegetable*'.[31] This concept combination uses the connective '*Or*,' that is formally defined in logic and probability. The question is here: Is it possible to apply the formal definition of the connective '*Or*' to build the structure of '*Fruit Or Vegetable*' from the structures of '*Fruit*' and '*Vegetable*'? Traditional approaches to cognitive phenomena assume that this question has a positive answer. This assumption is known as the principle of semantic compositionality,[32] and later called simply principle of compositionality. The principle of compositionality was first formulated with the aim to formalize how to assign meaning to a complex logical expression, and was later applied to linguistics,[33] and concept theory.[6] Cognitive psychologists have largely debated whether or not concepts are compositional.[14,34] They have performed several experiments measuring semantic estimations such as membership, typicality, and similarity, for concept combinations of diverse nature, including logical combinations such as '*Pet And Fish*', and '*Not Sport*',[35] and adjective-noun compounds such as '*Red Apple*',[29,36] among others. The evidence collected during two decades of research revealed that concept combinations are not compositional in general, at least in the sense suggested by classical logic, fuzzy logic, and probability theory.

One psychological situation that illustrates this is the so called 'borderline contradiction.' A borderline contradiction consist of a contradictory sentence that humans believe to be true. For example consider the sentence $S=$'*John Is Tall And John Is Not Tall.*' This sentence clearly entails a logical contradiction. However in,[37] an experiments in which participants were requested to estimate the truth value of the sentence S assuming that John's height is $h_1 = 5'4''$, $h_2 = 5'47''$, $h_3 = 5'11''$, $h_4 = 6'2''$, and $h_5 = 6'6''$ respectively,shows that participants rate S as 'true' in $14.5\%, 21.1\%, 44.6\%, 28.9\%$, and 5.3% of the cases respectively. Hence, a significant proportion of participants believes S is true for cases h_2, h_3 and h_4. These instances are called borderline contradiction cases. In consequence, this experiment proves that borderline contradiction occurs in human cognition.

For more general concept combinations, involving any two daily life concepts combined by conjunction or disjunction, the gradedness structure exhibits features that are even less obvious than what has been found in borderline contradiction research. For example, a large body of experimental evidence indicates that the membership and typicality of exemplars with respect to the conjunction of concepts is found to be larger than the memberships of at least one of the former concepts. This effect contradicts the principles of any logical or probabilistic theory, because the membership of an exemplar of a set cannot be larger than the membership of the exemplar with respect to the intersection of two sets.[38] For example, in[35] it is shown that the membership of the item *Coffee Table* with respect to concepts *Furniture, Household appliances* and their conjunction are 1, 0.15 and 0.38 respectively, and hence singly overextended, and the membership of the item *Tree house* with respect to concepts *Building, Dwelling,* and their conjunction are 0.5, 0.9 and 0.95 respectively, and hence doubly overextended.

Another famous example[c] states that 'guppy' is a typical '*Pet And Fish*' but neither a typical '*Pet*' nor a typical '*Fish*'.[39] In addition, estimations of the applicability of relevant properties of concepts and their conjunctions exhibit the same effect. For example, *talk* is not a relevant property for neither '*Pet*' or '*Bird*,' but it is for '*Pet and Bird*'.[40,41] Overextensions have also been observed in experiments considering negated concepts. For example, 'chess' is overextended with respect to the concepts '*Game*' and '*not Sport*' and their conjunction.[42] For the disjunction of two concepts, the analogous 'underextension' effect occurs correspondingly. Namely, the membership (typicality, property applicability) of an item with respect to the disjunction of two concepts is in general smaller than the membership of the item with respect to the former concepts in combination.[31] The following theorem from[18] summarizes the cases when a concept conjunction and a concept disjunction can be modeled using classical probability.

Theorem 2.1. *The membership weights $\mu(A)$, $\mu(B)$, and $\mu(A$ and $B)$ ($\mu(A$ or $B)$) of an item X with respect to concepts A and B and their conjunction (disjunction) 'A and B' ('A or B') can be modeled in a classical*

---

[c]Known as the guppy effect.

probabilistic space if and only if they satisfy the following inequalities

$$0 \leq \mu(A \text{ and } B) \leq \mu(A) \leq 1 \quad (0 \leq \mu(A) \leq \mu(A \text{ or } B) \leq 1)$$

$$0 \leq \mu(A \text{ and } B) \leq \mu(B) \leq 1 \quad (0 \leq \mu(B) \leq \mu(A \text{ or } B) \leq 1) \quad (1)$$

$$\mu(A) + \mu(B) - \mu(A \text{ and } B) \leq 1 \quad (0 \leq \mu(A) + \mu(B) - \mu(A \text{ or } B))$$

3. The Development of Quantum Models of Concepts

3.1. *The Discovery of Non-classical Structures in Cognition*

As we reviewed in section 2.1, semantic estimations of concepts are graded. Thus, when a phenomenon is observed a large number of times by some experimental procedure, this tendency is reflected in the relative frequencies, so called statistics, of the experimental outcomes. The statistics obtained from the observation of a phenomenon can be used to characterize its mathematical properties. For example, consider an urn with a large number of balls in it, and let us define the following questions $\mathcal{E}_1 =$'The ball is red', and $\mathcal{E}_2 =$'The ball is wooden.' The experimental situation consists of deciding one of this questions, extract a ball from the urn, and check whether the answer to the question is 'yes' or 'no.' Note that other questions $\mathcal{E}_1^c =$'The ball is not red,' $\mathcal{E}_2^c =$'The ball is not wooden,' $\mathcal{E}_1 \cap \mathcal{E}_2 =$'The ball is red and wooden,' $\mathcal{E}_1 \cup \mathcal{E}_2^c =$'The ball is red or not wooden,' etc., can be trivially defined using probability theory. A probabilistic model of the urn system must deliver a consistent description of the relative frequencies of the outcomes obtained for all these experimental situations. For example, let $P(red), P(wooden)$ and, $P(r\&w)$ be probabilities to obtain the answer 'yes' respectively for $\mathcal{E}_1, \mathcal{E}_2$ and $\mathcal{E}_1 \cap \mathcal{E}_2$. Note that in this situation, the following mathematical conditions must be satisfied[d]:

$$P(r\&w) \leq P(red)$$
$$P(r\&w) \leq P(wooden) \quad (2)$$
$$P(red) + P(wooden) - P(r\&w) \leq 1$$

The first two inequalities in (2) are trivial. The third inequality reflects that the probability of event $\mathcal{E}_1 \cup \mathcal{E}_2$ must be well defined, i.e. cannot be larger than the probability of the former events in the conjunction. Now, suppose that we repeat 100 times the experiment and we obtain that 60 balls are red, 75 balls are wooden and 32 are both red and wooden. Note that in this

---

[d]The results follows from theorem 2.1

case, the relative frequencies for $P(red), P(wooden)$ and, $P(r\&w)$ are 0.6, 0.75 and 0.32 respectively. However, note that the third condition in (2) is not hold by this statistical situation, as $0.6 + 0.75 - 0.32 > 1$. Hence, this example cannot occur for any real urn, as the proportion of red, wooden, and red and wooden balls chosen in the example entail a logical contradiction.[43]

The conditions (2) were first derived by George Boole to put forward particular examples of what can and cannot occur in the statistics of an experimental situation. These conditions were named conditions of possible experience.[44]

It is important to note that the conditions of possible experience cannot be violated when all properties can be jointly measured on a single sample. Therefore, when all measurements can be applied to a single sample, because the conditions of possible experience hold, we can ensure it is possible to build a classical probabilistic representation. However, not all systems allow all properties to be jointly measured in a single sample. Particularly most quantum systems require the measurement of incompatible properties, which are by definition measurements that cannot be measured in a single sample without non-deterministic disturbances.

When properties cannot be jointly measured on a single sample, we can still build a probabilistic model of the system from the marginal probabilities obtained from those measurements that can be jointly measured. However, in order for this probabilistic model to be classical, we require that the marginal probabilities we obtain do not lead to a violation of the conditions of possible experience. This is not a trivial matter, as the marginal probabilities of a system might seem consistent, but hide the violation of some condition of possible experience in an indirect way.[45] Hence, systems where measurements can be incompatible might deliver statistics that cannot be represented using classical probability theory.

Some scientists and philosophers, and remarkably among them the founding fathers of quantum mechanics[e], have suggested that the joint measurements are not always possible for cognitive phenomena. Therefore, non-classical probabilistic models, and particularly quantum probabilistic modeling, might be more realistic than classical probability.[47]

The first concrete proposal of a cognitive phenomena exhibiting non-classical probabilistic features was put forward in.[48] The example consists of an opinion poll that contains three questions, each question having only

---

[e]Although these remarks are not explicit they can be clearly understood in essays of Bohr about the nature of complementarity (see[46])

two possible answers ('yes' or 'no'). In particular, what brings the non-classicality to this situation is that it is assumed that some participants do not have a predefined answer to the questions in the opinion poll, but their answer is formed at the moment the question is posed. Note that if we try to draw an analogy between this cognitive situation and the urn example above, a participant 'forming its answer at the moment the question is posed' corresponds to 'a ball acquiring its color when the ball is extracted from the urn'. Clearly, balls do not acquire their color when they are extracted from an urn, but have a color before the experiment is performed. However, if we remind that quantum systems do acquire their properties when observed[f], it can be suggested that the formalism of quantum physics might be a sensible alternative to model cognitive phenomena. The opinion poll contained the following three questions:

(1) *Are you a proponent of the use of nuclear energy?* ('yes or 'no')
(2) *Do you think it would be a good idea to legalize soft-drugs?* ('yes' or 'no')
(3) *Do you think it is better for people to live in a capalistic system?* ('yes' or 'no')

Assuming the hypothetical case that for each question, 50% of the participants have answered 'yes', and only 30% of the total of persons where convinced of their choice before the question was actually posed (15% towards 'yes' and 15% towards 'no'), it was proven that it is not possible to provide a classical probabilistic account of the statistical situation. For reasons of space we do not elaborate on the mathematical details here, but for a detailed description of the incompatibility of this statistical situation with classical probability theory we refer to.[48]

3.2. *Quantum Modeling of Concepts*

The discovery of non-classical statistical situations explained in section 3.1 was followed by the application of the standard quantum formalism to the challenging conceptual phenomena reviewed in section 2. We first review the basics of how to model a quantum entity, and then show how by assuming that concepts are quantum entities we can successfully cope with the problems reviewed in section 2.

---

[f]This is known as the collapse of the wave function explained in section 1.3

3.2.1. *Quantum Modeling*

In quantum physics, the state of a quantum entity is described by a complex wave function. By complex we mean it assumes complex values. The complex wave function embodies the probability to find the entity in different configurations. In an abstract setting, this complex wave function corresponds to a vector of unit length. Such abstract setting is called the complex Hilbert space $\mathcal{H}$ of quantum mechanics, and is essentially the set of these (complex-valued) vectors[g]. Hence, each unit vector represents a possible state of the quantum entity under consideration.

Vectors are denoted using the bra-ket notation introduced by, one of the founding fathers of quantum mechanics, Paul Dirac. Dirac notation introduces two types of vectors: 'bra' vectors denoted by $\langle A|$, and 'ket' vectors denoted by $|A\rangle$. By convention, the state of a quantum entity is described by a 'ket' vector, hence the state of a concept $\mathcal{A}$ is represented by $|A\rangle$. The inner product of two vectors $|A\rangle$ and $|B\rangle$ is denoted as $\langle A|B\rangle$ and called a bra-ket, where the 'bra' vector $\langle A|$ is the complex conjugate of the 'ket' $|A\rangle$. The bra-ket notation is convenient in quantum mechanics to simplify formulas involving inner products.

Definition 3.1. A bra-ket is a complex inner product, and induces the norm $|| \cdot || = \sqrt{\langle \cdot | \cdot \rangle}$.

We say that $|A\rangle$ and $|B\rangle$ are orthogonal if and only if $\langle A|B\rangle = 0$. We denote it by $|A\rangle \perp |B\rangle$. Additionally, we say that $\langle A|B\rangle$ is the complex conjugate of $\langle B|A\rangle$. Therefore

$$\langle A|B\rangle^* = \langle A|B\rangle \tag{3}$$

The operation bra-ket $\langle \cdot | \cdot \rangle$ is linear in the ket and anti-linear in the bra. Therefore, for any $\alpha, \beta \in \mathbb{C}$, $\langle A|(\alpha|B\rangle + \beta|C\rangle) = \alpha\langle A|B\rangle + \beta\langle A|C\rangle$ and $(\alpha\langle A| + \beta\langle B|)|C\rangle = \alpha^*\langle A|C\rangle + \beta^*\langle B|C\rangle$.

Definition 3.2. Let $|A\rangle, |B\rangle \in \mathcal{H}$, and

$$\mathbf{M} : \mathcal{H} \to \mathcal{H},$$

$$|A\rangle \to \mathbf{M}|A\rangle.$$

---

[g]A Hilbert space is formally a vector space equipped with an inner product such that infinite sums of elements belong to the space (i.e. closed)

We say **M** is

(1) Linear: for $\alpha, \beta \in \mathbb{C}$ we have $\mathbf{M}(\alpha|A\rangle + \beta|B\rangle) = \alpha\mathbf{M}|A\rangle + \beta\mathbf{M}|B\rangle$
(2) Hermitian: $\langle A|\mathbf{M}|B\rangle = \langle B|\mathbf{M}|A\rangle$
(3) Idempotent: $\mathbf{M} \cdot \mathbf{M} = \mathbf{M}$

We say **M** is an orthogonal projector if it is linear, Hermitian, and idempotent.

Measurable quantities, known as observables in quantum physics, correspond to Hermitian linear functions on the Hilbert space. In particular, any two-valued observables corresponds to an orthogonal projector. Therefore, estimations such as the membership of an instance with respect to a concept, e.g. 'apple' is a member of '*Fruit*' versus 'apple' is not a member of '*Fruit*,' can be modeled by means of orthogonal projections on the Hilbert space. Note that a projector **M** can also be characterized by the subspace $\mathcal{H}_{\mathbf{M}} = \{\mathbf{M}|v\rangle, |v\rangle \in \mathcal{H}\}$ into which vectors are projected. Quantum theory associates probability to events by measuring the extent to which a vector lies in the subspace determined by the measurement operator. This is mathematically formalized by the Born rule of probability:

Definition 3.3. Let $|A\rangle$ be a state of an entity $\mathcal{A}$, and **M** be an orthogonal projector. The probability of an answer 'yes' to the question measured by **M** is given by

$$\mu(A) = \langle A|\mathbf{M}|A\rangle \tag{4}$$

Quantum theory has a particular form to model composite systems. Namely, consider a system $\mathcal{A}$ is formed by the composition of two sub-systems $\mathcal{A}_1$ and $\mathcal{A}_2$. Note that it is not a trivial question to decide how are the superposed states of each entity. In fact we have that, on the one hand, each sub-system should exist in its own superposed state, but on the other hand, the system as a whole must exist in a superposed state as well. Hence, although a quantum system exists physically as two separated entities, it behaves in some cases as a non-decomposable entity. These non-decomposable situations of the system are called entangled states, and are one of the most important cases of modern applications of quantum physics.[49] Consider for example two quantum systems $\mathcal{A}_1, \mathcal{A}_2$ forming a composite quantum system $\mathcal{A}$. If we model these entities separately, each entity exist in a superposed state $|A_i\rangle, i = 1, 2$. Hence, we might think that the ordered pair $|A\rangle = (|A_1\rangle, |A_2\rangle)$ is a proper representation of the composite system $\mathcal{A}$. In quantum mechanics however, composite systems cannot always be represented by an ordered pair of sub-systems. Formally, composition of quantum entities are not represented by vectors in the Cartesian product $\mathcal{H} \times \mathcal{H}$, but rather in the tensor product space[h] $\mathcal{H} \otimes \mathcal{H}$. A vector $|A\rangle$ in a tensor space is in general represented as a superposition of tensor products. For example, suppose that $\{|A_i\rangle\}_1^n$ is a basis of $\mathcal{H}$. Then $|C\rangle = \sum_{i,j}^n c_{ij} |A_i\rangle \otimes |A_j\rangle$.

Definition 3.4. Let $|A\rangle \in \mathcal{H} \otimes \mathcal{H}$. If $|A\rangle$ can be factorized as $|A\rangle = |A_1'\rangle \otimes |A_2'\rangle$, where $|A_1'\rangle \in \mathcal{H} \otimes \mathbf{1}$, and $|A_2'\rangle \in \mathbf{1} \otimes \mathcal{H}$, we say the $|A\rangle$ is a 'separable vector.' Otherwise, $|A\rangle$ is a 'non-separable vector', or also called entangled state.

Separable vectors can be represented as ordered pairs. This implies that when a measurement is performed in one of the sub-systems, the wave function collapse induced by the measurement only occurs at the measured sub-system, while the other sub-system remains in its original superposed state. Non-separable vectors however, cannot be factorized as a single tensor product. Then, they cannot be represented as an ordered pair. When a measurement is performed on a non-separable vector, the collapse of the wave function induced by the measurement will affect both sub-systems because they cannot be decomposed as an ordered pair. This is known as quantum entanglement. Namely, performing a measurement on a sub-system that is in an entangled state with other systems, induces a collapse in all the entangled sub-systems rather than in the measured sub-system only.

In the remaining of this section we will show how by assuming that a concept is a quantum entity it is possible to develop models for concept combinations that cope with the problems described in section 2.4.

3.3. Quantum Modeling of Concepts and their Combinations

A quantum model of a concept assumes that concepts exist in superposed states, and a semantic estimation corresponds to an orthogonal projector. Analogous to quantum physics, the probability to obtain a certain outcome of the semantic estimation is obtained by applying the Born rule (4).

In particular, consider an instance X, and the membership of X with respect to a certain concept to be represented by an orthogonal projection

---

[h]The remaining of this section requires that the reader is familiar with tensor product spaces, for an introduction see[50]

M. The probability $\mu(A)$ for a test subject to decide 'in favor of membership' of item X with respect to concepts $\mathcal{A}$ and $\mathcal{B}$ are

$$\mu(A) = \langle A|M|A\rangle \text{ and } \mu(B) = \langle B|M|B\rangle \tag{5}$$

respectively.

Next, consider a situation where two concepts $\mathcal{A}$ and $\mathcal{B}$ are elicited in conjunction. In this case, the conjunction of the concepts represented by '$\mathcal{AB}$' is modeled by means of the normalized superposition state $\frac{1}{\sqrt{2}}(|A\rangle + |B\rangle)$, for simplicity we also suppose that $|A\rangle$ and $|B\rangle$ are orthogonal. Hence $\langle A|B\rangle = 0$.

The non-classical membership estimates obtained for concept conjunction reviewed in section 2.4, and presented in the literature in,[35] can be modeled in accordance with the quantum rules by equation (5) for $\mu(A)$ and $\mu(B)$. Namely, the probability for the membership $\mu(AB)$ of item X with respect to the conjunction $\mathcal{AB}$ is given by

$$\mu(AB) = \frac{1}{2}\langle A + B|M|A + B\rangle$$
$$= \frac{1}{2}\left(\langle A|M|A\rangle + \langle B|M|B\rangle + \langle A|M|B\rangle\langle A|M|B\rangle\right) \tag{6}$$

Recalling that $\langle A|M|B\rangle$ is the conjugate of $\langle B|M|A\rangle$, then $\langle A|M|B\rangle + \langle B|M|A\rangle$ corresponds to two times the real part of $\langle A|M|B\rangle$. Then,

$$\mu(AB) = \frac{(\mu(A) + \mu(B))}{2} + \Re\langle A|M|B\rangle \tag{7}$$

Therefore, the membership probability $\mu(AB)$ corresponds to the sum of the average of $\mu(A)$ and $\mu(B)$, plus an interference term $\Re\langle A|M|B\rangle$. The interference term can assume positive or negative values, depending on the particular choices of the vectors representing the conceptual state. Positive interference implies a membership $\mu(AB)$ larger than average, and it normally accounts for overextended estimations of membership. Negative interference implies a membership $\mu(AB)$ smaller than average, and it normally accounts for underextended estimations of membership. It is interesting to note that in the absence of interference, i.e. when $\langle A|M|B\rangle = 0$, the probability formula is reduced to the average of the former probabilities, and hence we can say in the quantum model a concept combination is singly overextended and singly underextended by default.[51] Interestingly, the same formula can be used to model other semantic estimations (for example, see[18] for the modeling of typicality), other conceptual phenomena such as non-classical disjunction,[52] Borderline contradictions,[53] and other controversial phenomena in cognitive science such as the conjunction fallacy,[54] the Ellsberg and Machina paradoxes,[55] and others (for a review, see[47]).

3.4. *An Extended Quantum Model of Concept Combination based on Modes of Thought*

In section 3.3, we explained how the quantum approach to concepts is able to represent non-classical conjunctions. This quantum model is based on formula (7), whose mathematical form involves an average term plus an interference term. It is possible to prove that the absolute value of the interference term is bounded by $\sqrt{\mu(A)\mu(B)}$.[18] This implies that formula (7) cannot always account for classical conjunctions. For example, let $\mu(A) = 0.1, \mu(B) = 0.8$ and $\mu(AB) = 0.05$. This is a classical type of reasoning as conditions of theorem 2.1 are hold. However, for the quantum model the membership of conjunction cannot hence be smaller than $0.45 - 0.2828 = 0.1672 > 0.05 = \mu(AB)$. Hence, the quantum model presented in section 3.3, although explains the deviations from classical logical reasoning, it does not incorporate classical logical reasoning in general.

Aerts proposed in[18] that the classical logical structure of reasoning characterized in theorem 2.1 can be obtained applying the tensor product of Hilbert spaces to model concepts as compound entities rather than superposed entities. Namely, given the membership of an exemplar with respect to concepts $\mathcal{A}$, $\mathcal{B}$ and their combination, the simplest tensor product model represents the states and measurements of the former concepts of the combination. Namely, a unitary vector $|A\rangle \in \mathcal{H}$ represents the state of concept $\mathcal{A}$, a unitary vector $|B\rangle \in \mathcal{H}$ represents the state of concept $\mathcal{B}$, a projector $\mathbf{M} : \mathcal{H} \to \mathcal{H}$ represents the membership measurement. Hence,

$$\begin{aligned} \mu(A) &= \langle A|\mathbf{M}|A\rangle \\ \mu(B) &= \langle B|\mathbf{M}|B\rangle, \end{aligned} \tag{8}$$

and the state representing the combination of concepts is given by the tensor product of $|A\rangle$ and $|B\rangle$. Therefore, $|AB\rangle = |A\rangle \otimes |B\rangle$. Note that the membership operator $\mathbf{M}$ can be extended to the tensor product $\mathcal{H} \otimes \mathcal{H}$ by the operators $\mathbf{M}^A = \mathbf{M} \otimes 1$ and $\mathbf{M}^B = 1 \otimes \mathbf{M}$ respectively. Indeed, note that

$$\begin{aligned} \langle AB|\mathbf{M}^A|AB\rangle &= ((\langle A| \otimes \langle B|)\mathbf{M} \otimes 1(|A\rangle \otimes |B\rangle) \\ &= \langle A|\mathbf{M}|A\rangle \otimes \langle B|1|B\rangle = \mu(A) \cdot 1 \\ \langle AB|\mathbf{M}^B|AB\rangle &= ((\langle A| \otimes \langle B|)1 \otimes \mathbf{M}(|A\rangle \otimes |B\rangle) \\ &= \langle A|1|A\rangle \otimes \langle B|\mathbf{M}|B\rangle = \mu(B) \cdot 1 \end{aligned} \tag{9}$$

If we want to measure the membership of an exemplar with respect to the conjunction of concepts $\mathcal{A}$ and $\mathcal{B}$, we must determine whether the exemplar is a member of both concepts simultaneously. Then, the membership operator for the conjunction of two concepts is given in this case by $\mathbf{M}^\wedge = \mathbf{M} \otimes \mathbf{M}$. Applying this operator to the state $|AB\rangle$ we compute the membership of an exemplar with respect to the conjunction of concepts $\mathcal{A}$ and $\mathcal{B}$ and obtain

$$\langle AB|\mathbf{M}^\wedge|AB\rangle = ((\langle A| \otimes \langle B|)\mathbf{M} \otimes \mathbf{M}(|A\rangle \otimes |B\rangle))$$
$$= \langle A|\mathbf{M}|A\rangle \otimes \langle B|\mathbf{M}|B\rangle = \mu(A)\mu(B) \tag{10}$$

Similarly, if we want to measure the membership of the exemplar with respect to the disjunction of concepts $\mathcal{A}$ and $\mathcal{B}$, we introduce the operator $\mathbf{M}^\vee = \mathbf{M} \otimes \mathbf{M} + \mathbf{M} \otimes (1 - \mathbf{M}) + (1 - \mathbf{M}) \otimes \mathbf{M}$ which is shortly written as $1 \otimes 1 - (1 - \mathbf{M}) \otimes (1 - \mathbf{M})$. Hence, the membership of the exemplar with respect to the disjunction of the concepts $\mathcal{A}$ and $\mathcal{B}$ is given by

$$\langle AB|\mathbf{M}^\wedge|AB\rangle$$
$$= ((\langle A| \otimes \langle B|)\mathbf{M} \otimes \mathbf{M} + \mathbf{M} \otimes (1 - \mathbf{M}) + (1 - \mathbf{M}) \otimes \mathbf{M}(|A\rangle \otimes |B\rangle))$$
$$= \langle A|\mathbf{M}|A\rangle\langle B|\mathbf{M}|B\rangle + \langle A|\mathbf{M}|A\rangle\langle B|1 - \mathbf{M}|B\rangle + \langle A|1 - \mathbf{M}|A\rangle\langle B|\mathbf{M}|B\rangle$$
$$= \mu(A)\mu(B) + \mu(A)(1 - \mu(B)) + (1 - \mu(A))\mu(B)$$
$$= \mu(A) + \mu(B) - \mu(A)\mu(B) \tag{11}$$

Note that the formulas for the membership of the conjunction and disjunction of two concepts, given by (10) and (11) respectively, are equivalent to a classical probability formulas where the membership of concepts $\mathcal{A}$ and $\mathcal{B}$ are assumed to be a conjunction and disjunction of independent events. Namely, equations (10) and (11) refer to the membership estimations of an exemplar with respect to the conjunction and disjunction as classical probabilistic conjunction and disjunction of independent events. Interpreting this model, we assume that $\mathcal{A}$ and $\mathcal{B}$ are concepts whose meanings are independent from each other. Hence, the classical probabilistic rules governing independent events are applied to estimate the membership of concept conjunction and disjunction. In this sense, the tensor product model resembles a classical probabilistic version for concept combination. The resemblance between this simple tensor product model and classical probabilistic formulas can be formalized in a more general tensor product model that assumes the state of the conceptual situation not to be the tensor product $|A\rangle \otimes |B\rangle$, but a general tensor $|C\rangle \in \mathcal{H} \otimes \mathcal{H}$. Note that $|A\rangle \otimes |B\rangle$ is a separable tensor, but $|C\rangle$ is not necessarily separable, and hence can be an entangled state. For this general tensor model we can prove the following result[56]

Theorem 3.1. *Let* $\mu = \{\mu(A), \mu(B), \text{ and } \mu(AB)\}$ *be a triplet denoting the membership of concepts* $\mathcal{A}$, $\mathcal{B}$ *and their conjunction. The triplet* μ *can be modeled in a classical probabilistic space (see theorem 2.1) if and only if the triplet admits a representation in a tensor product space.*

This theorem proves that there are two quantum approaches to model conceptual combination. One approach is based on superposed states, and models non-classical behavior such as overextensions and underextensions. The other approach is based on tensor products, and models classical probabilistic behavior. These two approaches refer to two completely different types of reasoning. Hence, the question is how to reconcile these two approaches so we can have a unified quantum interpretation of the phenomena of concepts.

To answer this question, note that the human mind, when dealing with conjunctions and disjunctions of concepts, can operate in two profoundly different ways.[18,42] Namely, concepts are conceived in two different forms: In the first form a concept combination is viewed as a single entity, and in the second form the concept combination is viewed as two different entities being combined. To make this point clear, consider the concepts '*Fruit*' and '*Vegetable*' and suppose we want to estimate the membership of the instance 'olive' with respect to the conjunction of these two concepts. Now, the first type of reasoning considers the concept '*Fruit and Vegetable*' as a single concept, and then 'olive' would be estimated hence with respect to the meaning of '*Fruit and Vegetable*'. To do so, we conceive one olive in our mind, and ask to ourselves whether this olive is a '*Fruit and Vegetable*'. In the second type of reasoning, '*Fruit and Vegetable*' as a combination of concepts '*Fruit*' and '*Vegetable*', and hence 'olive' would be estimated separately with respect to the meaning of '*Fruit*', and with respect to the meaning of '*Vegetable*'. After these two steps, the logical connective '*and*' is applied to finalize the estimation of 'olive' with respect to the concept combination. The two modes of thought have been named '*Emergent*' and '*Logical*' modes of thought respectively. These names have been chosen because the first mode operates beyond logic and combines the concept in a way that a new entity is created out of the two concepts, i.e. '*Fruit and Vegetable*' is an emergent concept created from '*Fruit*' and '*Vegetable*', while the second mode operates within what is understood by classical logic to be a concept combination, i.e. '*Fruit and Vegetable*' is a compositional concept created from '*Fruit*' and '*Vegetable*'.

Surprisingly, there exist a mathematical construction in quantum physics that reconciles these two modes of thought in a very simple way. Namely, we must assume that human mind operates in none of the two modes of thought, but in a superposition of them.[18] This mathematical construction is called Fock space (for a review on concept combination models in Fock spaces see[57]). The Fock space model of concept combination provides a much more powerful model than the interference model of section 3.3, or than a classical probabilistic model. The state of the concept combination is given by

$$|AB\rangle = me^{i\theta_1}\left(\frac{|A\rangle + |B\rangle}{\sqrt{2}}\right) + \sqrt{1 - m^2}e^{i\theta_2}|C\rangle$$

where $m \in [0, 1]$.

This formula to represent a state of a concept combination based in the Fock space model is based on the superposition of the two modes of thought explained above. Namely, the first term entails the emergent mode of thought, while the second term entails the logical mode of thought. The parameter m determines what is the weight of each mode of thought in the conceptual combination. Indeed, these emergent mode of thought of model in section 3.3 is recovered choosing $m = 1$, and the classical mode of thought is recovered choosing $m = 0$.

It is important to notice that the Fock space model of concept combination proposes a radically new interpretation to how reasoning about conjunctions and disjunctions occurs. Contrary to all other models applied to reasoning, it does not propose fundamental principles that explain how people reason. Instead, it proposes the existence of multiple fundamental reasoning processes, but assumes that all these reasoning processes can exist in superposed states in the human mind.

In order to test the plausibility of this new interpretation to reasoning about concept combinations, an experiment that extends the scope of concept combination to conjunctions and disjunctions was proposed. Namely, experimental data about estimations of concepts, their negations and conjunctions using their negations, was collected and reported in.[42] As an example, the memebership of the instance 'olive' is measured with respect to the concepts $A = $'Fruit' and $B = $'Vegetable', their negations $\bar{A} = $'not Fruit', $\bar{B} = $'not Vegetable', and the possible combinations with these four concepts: $AB = $'Fruit and Vegetable', $A\bar{B} = $'Fruit and not Vegetable', $\bar{A}B = $'not Fruit and Vegetable', and $\bar{A}\bar{B} = $'not Fruit and not Vegetable'. In this example, the memberships estimations gave the following values:

$$\mu(A) = 0.53, \; \mu(\bar{A}) = 0.47$$

$$\mu(B) = 0.63, \; \mu(\bar{B}) = 0.44 \tag{12}$$

$$\mu(AB) = 0.65, \; \mu(A\bar{B}) = 0.34, \; \mu(\bar{A}B) = 0.51, \; \mu(\bar{A}\bar{B}) = 0.36$$

Note that AB is a doubly overextended estimation, $\bar{A}B$ is a singly overextended estimation, and $A\bar{B}$ and $\bar{A}\bar{B}$ are not overextended, and hold the conditions of theorem 2.1. Then, we confirm that some conceptual combinations lead to classical reasoning behavior but other combinations do not. In order for a pair of concepts, their negations and conjunctions to be put in correspondence with classical reasoning, we need that all the conjunctions are classical. This is formalized mathematically in the following theorem.

Theorem 3.2. *If the membership weights* $\mu(A), \mu(B), \mu(\bar{A}), \mu(\bar{B}), \mu(AB),$ $\mu(A\bar{B}), \mu(\bar{A}B)$ *and* $\mu(\bar{A}\bar{B})$ *of an exemplar* x *with respect to the concepts* A, B, *their negations* $\bar{A}$, *and* $\bar{B}$, *and the conjunctions* $AB, \bar{A}B, A\bar{B},$ *and* $\bar{A}\bar{B}$, *are all contained in the interval* $[0, 1]$, *they can be represented in a classical probabilitic space if and only if they satisfy the following conditions.*

$$I_A = \mu(A) - \mu(AB) + \mu(A\bar{B}) = 0 \tag{13}$$

$$I_B = \mu(B) - \mu(AB) + \mu(\bar{A}B) = 0 \tag{14}$$

$$I_{\bar{A}} = \mu(\bar{A}) - \mu(\bar{A}\bar{B}) + \mu(\bar{A}B) = 0 \tag{15}$$

$$I_{\bar{B}} = \mu(\bar{B}) - \mu(\bar{A}\bar{B}) + \mu(A\bar{B}) = 0 \tag{16}$$

$$I_{AB\bar{A}\bar{B}} = 1 - \mu(AB) + \mu(A\bar{B}) + \mu(\bar{A}B) + \mu(\bar{A}\bar{B}) = 0 \tag{17}$$

To clarify the application of this theorem, note that in our example we have that

$$I_A = -0.46, \; I_B = -0.53, \; I_{\bar{A}} = 0.41, \; I_{\bar{B}} = -0.26, I_{AB\bar{A}\bar{B}} = -0.86$$

Hence, the example above cannot be represented in a classical probabilistic space. However the Fock space model provides a faithful representation of the data. We do not elaborate on the mathematical details of the model but refer to.[42]

The analysis of this experimental situation provide two important results. First, it proved that the Fock space model of conceptual conjunctions and disjunctions can be naturally extended to model concept combinations involving negations of concepts. This means that the mathematical formalism, initially developed for conjunctions and disjunctions only, can be applied to other types of combinations and hence can be considered as a solid approach to model conceptual combinations, at least within the extent of logical connectives.[42]

Secondly, the conditions established in theorem 3.2 are violated in a strongly regular manner that is consistent with the structure of the Fock space model. Indeed, the 95%-confidence interval has been computed for the parameters of theorem 3.2, and obtained interval $(-0.51, -0.33)$ for I_A, interval $(-0.42, -0.28)$ for $I_{\bar{A}}$, interval $(-0.52, -0.34)$ for I_B, interval $(-0.40, -0.26)$ for $I_{\bar{B}}$, and interval $(-0.97, -0.64)$ for $I_{AB\bar{A}\bar{B}}$. Moreover, these values are consistent with the mathematical predictions of the Fock space model when the two modes of thought are present and the emergent thought is dominant.[58]

From an interpretational point of view, these two results confirm that the superposition of modes of thought is a sensible approach to the phenomena of concepts and their combinations, as it provides simple but concrete principles to build a concept combination from single concepts that models both the classical and non-classical membership estimations. Moreover, the regularities of parameters I_A, $I_{\bar{A}}$, I_B, $I_{\bar{B}}$, and $I_{AB\bar{A}\bar{B}}$ observed in[42] suggest that the deviations from the classical laws of thinking are not at all random, but quite well structured. This confronts the well established laws of thinking developed by logicians since the times of Aristotle, and known in modern logic as the De Morgan laws.[59]

4. Perspectives in the Quantum Modeling of Concepts

In section 3 we have illustrated how to perform quantum models of concepts and their combinations, and shown that quantum cognition provides a better modeling language than their classical and fuzzy set theoretical predecessors.

The latter results stimulated further investigations about the structural relations between concepts and quantum entities. In the remaining of this section, we will elaborate on two of these investigations carried out by the Brussels group in quantum cognition.

4.1. *Indistinguishability in Concept Combinations*

One of the most profound differences between quantum and classical physics is how identical particles behave statistically[i]. Classical particles are distinguishable because every particle can be identified with a unique position

---

[i]By 'behave statistically' we mean the proportion of particles that hold a certain combination of properties when a large collection of particles is considered

in space. On the contrary, quantum particles are indistinguishable. In fact, since quantum particles exist in superposed states prior to measurement, they cannot be associated with a particular position in space (or any other property). It is only after a measurement that particles acquire a particular position in space (or a particular property of interest), but prior to measurement they cannot be distinguished.

The key element to comprehend the difference between the statistics of distinguishable and non-distinguishable entities is that the exchange on two entities in the description matters for the former case while for the latter it does not. For example, suppose we define an entity $\mathcal{A}$ such that it can exist in two possible states (instantiations) s_1 and s_2 only. If we consider two entities $\mathcal{A}$ that are identical, i.e. two copies of $\mathcal{A}$, we can describe the state of the system differently depending whether we assume the entities are distinguishable or non-distinguishable. For the distinguishable case, we can write the state of the pair of identical entities using an ordered pair (x, y), where x denotes the state of one of the entities, and y denotes the state of the other. Note that it is important that each element of the ordered pair always refer to the same entity, hence we can say that the entity we refer on the first element of the ordered pair is our first entity, and the entity we refer by the second element of the ordered pair is our second entity. It is clear that there are four possible states in this case. Namely (s_1, s_1), (s_1, s_2), (s_2, s_1), and (s_2, s_2). Now suppose that each of these two states represent some quantitative property a such as energy or mass for a physical entity, or membership, typicality, etc. for a conceptual case. For simplicity let a_1 the value of a on state s_1 and a_2 the value of this quantity for state s_2. We can easily note that the total value for each of the four states are $2a_1, a_1 + a_2, a_1 + a_2$ and $2a_2$ respectively. To finalize our construction, assuming that the two states have the same probability to be elicited, we conclude that when observing the value of property a for a pair of entities, we find that the value $2a_1$ is equally likely than the value $2a_2$, and that $a_1 + a_2$ is twice as likely than $2a_1$ and $2a_2$. This is because two out of the four possible configurations have a quantity a equal to $a_1 + a_2$, while the other two quantities occur for only one of the possible configurations. For the non-distinguishable case, the state of the pair of concepts cannot be represented as an ordered pair. Hence, we cannot make explicit what entity is on what state. In particular, there is no distinction between the states (s_1, s_2) and $(s_2, s1)$. Both states are the same state for non-distinguishable particles. This is the key feature of non-distinguishable entities, and is what imply that they behave statistically different. Namely, it is therefore equally likely to observe the quantity a having a value $2a_1$, $a_1 + a_2$, and $2a_2$ for the case of two non-distinguishable entities. This might seem strange as our intuition tells us that there should always be an underlying mechanism to represent pairs of entities as ordered pairs. However, quantum mechanics tells us this is not always possible, and hence we must embrace the fact that nature does not always reveal itself in terms of distinguishable entities.

To make this construction more general, we can assume that there is a certain natural preference for $\mathcal{A}$ to be in one of the two states. We can model this by saying that there is a probability p for $\mathcal{A}$ to be in state s_1 and a probability $1-p$ to be in state s_2. Moreover, we can also assume that we have more than two entities, let's say n entities. These assumptions can be used to conclude that a binomial distribution describes the statistical behavior for the case of distinguishable entities, and that a linear distribution describes the statistical behavior for the case of non-distinguishable entities.

If combining concepts resembles the combination of quantum entities, one can wonder whether the analogy between concepts as primary elements of human thought and quantum particles as primary elements of matter is even deeper than what appears in the situations of combinations using logical connectives. In particular, one of the most important aspects where quantum particles exhibit their particular nature is by the way they statistically behave. In particular, a collection of identical and distinguishable entities (e.g. a collection of dices) are statistically described by the well-known Maxwell-Boltzmann (MB) distribution, while a collection of identical quantum entities are described by the Bose-Einstein (BE) distribution for quantum particles with integer spin (e.g. a collection of photons), and by the Fermi-Dirac (FD) distribution for quantum particles with semi-integer spin (e.g. a collection of electrons). These two quantum statistical distribution are primarily consequence of the non-distinguishable nature of quantum entities. Thus, the distinguishability of a collection of identical entities is revealed by the type of statistical distribution they obey in an experimental setting.

Inspired by this fact, in a collaborative work where one of the authors participated,[60] an experiment tested how collections of identical concepts are elicited by the human mind, with the aim of identifying whether or not conceptual entities came to existence following the statistics of classical or quantum particles. Interestingly, strong evidence showed that in a number of cases they behave as non-distinguishable entities. Particularly, the BE statistics seem to explain better the statistics of concept elicitation than the MB distribution.

Before explaining the experiment, consider for example the linguistic expression 'eleven animals.' This expression can be viewed as the combination of concepts '*Eleven*' and '*Animals*' into '*Eleven Animals.*' If we assume that each of these animals can be either a cat or a dog, we obtain twelve possible instantiations: the first case is 'zero cats and eleven dogs', the second case is 'one cat and ten dogs', and successively until the twelfth case 'eleven cats and zero dogs'. If the concept '*Eleven Animals*' elicits an abstract idea of 'eleven animals', they will not be understood as distinguishable from each other. Instead, if the concept '*Eleven Animals*' elicits a concrete idea of 'eleven animals[j]', they will be understood as distinguishable from each other.

In the experimental setup we obtained the statistical distribution of the possible states of various concept combinations such as 'eleven animals', 'nine humans' (instantiations corresponded to man and women), 'seven expressions of affection' (instantiations were hugs and kisses), and others, by asking 88 participants to rate how likely is the happening of each of the possible instantiations for the given collection of concepts.

We fitted the experimental distributions for both MB and BE statistics, and used various tests[k] to determine which of the two distributions provides a better account for the experimental distributions. The results showed that consistently for concepts related to emotions the statistical distributions behave as indistinguishable (quantum) entities while concepts related to objectual or physical instantiations behave as distinguishable (classical) entities.

4.2. *Psychological Evidence of Conceptual Entanglement*

Quantum entanglement is a phenomena that manifests 'connections' between two particles regardless of their distance (see section 3.2), and has been shown to be useful to develop a number of novel technologies[l]. These connections between two entangled particles can be verified by computing the correlations of certain measurements such as (spin or polarization) applied to them. In particular, in the existence of entanglement between two particles, the correlations in their measurements are incompatible with classical probabilistic models.[49]

---

[j] As objects existing in a particular space and time
[k] Bayesian Information Criterion and R^2 goodness of fit.
[l] We will not elaborate on the technicalities of entanglement in this section for reasons of space, but refer to[49] for details

From a mathematical perspective, the non-classicality of quantum particles shows that there are tests to identify whether a statistical situation can be described by a classical probabilistic model, or it needs a quantum probabilistic model involving quantum entanglement instead. The constrains that characterize the need of entanglement to describe a statistical situation are known as Bell-like inequalities[m]. We will not elaborate on the complex mathematical aspects involving Bell-like inequalities, but remark that Bell-like inequalities are in the same spirit than the conditions of possible experience presented in section 3.1.

The fact that entangled states entail non-trivial correlations, has been putted in correspondence with the, perhaps less formal assumption, that the meaning of certain concept combinations exhibit non-compositional connections. Fortunately, although the latter assumption does not seems formal at all, it can be tested by measuring the correlations in psychological experiments where participants estimate the likelihood of instantiations for certain concepts and their combinations.

For example in,[61] a cognitive experiment confirmed that the correlations of such estimates violate a special type of Bell-like inequality. Namely, let the entities $\mathcal{A}$ and $\mathcal{B}$ refer to the concepts *Animal* and *Acts*, respectively. Let $\mathbf{M}_A$, and $\mathbf{M}_{A'}$ be two measurements for concept $\mathcal{A}$, and $\mathbf{M}_B$ and $\mathbf{M}_{B'}$ be two measurements for concept $\mathcal{B}$. The outcomes of these measurements are

$$
\begin{aligned}
\mathbf{M}_A &= \{A_1 = \text{`horse'}, A_2 = \text{`bear'}\} \\
\mathbf{M}_{A'} &= \{A_1' = \text{`tiger'}, A_2' = \text{`cat'}\} \\
\mathbf{M}_B &= \{B_1 = \text{`growls'}, B_2 = \text{`whinnies'}\} \\
\mathbf{M}_{B'} &= \{B_1' = \text{`snorts'}, B_2' = \text{`meows'}\}.
\end{aligned}
\tag{18}
$$

Next, a psychological experiment where 81 participants estimate the likelihood that the different combinations are instantiations of the concepts $\mathcal{A}$, $\mathcal{B}$, and the conceptual combination '*The Animal Acts*,' with respect to outcomes of measurements $\mathbf{M}_A\mathbf{M}_B$, $\mathbf{M}_A\mathbf{M}_{B'}$, $\mathbf{M}_{A'}\mathbf{M}_B$, and $\mathbf{M}_{A'}\mathbf{M}_{B'}$ were performed. The expected values of these joint observables were calculated and used to estimate the value of a Bell-like inequality known as CHSH inequality. From the statistics of the experiment, it is concluded that the correlations observed between the concepts $\mathcal{A}$ and $\mathcal{B}$ are non-classical, and that in fact concepts $\mathcal{A}$ and $\mathcal{B}$ are entangled in the experiment. Several

---

[m]Because the first mathematical formulation of this was developed by John Bell

other tests have been carried in psychological experiments confirming the entanglement of conceptual entities.[62,63]

5. Conclusion and Perspectives

We have reviewed some of the most fundamental problems for the modeling of concepts, and shown how the quantum cognition approach handles these problems. In particular, the vagueness of concepts can be modeled using the notion of superposed state, the contextuality of semantic estimations is modeled using the quantum notion of measurement and the outcome probabilities are obtained using the Born rule, and for the non-compositionality of concept combinations we showed that a model that combines classical reasoning and quantum reasoning (modeled using entangled and superposed states respectively) is able to overcome classical and fuzzy theoretical modeling of concept combinations. Moreover, we have shown that two quantum features, indistinguishability and entanglement, reveal non-classical features in the statistics of experiments involving concept combinations.

Although we have followed as a connecting line the contributions of the Brussels group leaded by Diederik Aerts, we do not want to subtract merit to other groups that have been fundamental in the establishment of quantum cognition as a research field. Namely, the development of a quantum cognitive perspective has been fostered not only by the Brussels group at the Centre Leo Apostel (CLEA), but also by other contemporary researchers in the early 2000's.[64,65] These investigations attracted the attention of a number of research groups in different areas of cognitive science such as psychology, decision making, and artificial intelligence. For reasons of space we cannot cover all these developments, but we will try to mention the most remarkable advances on the areas where the quantum cognitive perspective has been applied.

Concerning the study of language, Bruza and others have presented evidences of quantum structures in experimental research on human memory and human lexicon.[66] These seminal explorations were followed by applications in natural language processing, where quantum notions such as orthogonal spaces, density matrices, among others, have been applied to improve the current methods.[67] In decision making, Busemeyer, Khrennikov, and others have shown that important cognitive phenomena such as the conjunction fallacy, and a number of cognitive bias can be successfully approached by quantum models.[47,68] These ideas have been later applied in psychology of human responses on psychological surveys.

Notably, a quantum-inspired model explaining the non-commutativity of human responses in surveys has been put forward in,[69] and later confirmed using a strong statistical analysis in.[70] In information retrieval, Van Rijsbergen, Zuccon, and others have shown that quantum probability is very useful for improving rankings of relevant web pages,[71] while Melucci, Frommholz and others have used quantum-based methods of clustering and web page representation.[72] Some authors have proposed the applications of quantum probability in a variety of fields including social sciences,[73] finances,[47] and the dynamics of political systems.[74] Moreover, the applications of quantum structures in cognition have proposed novel perspectives in the foundations of quantum probability.[75] For general introductions to the application of quantum structures in cognition we refer to the special issue of the journal of mathematical psychology,[76] and the survey presented in.[77]

Interstingly, although the recent years have witnessed an expansion in the form of conferences, special issues, books, and grants in areas related to it, the quantum modeling of concepts has still not become the focus of attention in any of these academic activities. The latter seem to be triggered by the fact that the phenomenology of concepts seems less interesting, from an applied point of view, than other areas such as decision-making, natural language processing, or information retrieval. However, if we recall that a model of concepts and their combinations aims at representing how meaning is structured in our ideas and thinking, we can understand that the modeling of concepts is more fundamental than the modeling of decisions or other cognitive situations, and thus we can conclude that foundational research on concepts should have a more central role than its applications in complex cognitive tasks.

In order to advance on the fundamental aspects of the quantum modeling of concepts, the authors support the position that it is necessary to explore cognitive analogues to the notions of physical quantities such as energy, mass, and momentum. In this perspective, we can for example consider the 'impact' of a concept. Indeed, when we are in a certain mental state and suddenly we receive some concept as external information (someone says something for example), we can feel a sort of impact in our mental state by the concept mentioned. This impact plays a crucial role in rhetoric and poetry. Impact could be understood as a measure of 'what makes the difference between a coherent and simply non-contradictory formation of a conceptual combination'. The conceptual impact we are referring here is a type of interaction between the 'semantic momentum' that concepts carry, and we believe that this direction of thought is worth exploring to understand how the meaning of concepts is structured and related to each other.

This last reflection leads us to the question of what is the relation between quantum systems and concepts at a philosophical level. As it is well-known, the debate concerning the physical interpretation of the quantum-theoretical entities (wave-function collapse, measurement problem, non-local realism, etc.) is still open despite the tremendous efforts and multiple interpretations that have been proposed.[78] Hence, it is striking that the analogous debate for concepts, i.e. the interpretation and description of the basic units that structure the meaning of our experiences, has the same problems than its quantum physical counterpart. Therefore, the quantum cognitive approach to concepts might not only be a radically new approach for concept modeling, but also might help to provide a unified description of both the physical and conceptual entities. A highly speculative attempt in this vein is the conceptual interpretation of quantum mechanics developed by Aerts.[79] In Aerts' view:

> *The basic hypothesis underlying this new framework is that quantum particles are conceptual entities. More concretely, we propose that quantum particles interact with ordinary matter, nuclei, atoms, molecules, macroscopic material entities, measuring apparatuses, . . . , in a similar way to how human concepts interact with memory structures, human minds or artificial memories.*

If quantum particles are conceptual entities, it follows that the entire physical realm is ruled by laws of semantic-like nature. Aerts proposes that the physical rules of the physical realm have emerged from the semantic interaction of basic physical entities, and has evolved for a very long time, leading to a language of communication (physical interaction) that is parsimonious and thus full of symmetries. The parsimony of the physical realm we observe has made possible that our physical experimental methods, combined with human intellect, have been sufficient to unveil some of the rules of nature with success. For the case of the semantic realm the situation is different. The semantic realm that we collectively create through our cognition has evolved for a short time in a cosmological scale. Hence, although we have evolved language capacities that allow us to interact semantically, these language capacities do not have (yet) the symmetries we observe in the language in which particles communicate. However, if the same structural principles, leaving symmetries out, apply for both semantic and physical entities, the quantum approach to concepts should at some point discover some of the fundamental laws that govern our cognitive experience. If this speculative scenario occurs, we will witness a new era of understanding of our reality.

Acknowledgments

We would like to thank Sylvie Desjardins, Diederik Aerts, Christian de Ronde, Sandro Sozzo, and Massimiliano Sassoli de Bianchi for fruitful discussions that have contributed to the creation of this material, and to the anonymous reviewers for their dedicated reading as well as useful comments on the manuscript.

References

1. J. R. Searle, *Mind: a brief introduction*, vol. 259. Oxford University Press Oxford, 2004.
2. M. Boden, *Mind as machine: A history of cognitive science*, vol. 1. Oxford University Press, USA, 2008.
3. U. Neisser, *Cognition and reality: Principles and implications of cognitive psychology*. WH Freeman/Times Books/Henry Holt & Co, 1976.
4. S. Russell and P. Norvig, *Artificial Intelligence: A Modern Approach*. Prentice Hall, 1995.
5. N. Maillot and M. Thonnat, "Ontology based complex object recognition," *Image and Vision Computing*, vol. 26, pp. 102–113, Jan. 01 2008.
6. E. Rosch, C. B. Mervis, W. Gray, D. Johnson, and P. Boyes-Braem, "Basic objects in natural categories," *Cognitive Psychology*, vol. 8, pp. 382–439, 1976.
7. P. Gärdenfors, *Conceptual Spaces: The Geometry of Thought*. Cambridge, Massachusetts: The MIT Press, 2000.
8. E. Rosch, "Natural Categories," *Cognitive Psychology*, vol. 4, pp. 328–350, 1973.
9. E. Machery, *Doing without concepts*. Oxford University Press, 2009.
10. M. Steyvers and T. Griffiths, "Probabilistic topic models," *Handbook of latent semantic analysis*, vol. 427, no. 7, pp. 424–440, 2007.
11. J. Hampton, "Typicality, graded membership, and vagueness," *Cognitive Science*, vol. 31, no. 3, pp. 355–384, 2007.
12. R. Belohlavek and G. Klir, eds., *Concepts and Fuzzy Logic*. MIT Press, 2011.
13. R. Nosofsky, "Attention, similarity, and the identification–categorization relationship.," *Journal of Experimental Psychology: General*, vol. 115, no. 1, pp. 39–57, 1986.
14. J. Fodor, *Concepts: Where Cognitive Science Went Wrong*. Oxford: Oxford University Press, 1998.
15. D. Aerts and M. Czachor, "Quantum aspects of semantic analysis and symbolic artificial intelligence," *Journal of Physics A: Mathematical and General*, vol. 37, no. 12, L123, 2004.

16. D. Aerts, "Quantum particles as conceptual entities: a possible explanatory framework for quantum theory," *Foundations of Science*, vol. 14, no. 4, pp. 361–411, 2009.

17. D. Aerts and S. Aerts, "When can a data set be described by quantum theory?," in *Proceedings of the Second Quantum Interaction Symposium, Oxford* (P. Bruza, W. Lawless, K. van Rijsbergen, D. Sofge, B. Coecke, and S. Clark, eds.), pp. 27–33, London: College Publications, 2008.

18. D. Aerts, "Quantum structure in cognition," *Journal of Mathematical Psychology*, vol. 53, pp. 314–348, 2009.

19. D. Aerts and B. D'Hooghe, "Classical logical versus quantum conceptual thought: Examples in economics, decision theory and concept theory," in *Quantum Interaction, Third International Symposium, QI 2009, Saarbrücken, Germany, March 25-27, 2009. Proceedings* (P. Bruza, D. A. Sofge, W. F. Lawless, K. van Rijsbergen, and M. Klusch, eds.), vol. 5494 of *Lecture Notes in Computer Science*, pp. 128–142, Springer, 2009.

20. D. Medin, "Concepts and conceptual structure," *American Psychologist*, vol. 44, no. 12, pp. 1469–1481, 1989.

21. L. Zadeh, "Fuzzy sets," *Information and Control*, vol. 8, pp. 338–353, 1965.

22. L. Gabora and D. Aerts, "A model of the emergence and evolution of integrated worldviews," *Journal of Mathematical Psychology*, vol. 53, pp. 434–451, 2009.

23. R. Nosofsky, "Attention and learning processes in the identification and categorization of integral stimuli," *Journal of Experimental Psychology: Learning, Memory, and Cognition*, vol. 13, pp. 87–108, jan 1987.

24. A. Tversky, "Features of similarity," *Psychological Review*, vol. 44, no. 4, 1977.

25. E. Rosch and B. Lloyd, eds., *Cognition and Categorization*. Lawrence Erlbaum Assoc., 1978.

26. J. Meibauer, "What is a context?," *What is a Context?: Linguistic Approaches and Challenges*, vol. 196, p. 9, 2012.

27. L. Gabora and D. Aerts, "Contextualizing concepts using a mathematical generalization of the quantum formalism," *JETAI: Journal of Experimental & Theoretical Artificial Intelligence*, vol. 14, 2002.

28. T. Veloz, L. Gabora, M. Eyjolfson, and D. a, "Toward a formal model of the shifting relationship between concepts and contexts during associative thought," in *Quantum Interaction - 5th International Symposium, QI 2011, Aberdeen, UK, June 26-29, 2011, Revised Selected Papers* (D. Song, M. Melucci, I. Frommholz, P. Zhang, L. Wang, and S. Arafat, eds.), vol. 7052 of *Lecture Notes in Computer Science*, pp. 25–34, Springer, 2011.

29. D. L. Medin and E. J. Shoben, "Context and structure in conceptual combination," *Cognitive Psychology*, vol. 20, no. 2, pp. 158–190, 1988.

30. L. J. Rips, "The current status of research on concept combination," *Mind & Language*, vol. 10, no. 1-2, pp. 72–104, 1995.

31. J. Hampton, "Disjunction of natural concepts," *Memory & Cognition*, vol. 16, no. 6, pp. 579–591, 1988.

32. F. J. Pelletier, "The principle of semantic compositionality," *Topoi*, vol. 13, no. 1, pp. 11–24, 1994.
33. R. E. Grandy, "Understanding and the principle of compositionality," *Philosophical Perspectives*, vol. 4, pp. 557–572, 1990.
34. J. A. Fodor and Z. W. Pylyshyn, "Connectionism and cognitive architecture: A critical analysis," *Cognition*, vol. 28, no. 1, pp. 3–71, 1988.
35. J. A. Hampton, "Overextension of conjunctive concepts: Evidence for a unitary model of concept typicality and class inclusion.," *Journal of Experimental Psychology: Learning, Memory, and Cognition*, vol. 14, no. 1, p. 12, 1988.
36. H. Kamp and B. Partee, "Prototype theory and compositionality," *Cognition*, vol. 57, no. 2, pp. 121–191, 1995.
37. S. Alxatib and J. Pelletier, "On the psychology of truth-gaps," in *Vagueness in communication* (R. Nouwen, van Rooij Robert, U. Sauerland, and H.-C. Schmitz, eds.), pp. 13–36, Springer, 2011.
38. E. Smith and D. Osherson, "On the adequacy of prototype theory as a theory of concepts," *Cognition*, vol. 9, pp. 35–38, 1981.
39. J. A. Hampton, "Conjunctions of visually based categories: Overextension and compensation.," *Journal of Experimental Psychology: Learning, Memory, and Cognition*, vol. 22, no. 2, p. 378, 1996.
40. J. Hampton, "Inheritance of attributes in natural concept conjunctions," *Memory & Cognition*, vol. 15, pp. 55–71, 1997.
41. J. Fodor and E. Lepore, "The red herring and the pet fish: why concepts still can't be prototypes," *Cognition*, vol. 58, no. 2, pp. 253–270, 1996.
42. D. Aerts, S. Sozzo, and T. Veloz, "Quantum structure of negation and conjunction in human thought," *Frontiers in psychology*, vol. 6, p. 1447, 2015.
43. I. Pitowsky, "George boole's 'conditions of possible experience'and the quantum puzzle," *The British Journal for the Philosophy of Science*, vol. 45, no. 1, pp. 95–125, 1994.
44. G. Boole, *An Investigation of the Laws of Thought on which are Founded the Mathematical Theories of Logic and Probabilities by George Boole*. Walton and Maberly, 1854.
45. N. Vorob'ev, "Consistent families of measures and their extensions," *Theory of Probability & Its Applications*, vol. 7, no. 2, pp. 147–163, 1962.
46. M. Bitbol and S. Osnaghi, "Bohr's complementarity and kant's epistemology," in *Niels Bohr, 1913-2013* (O. Darrigol, B. Duplantier, and J.-M. Raimond, eds.), pp. 199–221, Springer, 2016.
47. A. Y. Khrennikov, *Ubiquitous quantum structures: from psychology to finance*. Springer, 2010.
48. D. Aerts and S. Aerts, "Applications of quantum statistics in psychological studies of decision processes," *Foundations of Science*, vol. 1, no. 1, pp. 85–97, 1995.
49. R. Horodecki, P. Horodecki, M. Horodecki, and K. Horodecki, "Quantum entanglement," *Reviews of Modern Physics*, vol. 81, no. 2, p. 865, 2009.
50. I. Bengtsson and K. Zyczkowski, *Geometry of quantum states: an introduction to quantum entanglement*. Cambridge University Press, 2007.

51. D. Aerts, J. Broekaert, L. Gabora, and T. Veloz, "The guppy effect as interference," in *Quantum Interaction* (J. R. Busemeyer, F. Dubois, A. Lambert-Mogiliansky, and M. Melucci, eds.), pp. 36–47, Springer, 2012.

52. D. Aerts, L. Gabora, and S. Sozzo, "Concepts and their dynamics: A quantum-theoretic modeling of human thought," *Topics in Cognitive Science*, vol. 5, no. 4, pp. 737–772, 2013.

53. S. Sozzo, "A quantum probability explanation in fock space for borderline contradictions," *Journal of Mathematical Psychology*, vol. 58, pp. 1–12, 2014.

54. R. Franco, "The conjunction fallacy and interference effects," *Journal of Mathematical Psychology*, vol. 53, no. 5, pp. 415–422, 2009.

55. D. Aerts, S. Sozzo, and J. Tapia, "A quantum model for the ellsberg and machina paradoxes," in *Quantum Interaction* (J. R. Busemeyer, F. Dubois, A. Lambert-Mogiliansky, and M. Melucci, eds.), pp. 48–59, Springer, 2012.

56. T. Veloz, "Towards a quantum theory of cognition: History, development, and perspectives," *University of British Columbia, PhD thesis*, 2015.

57. D. Aerts, "General quantum modeling of combining concepts: A quantum field model in fock space," *arXiv preprint arXiv:0705.1740*, 2007.

58. D. Aerts, S. Sozzo, and T. Veloz, "New fundamental evidence of non-classical structure in the combination of natural concepts," *Phil. Trans. R. Soc. A*, vol. 374, no. 2058, p. 20150095, 2016.

59. A. Arnauld and P. Nicole, *Logic, or the Art of Thinking*. Cambridge, England: Cambridge University Press, 1996. First published in 1662. Translated and edited by J.V. Buroker.

60. D. Aerts, S. Sozzo, and T. Veloz, "The quantum nature of identity in human thought: Bose-einstein statistics for conceptual indistinguishability," *International Journal of Theoretical Physics*, vol. 54, no. 12, pp. 4430–4443, 2015.

61. D. Aerts and S. Sozzo, "Quantum structure in cognition: Why and how concepts are entangled," in *Quantum Interaction* (D. Song, M. Melucci, I. Frommholz, P. Zhang, L. Wang, and S. Arafat, eds.), pp. 116–127, Springer, 2011.

62. P. Bruza, K. Kitto, L. Ramm, L. Sitbon, D. Song, and S. Blomberg, "Quantum-like non-separability of concept combinations, emergent associates and abduction," *Logic Journal of the IGPL*, vol. 20, no. 2, pp. 445–457, 2012.

63. D. Aerts, J. A. Arguëlles, L. Beltran, S. Geriente, M. S. de Bianchi, S. Sozzo, and T. Veloz, "Spin and wind directions i: Identifying entanglement in nature and cognition," *Foundations of Science*, vol. 23, pp. 1–13, 2015.

64. E. Conte, M. Lopane, A. Khrennikov, O. Todarello, A. Federici, and F. Vitiello, "A preliminar evidence of quantum like behavior in measurements of mental states," *NeuroQuantol.*, vol. 6, no. quant-ph/0307201, pp. 126–139, 2003.

65. C. J. Van Rijsbergen, *The geometry of information retrieval*, vol. 157. Cambridge University Press Cambridge, 2004.

66. P. D. Bruza and R. Cole, "Quantum logic of semantic space: An exploratory investigation of context effects in practical reasoning," *We Will Show Them: Essays in Honour of Dov Gabbay*, pp. 339–361, 2005.

67. R. T. Oehrle, J. R. (editors, D. Widdows, D. Widdows, S. Peters, and S. Peters, "Word vectors and quantum logic experiments with negation and disjunction," 2003.

68. J. R. Busemeyer, E. M. Pothos, R. Franco, and J. S. Trueblood, "A quantum theoretical explanation for probability judgment errors.," *Psychological review*, vol. 118, no. 2, p. 193, 2011.

69. J. Trueblood and J. Busemeyer, "A quantum probability account of order effects in inference," *Cognitive Science*, vol. 35, no. 8, pp. 1518–1552, 2011.

70. Z. Wang and J. R. Busemeyer, "A quantum question order model supported by empirical tests of an a priori and precise prediction," *Topics in cognitive science*, vol. 5, no. 4, pp. 689–710, 2013.

71. G. Zuccon and L. Azzopardi, "Using the quantum probability ranking principle to rank interdependent documents," in *Advances in information retrieval*, pp. 357–369, Springer, 2010.

72. B. Piwowarski, M.-R. Amini, and M. Lalmas, "On using a quantum physics formalism for multidocument summarization," *Journal of the American Society for Information Science and Technology*, vol. 63, no. 5, pp. 865–888, 2012.

73. E. Haven and A. Khrennikov, *Quantum social science*. Cambridge University Press, 2013.

74. P. Khrennikova, E. Haven, and A. Khrennikov, "An application of the theory of open quantum systems to model the dynamics of party governance in the US political system," *International journal of theoretical physics*, vol. 53, no. 4, pp. 1346–1360, 2014.

75. D. Aerts and M. S. de Bianchi, "The extended bloch representation of quantum mechanics. Explaining superposition, interference and entanglement," *arXiv preprint arXiv:1504.04781*, 2015.

76. P. Bruza, J. Busemeyer, and L. Gabora, "Introduction to the special issue on quantum cognition," *arXiv preprint arXiv:1309.5673*, 2013.

77. E. M. Pothos and J. R. Busemeyer, "Can quantum probability provide a new direction for cognitive modeling?," *Behavioral and Brain Sciences*, vol. 36, no. 03, pp. 255–274, 2013.

78. M. Schlosshauer, "Decoherence, the measurement problem, and interpretations of quantum mechanics," *Reviews of Modern physics*, vol. 76, no. 4, p. 1267, 2005.

79. D. Aerts, M. S. de Bianchi, S. Sozzo, and T. Veloz, "On the conceptuality interpretation of quantum and relativity theories," *arXiv preprint arXiv:1711.09668*, 2017.

QUANTUM COGNITION GOES BEYOND-QUANTUM: MODELING THE COLLECTIVE PARTICIPANT IN PSYCHOLOGICAL MEASUREMENTS

DIEDERIK AERTS

Center Leo Apostel for Interdisciplinary Studies and Department of Mathematics, Brussels Free University, 1050 Brussels, Belgium
E-mail: diraerts@vub.ac.be

MASSIMILIANO SASSOLI DE BIANCHI

Center Leo Apostel for Interdisciplinary Studies, Brussels Free University, 1050 Brussels, Belgium and Laboratorio di Autoricerca di Base, via Cadepiano 18, 6917 Barbengo, Switzerland.
E-mail: msassoli@vub.ac.be

SANDRO SOZZO

School of Business and Research Centre "IQSCS", University Road, LE1 7RH Leicester, United Kingdom.
E-mail: ss831@le.ac.uk

TOMAS VELOZ

Universidad Andres Bello, Departamento Ciencias Biológicas, Facultad Ciencias de la Vida, 8370146 Santiago, Chile and Instituto de Filosofía y Ciencias de la Complejidad. Los Alerces 3024, Ñuñoa, Santiago, Chile
E-mail: tveloz@gmail.com

In psychological measurements, two levels should be distinguished: the *individual level*, relative to the different participants in a given cognitive situation, and the *collective level*, relative to the overall statistics of their outcomes, which we propose to associate with a notion of *collective participant*. When the distinction between these two levels is properly formalized, it reveals why the modeling of the collective participant generally requires not only non-classical (*non-Kolmogorovian*) probabilistic models, but also non-quantum (*non-Bornian*) probabilistic models, when sequential measurements at the individual level are considered, and this though a classical or pure quantum description might remain valid for single measurement situations.

Keywords: Quantum structures; human cognition; cognitive modeling; probability models; universal measurements; extended Bloch representation.

1. Introduction

The aim of mathematical psychology is to develop theoretical models of cognitive (and more generally psychological) processes [20]. Its methodology comprises two main aspects. The first one is to establish sensible assumptions about how humans behave in the psychological/conceptual situations under study, which are then translated into specific mathematical descriptions. The second aspect is about confronting human subjects to these conceptual situations in controlled experimental settings, usually consisting of preliminary sessions where participants are instructed about what to do in the experiments, followed by the real situations where the data of their responses is collected, analyzed and then compared with the models. Generally speaking, a mathematical model will be considered satisfactory if capable of explaining the collected data (like the relative frequencies of outcomes, interpreted as probabilities) and possibly predict their structure (the relations which data obey [18,31].

A paradigmatic case is the study of the *conjunction fallacy* [26], where participants are observed to overall violate the rules of classical probability theory, when asked to judge likelihood of events. Numerous competing mathematical models exist to explain the observed non-classicality of their estimations, including models based on Bayesian probability [25, various types of heuristics [27] and more recently also quantum probability structures [19], and the success of these models is always contrasted in terms of how well they are able to account and explain the experimental probabilities [29].

A subtle element that is usually overlooked, or not discussed, in the construction of models, is that in a cognitive psychology experiment participants may either all respond (actualize potential outcomes) according to a same mathematical model, or each one according to a different mathematical model. More precisely, in experiments participants may all actualize an outcome in the *same way*, or in a *different way*. Also, depending on the situation they are confronted with, they can do this either in a *deterministic way*, or in a *indeterministic way*. By 'deterministic' we mean a process whose outcome is in principle predictable in advance, whereas by 'indeterministic' we mean a failure of predictability, i.e., a process where different possible outcomes are truly available, so that there is a situation of genuine unpredictability (the outcome cannot be determined in advance, not even in principle). When participants behave all deterministic, they can all choose the same outcome, or possibly different outcomes, and when they are all indeterministic, they can all actualize outcomes according to the same probabilistic model (which can be classical or pure quantum) or according to different probabilistic models, describing more general (beyond-classical and beyond-quantum, that is non-Kolmogorovian and non-Bornian) ways to respond or take a decision, that is, to actualize an outcome. Also, in the situation where they do not all behave in a statistically equivalent way, there can be a mixture of both deterministic and indeterministic ways of selecting an outcome.

The main goal of the present article is to show that, when the above possibilities are considered, fundamentally different types of probabilistic models might be required to account for the aggregated experimental data. In other words, the modeling of the *collective level* depends on the specificities of the *individual level*, so that a comprehensive theory of psychological experiments/measurements must necessarily take into account the latter. This does not mean that one has to possess information about the individual processes of actualization of potential outcomes (the individual way of producing responses, or decisions), but the modeling of the situation at the individual level must be compatible with the modeling at the collective level. This highlights the importance of making a clear distinction between these two levels, the individual being associable with the participant's decisions, and the collective with a notion of *collective participant*, whose behavior – its way of actualizing outcomes – is precisely described by the statistical indicators representing the entire collection of individual responses, in a unitary way.

In other words, in this article we pursue a new and definite approach to the foundations of cognitive psychology, and more specifically to the nature of psychological measurements, following our studies on the foundations of physical theories and measurement processes. By introducing the well-defined notion of collective participant, which was already implied in the *Brussels operational-realistic approach to cognition* (see [13] and the references cited therein), but was never made fully explicit until now (and which we think is a new theoretical notion in the psychological landscape), we draw general and fundamental conclusions about the probabilistic structure characterizing cognitive phenomena. To this end, we start in the next section by listing the basic elements of typical psychological measurements, explaining how they can be properly formalized.

2. The basic elements of psychological measurements

We introduce the following basic elements characterizing psychological measurements:

Number of participants. A measurement always comprises a given number n of participants (also called subjects, respondents, etc.) To obtain from them a significant statistics of outcomes, ideally n will be a large number (in certain measurements it can just be several tens, in others it can reach a few thousands).

Uniform average. The n individuals participating in a measurement are all confronted with the same conceptual situation, from whom the experimenter obtains individual outcomes (e.g., the selection of items, yes-no answers, the assessment of rankings about given exemplars, etc.). These individual outcomes are then counted to calculate their relative frequencies, interpreted as probabilities (for n sufficiently large), which are the main object of the mathematical modeling.

Individual level. The description of the conceptual situations and changes happening at the level of each individual participating in the experiment.

Collective level. The description of the conceptual situations and changes happening at the level where the different cognitive activities of the participants are considered as a whole, in particular when their outcomes are averaged out in the final statistics.

Intersubjectivity. Each participant is submitted to the same experimental situation, which can be described as a *conceptual entity S prepared in a given state* [13]. Such state is objective, or intersubjective, in the sense that each participant interacts with the same conceptual entity, prepared in the same initial state, describing the reality, or state of affair, of said entity.

State-space representation. For all participants, a same *state-space* Σ can be used to model the (conceptual entity describing the) cognitive situation. It can be a σ-algebra, a Hilbert space, a generalized Bloch sphere [6], or any other suitable mathematical structure [7–9]. Consequently, the same element of Σ is used to represent the state p_{in} of the conceptual entity S, describing the *initial* conceptual situation to which all participants are equally submitted to.

Outcomes. A psychological measurement is characterized by a given number N of possible outcomes, generally understood as the states to which the conceptual entity can transition to, when submitted to the action of a participant's mind subjected to an interrogation, or a decision-making process, specifying the outcomes (i.e., the answers, or the decisions) that are available to be actualized. Each participant is submitted to the same set of possible outcomes, as is clear that they are all part of the same measurement (or sub-measurement), and therefore have to follow the exact same protocol. So, in the same way a same state p_{in} is used to describe the initial condition of the measured conceptual entity S, also the outcome states will be represented by a same set $\{q_1, \ldots, q_N\}$ of elements of Σ, for all participants. In the special case where the state space is a Hilbert space, this means that each participant is associated with the same spectral family.[a]

State change. When submitted to a measurement (interrogative) context consisting of a conceptual entity S in the initial state p_{in}, with N possible outcomes $q_j \in \Sigma$, $j = 1, \ldots, N$, each participant, by providing one (and only one) of these outcomes, changes the entity's initial state to the state corresponding to the selected outcome. In other words, a participant's cognitive action produces one of the following N transitions: $p_{\text{in}} \to q_j$, $j = 1, \ldots, N$. This action will be assumed to happen in a *two-step process*, where the first step, if any, is deterministic, whereas the second step, if any, is indeterministic.

Deterministic context. Each participant, say the i-th one, starts her/his cognitive action by possibly changing the entity's initial state p_{in} in a deterministic way, producing a transition: $e_i^{\text{det}} : p_{\text{in}} \to p_i'$. One can for instance think of the deterministic context e_i^{det} as resulting from some kind of *information supply*. It can be some externally obtained background information, directly supplied by the experimenter in the ambit of the experimental protocol, or some internal information retrieval, resulting from the participants' *thinking activity*. The process could also correspond to the evocation of that portion of the persons' memory that need to be accessed in order to respond to the situation in question. In the Hilbert space language, e_i^{det} will be typically modeled by a unitary matrix, or by an orthogonal projection operator.

Indeterministic context. Following the possible information supply that produces the deterministic transition $p_{\text{in}} \to p_i'$, and assuming that p_i' is not already one of the outcome-states (i.e., an eigenstate of the measurement), the i-th participant will operate a genuinely indeterministic transition e_i^{ind}:

---

[a]This is clearly an idealization. But idealizations are what one tries to obtain when modeling experimental situations, as the scope of an idealization is precisely that of capturing the essential aspects of what a situation is all about, neglecting those aspects that are considered not to be essential. This is what physicists also do, for instance when modeling physical quantities (observables) by means of self-adjoint operators, or when describing state transitions by means of the Lüders-von Neumann projection formula. Another useful idealization here, in analogy with so-called measurements of the first kind in physics, is to assume that measurement outcomes can be associated with well-defined outcome states. i.e., that after the measurement the measured entity is always left in a well-defined (eigen) state.

$p'_i \to p''_i$, where $p''_i \in \{q_1, \ldots, q_N\}$ is one of the available outcome-states of the psychological measurement. One can think of the indeterministic context e_i^{ind} as being the result of some *subconscious mental activity*,[b] during which the i-th participant builds a mental condition of unstable equilibrium, resulting from the balancing of the different tensions between the state p'_i of the conceptual entity and the available mutually excluding answers $q_j \in \Sigma$, $j = 1, \ldots, N$, competing with each other. This unstable equilibrium, by spontaneously breaking at some unpredictable moment, in an unpredictable way, then actualizes one of the possible outcome-states, in what can be described as a *weighted symmetry breaking process* [11].

Note that if there are no deterministic context effects, like (external or internal) information supply, then e_i^{det} is just the trivial context, not affecting the initial state p_{in}, so that e_i^{ind} will directly operate on the initial state p_{in}. On the other hand, if the i-th participant knows in advance the answer to the interrogation, e_i^{ind} will be the trivial context, not affecting the state p'_i obtained through the previous information retrieval activity (consisting in the subject looking into her/his memory, to discover the already existing answer to the question addressed, which s/he will simply select as the outcome). In general, we can think of e_i^{ind} as a context triggering a process that takes a very short time (something like a sudden collapse). On the other hand, the transition induced by e_i^{det} will generally require a longer amount of time to be produced, corresponding to the time the i-th participant needs to obtain and assimilate the background information, before selecting an outcome. So, ideally and generally speaking, we will assume that to each individual participating in a measurement we can attach a context e_i, which can be understood as the composition of two contexts, one deterministic and the other indeterministic:

$$e_i = e_i^{\text{ind}} \circ e_i^{\text{det}}, \quad i \in \{1, \ldots, n\}. \tag{1}$$

It is important to specify that although the above decomposition, and the previous specifications of deterministic and indeterministic contexts, are meant to express the generality of our approach, they should not be misunderstood as a claim that all measurement situations should or could be modeled in this way. That said, we think it may be useful at this point to give a couple of examples. Consider a survey including the question: "Are

---

[b] One should not deduce from this that the deterministic contexts e_i^{det} would therefore be necessarily associated, or exclusively associated, with conscious/controlled mental activities. In other words, our distinction between deterministic and indeterministic contexts is different from the distinction of so-called *dual process theory*.

you a smoker or a non-smoker?". Obviously, e_i^{ind} will then be all trivial contexts, as each participant knows in advance if s/he is a smoker or a non-smoker, so only non-trivial deterministic contexts e_i^{det} will be present, whose actions will simply be the retrieval of the already existing answer about the participants' smoking or non-smoking habits. Consider now the question [3,5]: "Are you for or against the use of nuclear energy?" In this case, non-trivial indeterministic contexts will be present, being very plausible that some of the participants have no prior opinion about nuclear energy, hence for them the outcome will be literally created (actualized) when answering the question, in a way that cannot be predicted in advance (as in a symmetry breaking process). Imagine also that before the question is asked some literature is given to the participants to read, explaining how greenhouse gas emissions can be reduced through the use of nuclear energy. This will obviously influence in a determined way how decisions will be taken at the individual level, increasing the percentage of those who will answer by favoring nuclear energy (with some of them even possibly becoming fully deterministic in the way they will answer). So, we are in a situation where both non-trivial deterministic and non-trivial indeterministic contexts will both be possibly present and operate in a sequential way, in accordance with the decomposition (1).

3. Ways of choosing and background information

For the sake of clarity, in the following we shall simply call e_i^{det} the *background information* and e_i^{ind} the *way of choosing* of the i-th subject, respectively. Let then $\mu(q_j, e_i, p_{\text{in}}) \geq 0$ be the probability that the i-th participant produces the outcome q_j, via the individual context e_i, when submitted to the conceptual entity in the initial state p_{in}. We clearly have: $\sum_{j=1}^{N} \mu(q_j, e_i, p_{\text{in}}) = 1$, for every individual context e_i and every initial state $p_{\text{in}} \in \Sigma$. It is at this stage useful to distinguish the following three situations.

Situation 1: The way of choosing is trivial. This corresponds to the situation where all participants choose in a (in principle) predictable way, i.e., $e_i = e_i^{\text{det}}$, for every i, which however can be different for each participant. This means that the values of the probabilities $\mu(q_j, e_i, p_{\text{in}})$ are all either 0 or 1. However, if the number of individuals deterministically selecting the outcome q_j is n_j, then at the collective level the experimental probability $\mu(q_j, e, p_{\text{in}})$ for the transition $p_{\text{in}} \to q_j$ is simply given by the ratio $\frac{n_j}{n}$, where e denotes the context of the *collective participant* associated with the collection of n individual participants. Here we can distinguish two cases:

(a) $e_i^{\text{det}} = e^{\text{det}}$, for every $i = 1, \ldots, n$; (b) for some i and j, we can have $e_i^{\text{det}} \neq e_j^{\text{det}}$. In case (a), the process is deterministic also at the collective level. In case (b), since some of the probabilities $\mu(q_j, e, p_{\text{in}})$ are different from 0 or 1, the process becomes indeterministic at the collective level.

Situation 2: The background information is trivial. This corresponds to the situation where there is no information supply (or other deterministic context effects) prior to the selection of an outcome. We then have $e_i = e_i^{\text{ind}}$, for every $i = 1, \ldots, n$, so that the different participants only produce an indeterministic (quantum-like) transition. Again, we can distinguish two cases: (a) participants all choose in the same way, which of course does not mean they all choose the same outcome, i.e., $e_i^{\text{ind}} = e^{\text{ind}}$, for every $i = 1, \ldots, n$; (b) participants possibly choose in different ways, i.e., for some i and j, we can have $e_i^{\text{ind}} \neq e_j^{\text{ind}}$.

Situation 3: The way of choosing and the background information are both non-trivial. This is the most general and complex situation, and the following two cases can be distinguished: (a) $e_i^{\text{det}} = e^{\text{det}}$ and $e_i^{\text{ind}} = e^{\text{ind}}$, so that $e_i = e$, for every $i = 1, \ldots, n$, with $e = e^{\text{ind}} \circ e^{\text{det}}$; (b) for some i and j, we can have $e_i^{\text{det}} \neq e_j^{\text{det}}$ and/or $e_i^{\text{ind}} \neq e_j^{\text{ind}}$.

Coming back to the examples given at the end of the previous section, Situation 1 corresponds to participants being for instance asked if they are smokers. If they are all, then we are in case (a), otherwise, if there is a mix of smokers and non-smokers, we are in case (b). Situation 2 corresponds to participants being for instance asked if they are in favor of nuclear energy, assuming that none of them ever reflected or took a position on the matter before having to answer such question. If they are a very homogeneous group, say of same sex, cultural background, age group, etc., then we can assume that (at least in first approximation) we are in case (a), whereas if there are relevant differences among the participants, for instance because a portion of them do not even know what nuclear energy is, then we are in principle in case (b). Situation 3 corresponds to participants being for instance asked to answer a preliminary question (before being asked the nuclear energy question), to which all of them can respond in a predictable way. For example, the question could be: "Have you ever heard of Chernobyl's disaster?" Clearly, either participants have heard about it, or not. If they are a homogeneous group, we can assume they will all answer in the same way, say in an affirmative way, and that they will subsequently answer the nuclear energy question with same individual probabilities (which will have been altered in a deterministic way by the previous question, for instance because it contains the word "disaster").

So, this would be an example of case (a) in Situation 3, whereas case (b) would be when either participants are asked different preliminary questions, and/or when the group is non-homogeneous.

A few remarks are in order. Note that for all three situations above, in case (a) there is no difference between the individual and collective level, i.e., all participants are "cognitive clones," behaving exactly in the same way, whereas in the general case (b), they can produce distinct individual behaviors.

Different individual background information can be easily modeled in a Hilbert space representation. Indeed, it is sufficient to associate to each participant (or group of participants) a different unitary evolution [12], or a different projection operator [2]. On the other hand, different individual ways of choosing cannot be modeled by remaining within the confines of a single probability model, like that of standard quantum mechanics, as the latter only admits the *Born rule* way of choosing, apparently imposed by Gleason's theorem. We say "apparently" because Gleason's theorem only tells us that if the transition probabilities only have to depend on the state before the measurement and on the eigenstate actualized after the measurement, then they must be given by the Born rule. However, if we relax this constraint, one can introduce different parameter-dependent probability measures within a same Hilbert (state) space representation, and use them to describe the different individual ways of choosing of the participants. In other words, by extending the standard Hilbertian formalism (see for instance [6–8,10]), it becomes possible, while maintaining the Hilbertian structure for the states, to define different rules of probabilistic assignment, characterizing the participants' different ways of choosing (see also Sec. 5).

One may wonder what could be the meaning of an individual statistics of outcomes. We know that from the outcomes provided by all the participants we can determine the experimental probabilities, by calculating their relative frequencies. This is the description of what we have called the collective level, associated with the notion of collective participant. However, if we assume that the individual contexts e_i are generally non-deterministic, it is natural to also associate outcome probabilities $\mu(q_j, e_i, p_{\text{in}})$, $j = 1, \ldots N$, $i = 1, \ldots, n$, to each one of the n individuals participating in the experiment. Of course, this does not mean that these probabilities can be directly or easily determined. However, we also know that individuals have the ability to provide not only a specific answer to a given question, but also to estimate the probabilities for the available answers, which is a strong indication that it is correct to associate each individual with a specific statistics of outcomes. So, when we calculate the relative frequencies $\frac{n_j}{n}$ of the different outcomes, what we are in fact estimating is the probabilistic average:

$$\mu(q_j, e, p_{\text{in}}) = \frac{1}{n} \sum_{i=1}^{n} \mu(q_j, e_i, p_{\text{in}}), \quad j = 1, \ldots N, \tag{2}$$

where e, as we already mentioned above, denotes the context of the *collective participant*.

4. The collective participant

At this stage of our analysis, it is important to emphasize a difference between the present operational-realistic description of psychological measurements and the today most commonly adopted (subjectivist) view in quantum cognition (and cognitive modeling in general), according to which the initial state would describe the *belief system* of a participant about the cognitive situation under consideration. Here a question arises: Which participant? Consider for a moment the typical situation of a quantum measurement in a physics laboratory: a same measurement apparatus (and usually a same agent operating on it) is used n times, with the physical entity always prepared in the same state, to obtain the final statistics of outcomes. On the other hand, in a psychological laboratory n different participants play the role of the apparatus, i.e., we actually have n different measurement apparatuses used in a same experiment, and each one is typically used only once. So, if we want psychological measurements to be interpretable in a manner analogous to physics measurements, the n participants must be considered to be like clones, i.e., like measuring entities all having the same way of accessing the available background information, and the same way of choosing an outcome, the latter being described for instance by the Born rule. Also, one must assume that they all have the same way of updating their probabilities for subsequent measurements, by associating to their answers the same outcome states.

A first difficulty is that there are no reasons to think that all participants will necessarily share the exact same belief regarding the cognitive situation they are subjected to, i.e., that they will represent such situation by using the same vector in Hilbert space, and the same is true regarding the choice of the outcome states. On the other hand, as we mentioned already, in our approach the state describes an aspect of the reality of the conceptual entity under consideration, in a given moment, i.e., an intersubjective element of the cognitive domain shared by all the participants [13]. Therefore, the above difficulty does not arise in a realistic approach. However, the fact remains that a psychological measurement is performed by n participants, with different mindsets, and not just by one participant, i.e., by a single mind. Each of them will behave in a different way when confronted with the cognitive situation, i.e., will elicit one of the outcomes by pondering and choosing in possibly different ways. In other words, if the n participants are not assumed to behave as equivalent measurement apparatuses, i.e., are not assumed to be statistically equivalents, then we certainly cannot consider the experimental probabilities to be descriptive of their individual actions (or, in the subjectivist view, of their individual beliefs and judgments).

What we mean to say is that the states and probabilities describing the overall statistics generated by the collection of n individuals can only be associated with a *memoryless collective participant*, such that if it would be submitted n times to the interrogative context in question, the statistics of outcomes it would produce would be equivalent (for n large enough) to the statistics of outcomes generated by the n participants in the experiment. The collective mind of such collective participant has therefore to be understood as a composite entity formed by n separate sub-minds. When interrogated, it provides an answer by operating in the following way: first, it selects one of its internal sub-minds, say the i-th one (which works exactly as the mind of the i-th participant), then, it uses it to answer the question and to produce an outcome, and if asked again the question, it selects another of its internal sub-minds, among the $n-1$ that haven't yet provided an answer, and so on. This means that, for as long as the same question is asked no more than n times, the mind of the collective participant will show no memory effects (for instance, no response replicability effects). This memoryless property of the collective participant, associated with the overall statistics of outcomes, is what fundamentally distinguishes it from the individual participants. But apart from that, in many situations one can certainly describe the cognitive action of the collective participant in a way that is analogous to the description of the individual ones.

Now, when referring to the great success of quantum cognition, one usually points to the success of the standard quantum formalism in modeling the cognitive action of the collective participant. So, a natural question arises: Why this success? Many answers can of course be given (see for instance the first chapter of [17], as well as the introductory article [29] of a *TopiCS* volume dedicated to *quantum statistics*). Let us briefly explain the answer to which we arrived in [7,8]. For this, we consider *Situation 2* of Sec. 3, where the participants select an answer without acquiring any prior information. Then, the question is: Considering that the individual contexts e_i (here equal to e_i^{ind}) are in principle all different, and therefore not describable by means of the Born rule, why is it nevertheless possible, in general, to describe the context e, associated with the collective participant, by means of the latter? In other words: Why the averages (2) are usually well described by the quantum mechanical transition probabilities?

To answer this question, one has first to find a way to describe all possible ways of choosing an outcome. This can be done by exploiting the *general tension-reduction* (GTR) model [7–9], or its more specific implementation called the *extended Bloch representation* (EBR) of quantum mechanics [6], where the set of states is taken to be Hilbertian. Then, one has to find a way to calculate the average over all possible ways of choosing an outcome (called a *universal average*). This can be done by following a strategy similar to that used in physics in the definition of the *Wiener measure*, and the remarkable result is that the average probabilities so obtained are precisely those predicted by the Born rule (if the state space is Hilbertian), thus explaining why the latter generally appears as an optimal approximation in numerous experimental situations [6,8]. In other words, as n increases, the average (2) can be expected to become better and better approximated by the Born rule, i.e., the context e, characterizing the action of the collective participant, is expected to tend towards that context that is described by the Born rule (and the associated projection postulate).

5. Two-outcome measurements

Let us more specifically explain, in the simple situation of two-outcome processes, how different kinds of measurements, associated with a same initial state and pair of outcomes, can be modeled within the EBR. We consider the 3-dimensional Bloch sphere representation of states, with the initial state p_{in} described by a unit 3-dimensional real vector $\mathbf{x}_{\mathrm{in}}$. To fix ideas, one can imagine a virtual point-particle associated with it, i.e., positioned exactly at point $\mathbf{x}_{\mathrm{in}}$ on the surface of the unit sphere, which should also be imagined as an hollow structure in which the particle can penetrate. A measurement having two outcome-states q_1 and q_2 can then be represented as a one-dimensional simplex $\triangle_1$ with apex-points $\mathbf{a}_1$ and $\mathbf{a}_2$, corresponding to the two Bloch vectors representative of states q_1 and q_2, respectively (see Figure 1). One can think of $\triangle_1$ as an abstract elastic and breakable structure, extended between the two points $\mathbf{a}_1$ and $\mathbf{a}_2$. A measurement process can then be described as a two-phase process. During the first phase, the point particle enters the sphere, following a path orthogonal to $\triangle_1$, thus reaching a point $\mathbf{x}_e = (\mathbf{x}_{in} \cdot \mathbf{a}_1)\,\mathbf{a}_1$. Cognitively speaking, one can think of this first phase as a process during which the mind of an individual brings the situation described by p_{in} into the context of the two possible answers q_1 and q_2, i.e., as a (deterministic) preparation process during which the meaning of the situation is brought as close as possible to the meaning of the possible answers.

The second phase consists in the elastic breaking at some unpredictable point $\boldsymbol{\lambda}$, so that its subsequent collapse can bring the abstract point particle either towards point $\mathbf{a}_1$ or point $\mathbf{a}_2$, depending on whether $\boldsymbol{\lambda}$ belongs to the segment A_1, between $\mathbf{x}_e$ and $\mathbf{a}_2$, or to the segment A_2, between $\mathbf{a}_1$ and $\mathbf{x}_e$, respectively. Cognitively speaking, this second (typically indeterministic) phase corresponds to the reduction of the tensional equilibrium previously built, due to fluctuations causing the breaking of such equilibrium and consequent selection of only one of the two possible answers.

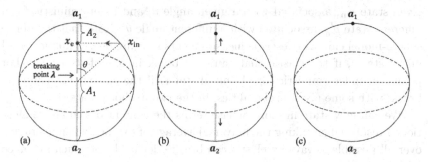

Fig. 1. A schematic description of the two phases of a two-outcome measurement, in the EPR of quantum mechanics. Here the breaking of the abstract elastic structure happens in A_1, so that the outcome is $\mathbf{a}_1$.

Being $\triangle_1$ of length 2, we can parametrize its points using the interval $[-1, 1]$, where the values $x = -1$ and $x = 1$ correspond to vectors $\mathbf{a}_2$ and $\mathbf{a}_1$, respectively, and $x_e = \mathbf{x}_{in} \cdot \mathbf{a}_1 = \cos\theta$ to the position $\mathbf{x}_e$ of the particle once it has reached the elastic, so that $A_1 = [-1, \cos\theta]$ and $A_2 = [\cos\theta, 1]$ (see Figure 1). If $\rho(y)$ is the probability density describing the way the elastic can break, then the probability for the transition $p_{in} \to q_1$ (i.e., the probability for the abstract point particle to go from the initial position $\mathbf{x}_{in}$ to the final position $\mathbf{a}_1$, passing through the equilibrium point $\mathbf{x}_e$) is given by:

$$\mu(q_1, e_\rho, p_{in}) = \int_{-1}^{\cos\theta} \rho(y)\,dy, \tag{3}$$

where e_ρ is the context associated with the ρ-way of breaking of the elastic, and of course $\mu(q_2, e_\rho, p_{\mathrm{in}}) = 1 - \mu(q_1, e_\rho, p_{\mathrm{in}})$.

It is worth observing that, depending on the breakability of the elastic, e_ρ will either be a deterministic or indeterministic context [1,4,5]. For instance, if the elastic can only break in the segment $[-1, x]$, with $x < \cos\theta$, it immediately follows from (3) that $\mu(q_1, e_\rho, p_{\mathrm{in}}) = 1$, so we are in the situation of a deterministic context. Note that such situation cannot be described by standard quantum mechanics, since a probability equal to 1, in a first kind measurement, is only possible if the initial state is an eigenstate, which is not necessarily the case here. Of course, if the elastic can instead only break in $[x, 1]$, with $x > \cos\theta$, then $\mu(q_1, e_\rho, p_{\mathrm{in}}) = 0$, so e_ρ still describes a deterministic process, but this time with the predetermined outcome being q_2. On the other hand, if the elastic has breakable parts both in A_1 and A_2, the outcome cannot be predicted in advance and e_ρ describes a genuine indeterministic context. Note that a context e_ρ can be deterministic for a given state p_{in}, associated with a given angle θ, and indeterministic for a different state p'_{in}, associated with a different angle θ'. Note that the above sphere-model can also describe measurements that are deterministic for all initial states, if the elastic that can only break in one of its two anchor points, or indeterministic for all initial states, if the elastic that can only break, with some given probabilities, in the two anchor points [1,7].

We can now state in more precise terms the content of the above mentioned result: when taking the universal average of (3), i.e., when averaging over all possible ρ-ways an elastic can break – let us denote such averaged probability $\langle\mu(q_1, e_\rho, p_{\mathrm{in}})\rangle^{\mathrm{univ}}$ – one can show that the latter is identical to the probability associated with a uniformly breaking elastic, characterized by the constant probability density $\rho_{\mathrm{u}}(y) = \frac{1}{2}$. In other words [6,8]:

$$\langle\mu(q_1, e_\rho, p_{\mathrm{in}})\rangle^{\mathrm{univ}} = \mu(q_1, e_{\rho_{\mathrm{u}}}, p_{\mathrm{in}}) = \frac{1}{2}\int_{-1}^{\cos\theta} dy = \frac{1}{2}(1 + \cos\theta). \qquad (4)$$

We observe that the universal average (4) is identical to the Born quantum probability.[c] In other words, when preforming a universal average, one recovers the quantum mechanical *Born rule*, if the state space has an Hilbertian structure (i.e., is a Blochean representation derived from the

---

[c]The average (4) can also be interpreted as the outcome probability of a so-called *universal measurement*, characterizing by a two-level "actualization of potential properties" process, i.e., such that there is not only the actualization of a measurement-interaction (the breaking point λ), but also of a way to actualize a measurement-interaction (the probabilty density ρ). At the present state of our knowledge, it is an open question to know if quantum measurements are universal measurements.

Hilbert space geometry), which in part explains why Hilbert-models based on the Born rule can be used to efficaciously account for many experimental situations, also beyond the domain of microphysics [7,14,15].

So, if the average (2) is performed on a sufficiently large sample of persons, each one describing a different way of selecting an outcome, one can expect $\mu(q_j, e, p_{in}) \approx \langle \mu(q_2, e_\rho, p_{in}) \rangle^{univ}$, i.e., one can expect the collective participant to behave as a *universal participant*, described by the Born probabilities. The above, however, can only work for as long as the averages are performed on single measurements. Indeed, if sequential measurements are considered at the level of the individual participants, then the situation becomes much more complex (the average defining the collective participant's probabilities becomes much more involved) and one cannot expect anymore the standard quantum formalism (or a classical probability structure[d]) to be able to model all possible experimental situations, as we will show by means of a simple example in Section 9. But before that, let us describe the experimental situations where sequential measurements are performed (Sec. 6), and what are the different modeling options (Sec. 7). We will then derive a well-known quantum equality (Sec. 8) and show that it can be easily violated if the individual participants are not all "quantum clones."

6. Sequential measurements

We consider two psychological measurements, which we denote A and B, and we assume that the M_A outcome states of A are $\{a_1, \ldots, a_{M_A}\}$, whereas the M_B outcome states for B are $\{b_1, \ldots, b_{M_B}\}$. In many experimental situations, it is observed that the outcomes probabilities obtained when A is performed first, and then B, are not the same as those obtained when B is performed first, and then A. More precisely, *question order effects* manifest in the fact that, in general, the probability for the sequential outcome 'a_j then b_k' (denoted $a_j b_k$ in the following), when the sequence of measurements 'A then B' (denoted AB) is considered, is different from the probability for the sequential outcome 'b_k then a_j' (denoted $b_k a_j$ in the following), when the sequence 'B then A' is considered (denoted BA) [10,17,21,28]. Clearly, the collective participant to whom the sequential outcome probabilities are to be associated with, cannot be the one we previously described, as its main characteristic was that of being memoryless, whereas to produce order

---

[d]Note that when only single measurement situations are considered, the Hilbert probability model, based on the Born rule, is structurally equivalent to a classical Kolmogorovian model.

effects some short-term memory is needed. So, the situation is now more complex, which is the reason why, as we are going to see, a non-Hilbertian probability structure will generally emerge.

In the practice, sequential measurements are executed on a same participant. For instance, assuming that we have $2n$ participants, half of them will be subjected to the sequence AB and the other half to the sequence BA. More precisely, if $i \in \{1, \ldots n\}$, then the i-th individual will first be submitted to measurement A and then, immediately after, to measurement B (which for instance are two sequential questions in an opinion poll). The outcome of the first measurement (the answer to the first question) can change the initial state of the conceptual entity for the second one, as the two measurements are assumed to be performed in a rapid succession (i.e., the two questions are asked one after the other), so that the outcome of the first will remain in the sphere of consciousness of the i-th participant when submitted to the second measurement (i.e., when answering the second question). Similarly, if $i \in \{n+1, \ldots 2n\}$, then the i-th participant will first be submitted to measurement/question B followed by measurement/question A. If $n(a_j b_k)$ and $n(b_k a_j)$ are the total counts for the sequential outcomes $a_j b_k$ and $b_k a_j$, respectively, we have the experimental probabilities (relative frequencies):

$$p(a_j b_k) = \frac{n(a_j b_k)}{n}, \quad p(b_k a_j) = \frac{n(b_k a_j)}{n}, \tag{5}$$

which in general will exhibit order effects, i.e., $p(a_j b_k) \neq p(b_k a_j)$, for some j and k. These are the probabilities one will typically attempt to model using the quantum formalism.

Since the sequential measurements are performed at the level of the individual participants, we can generally write:

$$p(a_j b_k) = \frac{1}{n} \sum_{i=1}^{n} \mu(a_j, e_{A,i}, p_{\text{in}}) \mu(b_k, e_{B,i}, a_j),$$

$$p(b_k a_j) = \frac{1}{n} \sum_{i=n+1}^{2n} \mu(b_k, e_{B,i}, p_{\text{in}}) \mu(a_j, e_{A,i}, b_k), \tag{6}$$

where $e_{B,i}$ denotes the context associated with the i-th individual, when submitted to the B-measurement, which needs not to be equal to the context $e_{A,i}$ responsible for the outcomes of the A-measurement. Now, it is clear that the averages (6) are very different from (2), as they do not consist in a sum of probabilities (characterizing the different individual contexts) all associated with the same initial and final state transition. Here we have a much more intricate sum of products of probabilities, associated with different state transitions, where the different contexts and states get mixed in a complicate way. This is a very different situation than that of a single measurement universal average, and one cannot expect anymore the overall statistics of outcomes to be well approximated by the Born rule, even when n becomes very large.

Consider once more a collective participant to be associated with the experimental probabilities (5), i.e., such that, if submitted to the sequential measurements in question, it would deliver those same probabilities. We can write:

$$p(a_j b_k) = \mu(a_j, e_A, p_{\text{in}})\mu(b_k, e_B, a_j),$$
$$p(b_k a_j) = \mu(b_k, e_B, p_{\text{in}})\mu(a_j, e_A, b_k). \tag{7}$$

This time, however, it is not the collective mind of the collective participant that is subjected to the sequential measurements. Indeed, if this would be the case, no relevant order effects would be observed, as the sub-minds selected to answer the first question would not be the same as those answering the second question. So, the collective participant, in this case, turns out to be a much more artificial construct, as it would operate differently than how it does when subjected to a single measurement situation. Indeed, in the latter case each repetition of the measurement is performed by a different sub-mind (a different individual participant), whereas in the case of sequential measurements the same sub-mind is used to answer the two questions in the sequence.[e] Also, as we said already, even if n is large, there are no reasons to expect that the two contexts e_A and e_B, characterizing the collective participant in relation to the A and B measurements, would be both describable by the Born rule, hence be also equal. In fact, because of the mixing between states and contexts in the sums (6), we cannot even expect in this case the collective participant to use the same vectors as the individual participants to describe the initial and final states, in the sense that, to be able to find two contexts e_A and e_B modeling the experimental probabilities as per (7), one will generally need to use an effective description for the states that is different from the one that is inter-subjectively employed by the individual participants (see the penultimate section of [10] for a discussion of this point).

---

[e]One could imagine here that the sequential measurements are performed so swiftly one after the other that the collective participant does not have the time to activate a different sub-mind to answer the two questions in the sequence, whereas enough time would be available in-between the sequences.

7. Different modeling options

Let us more specifically consider *Situation 3* of Sec. 3, where participants not only can take decisions in different ways, but are also supplied with some information before doing so. We emphasize again that if one introduces deterministic processes that can change the initial state into a state which is possibly different for each participant, these have to be associated with clearly identifiable processes, which should be in principle predictable in advance (for instance, by having access to sufficient information about the cultural background and psychological profile of each individual). This because the contexts e_i^{det} are defined to be deterministic not only for the reason that they change the initial state in a predetermined way, but also because they are given in advance. If this would not be the case, i.e., if they would be actualized at the moment, when the participants are submitted to the conceptual situation, then they would be fundamentally indeterministic and their description should be included in e_i^{ind}.

What about the collective participant in this case? Should we also associate it with a process of information supply, i.e., with a deterministic context e^{det}, if this happens at the level of the individual participants? To answer this question, we assume for simplicity that there are only two intakes of information during the measurement, and more precisely that n_a among the n participants receive some information characterized by the context e_a^{det}, whereas the remaining $n_b = n - n_a$ receive some information characterized by the context e_b^{det}. Then, if p_a is the state obtained when the initial state p_{in} is submitted to context e_a^{det}, and we simply write $p_a = e_a^{\text{det}} p_{\text{in}}$, and similarly $p_b = e_b^{\text{det}} p_{\text{in}}$ is the state obtained when the initial state p_{in} is submitted to context e_b^{det}, (2) becomes:

$$\mu(q_j, e, p_{\text{in}}) = \frac{1}{n} \sum_{i=1}^{n_a} \mu(q_j, e_i^{\text{ind}}, p_a) + \frac{1}{n} \sum_{i=n_a+1}^{n} \mu(q_j, e_i^{\text{ind}}, p_b). \quad (8)$$

To further simplify the discussion, we also assume that all participants choose in a trivial way (*Situation 1* of Sec. 3) and that, say, $p_a = q_1$ and $p_b = q_2$. Then, $\mu(q_1, e, p_{\text{in}}) = \frac{n_a}{n}$ and $\mu(q_2, e, p_{\text{in}}) = \frac{n_b}{n}$. This means that, as we observed already, different deterministic cognitive actions performed by the individual participants translate, at the level of the collective participant, in an indeterministic action, to be described as a 'way of choosing' and not as an 'information supply'. Of course, the situation is different if all participants would access exactly the same information, as in this case one can also do as if the same would happen at the level of the collective participant.

In the situation of sequential measurements, however, it is much less clear if the information supply process should also be included in the modeling of the collective participant, even though all (real) individual participants access the same information. Assume for instance that we are in the situation where before measurement A some preliminary background information is given, which is the same for all participants, changing the initial state according to the deterministic context e_A^{det}, and that following measurement A, and before measurement B, some further information is given, changing the outcome state of measurement A according to the deterministic context e_B^{det}, and same thing when the order of the measurements is changed (an example of this kind of situation is the so-called *Rose/Jackson experiment* [10,22,28,30]). Then we can write:

$$p(a_j b_k) = \frac{1}{n} \sum_{i=1}^{n} \mu(a_j, e_{A,i}^{\text{ind}}, e_A^{\text{det}} p_{\text{in}}) \mu(b_k, e_{B,i}^{\text{ind}}, e_B^{\text{det}} a_j),$$

$$p(b_k a_j) = \frac{1}{n} \sum_{i=1}^{n} \mu(b_k, e_{B,i}^{\text{ind}}, e_B^{\text{det}} p_{\text{in}}) \mu(a_j, e_{A,i}^{\text{ind}}, e_A^{\text{det}} b_k), \tag{9}$$

and the question is: At the level of the collective participant, should we model the probabilities (9) by writing:

$$p(a_j b_k) = \mu(a_j, e_A^{\text{ind}}, e_A^{\text{det}} p_{\text{in}}) \mu(b_k, e_B^{\text{ind}}, e_B^{\text{det}} a_j),$$

$$p(b_k a_j) = \mu(b_k, e_B^{\text{ind}}, e_B^{\text{det}} p_{\text{in}}) \mu(a_j, e_A^{\text{ind}}, e_A^{\text{det}} b_k), \tag{10}$$

i.e., by also associating the collective participant with information supply deterministic contexts, or should we instead write:

$$p(a_j b_k) = \mu(a_j, e_A^{\text{ind}}, p_{\text{in}}) \mu(b_k, e_B^{\text{ind}}, a_j),$$

$$p(b_k a_j) = \mu(b_k, e_B^{\text{ind}}, p_{\text{in}}) \mu(a_j, e_A^{\text{ind}}, b_k), \tag{11}$$

without including deterministic contexts at the collective level?

The modeling option (10) can be defended by saying that since the same information is given to all participants, and that the collective participant is meant to represent their overall behavior, the description of its virtual cognitive action should also include the deterministic contexts e_A^{det} and e_B^{det}, in an explicit way. On the other hand, the modeling option (11) can be defended by saying that we can consider the information accessed by the participants before answering the questions to be part of the questions themselves. Indeed, questions always have some built-in context, i.e., some background information, appearing in more or less explicit terms in the way they are formulated. So, these deterministic contexts should be integrated in the indeterministic ones (see [10] for a discussion of this point in the ambit of the Rose/Jackson measurement).

Another possible argument in favor of the modeling option (11) is the following. Since the averages (9) result from a sum of products of probabilities, so that the effects of the deterministic and indeterministic contexts are mixed in a complicate way, it is questionable if one should really attribute to the collective participant the same 'information supply' processes of the individual participants (also considering that the former will not generally use the same representation for the initial and outcome states). In fact, one can go even further and question if it is really meaningful to model the experimental probabilities $p(a_j b_k)$ and $p(b_k a_j)$ as the products (10) or (11). Indeed, being the same individual who answers the sequence of questions, one could object that the correct way to interpret the experimental situation is to say that the AB and BA measurements are in fact single measurements with $M_A M_B$ outcomes each. This because the fact that a same participant (a same sub-mind of the collective participant) answers both questions should maybe be considered more relevant than the fact that they provide the answers in a sequential way.

According to this last viewpoint, it would not be (or only be) the sequentiality of the answers at the origin of the observed order effects, in the sense that we can imagine a slightly different experimental context where each participant would be jointly submitted to both questions, and jointly provide a couple of answers, and it is not unreasonable to expect that it is the very fact that in the AB and BA measurements the couples of questions are presented in a different order that would be at the origin of the difference (or part of the difference) between the probabilities for the outcomes $a_j b_k$ and $b_k a_j$. In other words, the order effects would originate (or in part originate) at the level of how questions are formulated, as is clear that the order of the different statements contained in a sentence can be relevant for what concerns its perceived meaning. Take the following example [24]:

A novice asked the prior: "Father, can I smoke when I pray?" And he was severely reprimanded. A second novice asked the prior: "Father, can I pray when I smoke?" And he was praised for his devotion.

We see that *pray & smoke* does not elicit the same meanings as *smoke & pray*. In the same way, the perceived meaning of the joint question *Are Clinton & Gore honest?* is not exactly the same as that of the question

Are Gore & Clinton honest? Accordingly, the perceived meaning of the answer *Clinton is honest & Gore is honest* is not the same as that of the answer *Gore is honest & Clinton is honest,* i.e, they do not correspond to the same state. This means that the two measurements AB and BA, when interpreted as single measurements, their outcome states will be in general different, i.e., AB and BA will be described in the quantum formalism by two different non-commuting Hermitian operators. So, even though AB and BA are in practice executed as two-step processes, i.e., as processes during which an outcome state is created in a sequential way, one can wonder to which extent one is allowed to experimentally disentangle the sequence into two distinct measurements. In other words, in general, measurements A and B are to be considered entangled in the combinations AB and BA (for the notion of entangled measurements in cognition, see [16]).

8. A quantum equality

In this section, we consider measurements only having two (possibly degenerate) outcomes, and the following quantity [23,28]:

$$q = p(a_1b_1) - p(b_1a_1) + p(a_2b_2) - p(b_2a_2). \tag{12}$$

If we assume that probabilities have to be modeled as sequential processes, and that background information is also provided, which we also want it to be modeled at the level of the collective participant, then according to (10) for the first term of (12) we can write:

$$p(a_1b_1) = \mu(a_1, e_A^{\text{ind}}, e_A^{\text{det}} p_{\text{in}})\mu(b_1, e_B^{\text{ind}}, e_B^{\text{det}} a_1), \tag{13}$$

and similarly for the other terms. Let us model the above using the standard quantum formalism. The initial state p_{in} is then described by a ket $|\psi_{\text{in}}\rangle \in \mathcal{H}$, where $\mathcal{H}$ denotes a Hilbert space of arbitrary dimension. Also, the two indeterministic contexts e_A^{ind} and e_B^{ind} are necessarily the same and their action is described by the Born rule and corresponding Lüders-von Neumann projection formula. Finally, the two deterministic contexts e_A^{det} and e_B^{det} can be associated with two unitary operators, which we denote U_A and U_B, respectively. For the first factor on the r.h.s. of (13), we can write: $\mu(a_1, e_A^{\text{ind}}, e_A^{\text{det}} p_{\text{in}}) = \|P_A U_A|\psi_{\text{in}}\rangle\|^2$, where P_A denotes the orthogonal projection operator onto the subspace associated with the outcome-state a_1 of observable A, described by the vector $|a_1\rangle = \frac{P_A U_A|\psi_{\text{in}}\rangle}{\|P_A U_A|\psi_{\text{in}}\rangle\|}$ (assuming that $P_A U_A|\psi_{\text{in}}\rangle \neq 0$). Therefore, the second factor on the r.h.s. of (13) can be written: $\mu(b_1, e_B^{\text{ind}}, e_B^{\text{det}} a_1) = \|P_B U_B|a_1\rangle\|^2 = \frac{\|P_B U_B P_A U_A|\psi_{\text{in}}\rangle\|^2}{\|P_A U_A|\psi_{\text{in}}\rangle\|^2}$, where P_B

denotes the projection onto the subspace associated with the outcome-state b_1 of observable B. Multiplying these two factors, we thus find:

$$p(a_1 b_1) = \|P_B U_B P_A U_A |\psi_{\text{in}}\rangle\|^2 = \langle \psi_{\text{in}} | U_A^\dagger P_A U_B^\dagger P_B U_B P_A U_A |\psi_{\text{in}}\rangle. \quad (14)$$

Proceeding in the same way with the other terms in (12), one obtains that $q = \langle \psi_{\text{in}} | Q | \psi_{\text{in}} \rangle$, with the self-adjoint operator Q given by:

$$Q = U_A^\dagger P_A P_B' P_A U_A - U_B^\dagger P_B P_A' P_B U_B + U_A^\dagger \bar{P}_A \bar{P}_B' \bar{P}_A U_A - U_B^\dagger \bar{P}_B \bar{P}_A' \bar{P}_B U_B, \quad (15)$$

where we have defined the orthogonal projectors: $P_A' \equiv U_A^\dagger P_A U_A$, $P_B' \equiv U_B^\dagger P_B U_B$, $\bar{P}_A = \mathbb{I} - P_A$, $\bar{P}_B = \mathbb{I} - P_B$, $\bar{P}_B' = \mathbb{I} - P_B'$ and $\bar{P}_A' = \mathbb{I} - P_A'$. We have:

$$\begin{aligned}
U_A^\dagger \bar{P}_A \bar{P}_B' \bar{P}_A U_A &= U_A^\dagger (\mathbb{I} - P_A) \bar{P}_B' \bar{P}_A U_A = U_A^\dagger \bar{P}_B' \bar{P}_A U_A - U_A^\dagger P_A \bar{P}_B' \bar{P}_A U_A \\
&= U_A^\dagger \bar{P}_B' (\mathbb{I} - P_A) U_A - U_A^\dagger P_A \bar{P}_B' (\mathbb{I} - P_A) U_A \\
&= U_A^\dagger \bar{P}_B' U_A - U_A^\dagger \bar{P}_B' P_A U_A - U_A^\dagger P_A (\mathbb{I} - P_B') U_A + U_A^\dagger P_A (\mathbb{I} - P_B') P_A U_A \\
&= U_A^\dagger (\mathbb{I} - P_B') U_A - U_A^\dagger (\mathbb{I} - P_B') P_A U_A - U_A^\dagger P_A (\mathbb{I} - P_B') U_A \\
&\quad + U_A^\dagger P_A (\mathbb{I} - P_B') P_A U_A \\
&= \mathbb{I} - U_A^\dagger P_B' U_A - U_A^\dagger P_A U_A + U_A^\dagger P_B' P_A U_A - U_A^\dagger P_A U_A + U_A^\dagger P_A P_B' U_A \\
&\quad + U_A^\dagger P_A P_A U_A - U_A^\dagger P_A P_B' P_A U_A \\
&= \mathbb{I} - U_A^\dagger P_B' U_A - U_A^\dagger P_A U_A + U_A^\dagger P_B' P_A U_A + U_A^\dagger P_A P_B' U_A \\
&\quad - U_A^\dagger P_A P_B' P_A U_A. \quad (16)
\end{aligned}$$

Therefore:

$$\begin{aligned}
U_A^\dagger P_A P_B' P_A U_A + U_A^\dagger \bar{P}_A \bar{P}_B' \bar{P}_A U_A &= \mathbb{I} - U_A^\dagger P_B' U_A - U_A^\dagger P_A U_A \\
&\quad + U_A^\dagger P_B' P_A U_A + U_A^\dagger P_A P_B' U_A = \mathbb{I} - U_A^\dagger P_B' U_A - P_A' + U_A^\dagger P_B' U_A P_A' \\
&\quad + P_A' U_A^\dagger P_B' U_A. \quad (17)
\end{aligned}$$

Similarly, exchanging the roles of A and B, we obtain:

$$\begin{aligned}
U_B^\dagger P_B P_A' P_B U_B + U_B^\dagger \bar{P}_B \bar{P}_A' \bar{P}_B U_B &= \mathbb{I} - U_B^\dagger P_A' U_B - U_B^\dagger P_B U_B \\
&\quad + U_B^\dagger P_A' P_B U_B + U_B^\dagger P_B P_A' U_B = \mathbb{I} - U_B^\dagger P_A' U_B - P_B' + U_B^\dagger P_A' U_B P_B' \\
&\quad + P_B' U_B^\dagger P_A' U_B. \quad (18)
\end{aligned}$$

The difference of the above two expressions then gives [12]:

$$\begin{aligned}
Q &= (P_B' - U_A^\dagger P_B' U_A) + (U_B^\dagger P_A' U_B - P_A') + (U_A^\dagger P_B' U_A P_A' - P_B' U_B^\dagger P_A' U_B) \\
&\quad + (P_A' U_A^\dagger P_B' U_A - U_B^\dagger P_A' U_B P_B'). \quad (19)
\end{aligned}$$

We see that in general $Q \neq 0$, so that the average $q = \langle \psi_{in} | Q | \psi_{in} \rangle$ can in principle take any value inside the interval $[-1, 1]$. However, if we consider that the modeling should not explicitly include deterministic contexts, or that they would be trivial, then we can set $U_A = U_B = \mathbb{I}$ in (19), and we clearly obtain $Q = 0$, so that for every initial state we have the remarkable equality $q = 0$. Clearly, if experimental data obey the latter, they possibly (although not necessarily) have a pure quantum structure, whereas if the $q = 0$ equality is disobeyed, one has to use beyond-quantum (and of course also beyond-classical) probabilistic models to fit the data.

9. Testing the collective participant

In this section, we submit the collective participant to the q-test derived in the previous section. More precisely, we show by means of a simple example that although the collective participant can behave in a pure quantum way on single measurements, when sequential measurements are performed at the individual level, the modeling of the obtained statistics of outcomes at the collective level will generally be non-quantum (i.e., non-Bornian), as it will violate the $q = 0$ equality. To do so, we place ourselves in the simplest possible situation: that of an experiment only using two participants ($n = 2$). We also assume that they both select outcomes in a way that is independent of the measurement considered, characterized by the probability distributions $\rho_1(y)$ for the first participant, and $\rho_2(y)$ for the second participant, which are such that $\rho_1(y) + \rho_2(y) = 1$. This means that the average probability (2) exactly corresponds to the Born quantum probability, characterized by the uniform probability density $\rho_u(y) = \frac{1}{2}$. In other words, The collective participant, describing the average behavior of these two individuals, is associated with a pure quantum context e_{ρ_u}, described by a uniform ρ_u-way of selecting an outcome.

In the following, we also consider for simplicity that no processes of information supply are to be considered. Eqs. (9) then become:

$$p(a_j b_k) = \frac{1}{2} \sum_{i=1}^{2} \mu(a_j, e_{\rho_i}, p_{in}) \mu(b_k, e_{\rho_i}, a_j),$$

$$p(b_k a_j) = \frac{1}{2} \sum_{i=1}^{2} \mu(b_k, e_{\rho_i}, p_{in}) \mu(a_j, e_{\rho_i}, b_k), \tag{20}$$

so that (12) becomes:

$$2q = 2[p(a_1b_1) - p(b_1a_1) + p(a_2b_2) - p(b_2a_2)]$$
$$+ \ \mu(a_1, e_{\rho_1}, p_{\text{in}})\mu(b_1, e_{\rho_1}, a_1) + \mu(a_1, e_{\rho_2}, p_{\text{in}})\mu(b_1, e_{\rho_2}, a_1)$$
$$- \ \mu(b_1, e_{\rho_1}, p_{\text{in}})\mu(a_1, e_{\rho_1}, b_1) - \mu(b_1, e_{\rho_2}, p_{\text{in}})\mu(a_1, e_{\rho_2}, b_1) \qquad (21)$$
$$+ \ \mu(a_2, e_{\rho_1}, p_{\text{in}})\mu(b_2, e_{\rho_1}, a_2) + \mu(a_2, e_{\rho_2}, p_{\text{in}})\mu(b_2, e_{\rho_2}, a_2)$$
$$- \ \mu(b_2, e_{\rho_1}, p_{\text{in}})\mu(a_2, e_{\rho_1}, b_2) - \mu(b_2, e_{\rho_2}, p_{\text{in}})\mu(a_2, e_{\rho_2}, b_2).$$

Using $\mu(a_2, e_{\rho_1}, p_{\text{in}}) = 1 - \mu(a_1, e_{\rho_1}, p_{\text{in}})$, as well as $\mu(b_2, e_{\rho_1}, p_{\text{in}}) = 1 - \mu(b_1, e_{\rho_1}, p_{\text{in}})$, we obtain:

$$2q = + \ \mu(a_1, e_{\rho_1}, p_{\text{in}})[\mu(b_1, e_{\rho_1}, a_1) - \mu(b_2, e_{\rho_1}, a_2)]$$
$$+ \ \mu(a_1, e_{\rho_2}, p_{\text{in}})[\mu(b_1, e_{\rho_2}, a_1) - \mu(b_2, e_{\rho_2}, a_2)]$$
$$- \ \mu(b_1, e_{\rho_1}, p_{\text{in}})[\mu(a_1, e_{\rho_1}, b_1) - \mu(a_2, e_{\rho_1}, b_2)] \qquad (22)$$
$$- \ \mu(b_1, e_{\rho_2}, p_{\text{in}})[\mu(a_1, e_{\rho_2}, b_1) - \mu(a_2, e_{\rho_2}, b_2)]$$
$$+ \ \mu(b_2, e_{\rho_1}, a_2) - \mu(a_2, e_{\rho_1}, b_2) + \mu(b_2, e_{\rho_2}, a_2) - \mu(a_2, e_{\rho_2}, b_2).$$

We can observe that the third line of (22) is zero if $\mu(b_2, e_{\rho_1}, a_2) = \mu(a_2, e_{\rho_1}, b_2)$ and $\mu(b_2, e_{\rho_2}, a_2) = \mu(a_2, e_{\rho_2}, b_2)$, which will generally be the case for quantum transition probabilities. In the more general situation where measurements are not characterized by a globally uniform probability density ρ_{u}, this can still be the case if individuals select outcomes in the same way in A and B measurements, which is what we have also previously assumed, for simplicity. Concerning the first line of (22), we observe it is zero if $\mu(b_1, e_{\rho_1}, a_1) = \mu(b_2, e_{\rho_1}, a_2)$ and $\mu(b_1, e_{\rho_2}, a_1) = \mu(b_2, e_{\rho_2}, a_2)$. For this to be so, we need ρ_1 and ρ_2 to be symmetric with respect to the origin of the Bloch sphere, i.e., $\rho_1(y) = \rho_1(-y)$ and $\rho_2(y) = \rho_2(-y)$, which of course will not be true in general. Similarly, the second line of (22) is zero if $\mu(a_1, e_{\rho_1}, b_1) = \mu(a_2, e_{\rho_1}, b_2)$ and $\mu(a_1, e_{\rho_2}, b_1) = \mu(a_2, e_{\rho_2}, b_2)$, which again requires ρ_1 and ρ_2 to be symmetric.

So, if ρ_1 and ρ_2 are symmetric, and if we assume that the two individuals select outcomes in the same way (although not necessarily as per the Born rule) in the two measurements, then the $q = 0$ equality is obeyed. But in general situations this will not be the case, so that the q-test will not be passed. As a simple example, consider $\rho_1(y) = \chi_{[-1,0]}(y)$ and $\rho_2(y) = \chi_{[0,1]}(y)$, where $\chi_I(y)$ denotes the characteristic function of the interval I. If α denotes the the angle between the ρ_1-elastic and the ρ_2-elastic, i.e., $\mathbf{a}_1 \cdot \mathbf{b}_1 = \cos\alpha$, and assuming for simplicity that $p_{\text{in}} = a_1$, we have: $\mu(a_1, e_{\rho_1}, a_1) = \mu(a_1, e_{\rho_2}, a_1) = \mu(b_2, e_{\rho_2}, a_2) = \mu(a_2, e_{\rho_2}, b_2) = \mu(b_1, e_{\rho_1}, a_1) = \mu(a_1, e_{\rho_1}, b_1) = 1$, $\mu(b_2, e_{\rho_1}, a_2) = \mu(a_2, e_{\rho_1}, b_2) =$

$\mu(b_1, e_{\rho_2}, a_1) = \mu(a_1, e_{\rho_2}, b_1) = \cos\alpha$. Inserting these values in (22), we obtain: $q = -\frac{1}{2}(1 - \cos\alpha)^2$, which is clearly different from zero for $\alpha \neq 0$.

Note that in all known sequential measurements the $q = 0$ equality is violated: in some of them only very weakly, in others quite strongly, showing that the underlying probability model is intrinsically non-Hilbertian. The measurements where the violation is stronger appear to be those where some background information is provided to the participants, before answering the questions [10,12,17,28,30]. Since in this case Q, given by (19), is different from zero, apparently the above description in terms of pure Bornian sequential measurements seems to provide an interesting modeling of the measurements, so much so that it is even able to predict, with good approximation, the $q = 0$ remarkable relation. However, it can be objected that what is really important is not if the q value is small, because even a small value is, strictly speaking, a violation, but if such value goes to zero as the number n of participants increases, which as far as we know is something that has not been studied yet. Note also, as we mentioned already, that even an exact obedience of the $q = 0$ equality would be insufficient to deduce that the underlying probability model is purely quantum, as probabilities that are more general than the Born probabilities are also able to obey the $q = 0$ identity; see [10] for a more specific analysis of this aspect.

10. Conclusion

In this article, we emphasized the importance of distinguishing between the individual level of the participants in a psychological experiment, and their collective level, which we have associated with a notion of collective participant. When the latter only describes single measurement situations, one can generally expect the standard quantum formalism to provide an effective modeling tool for the data. This because when averaging over all possible ways of selecting an outcome, one recovers the Born quantum rule, so that a measurement associated with a collective participant who is the expression of a sufficiently representative collection of individuals can be expected to be structurally close to a pure quantum measurement (and of course, for as long as single measurements are considered, also classical probability theory will do the job).

The modeling becomes however much more involved when individuals perform more than a single measurement, in a sequential way. Then, one should not expect anymore the standard quantum formalism (or the use of classical stochastic processes, like Markov chains) to be sufficient to model typical data. In the article, we have shown this by considering the simplest possible situation of a collective participant formed by only two individuals, whose collective action is pure-quantum when only single (non-sequential) measurements are considered, but irreducibly beyond-quantum, that is non-Born (and of course also beyond-classical, considering that measurements are incompatible[f]) when at the individual level outcomes are selected in a sequential way, so that a more general mathematical framework is needed to model the obtained experimental probabilities [10,12].

In our analysis, we have also proposed a distinction between deterministic and indeterministic cognitive processes/contexts, as formalized in the decomposition (1) of the (possibly different) individual contexts. As far as this distinction is concerned, we observed that different modeling options are possible, putting the cognitive action more in the deterministic evolution of the initial state or in the indeterministic collapse of it, or in some combination of the two. Concerning the individual cognitive processes, it is however clear that the different possible ways of modeling them should correspond to objective components of the cognitive situations, possibly testable in well-designed experiments.

References

1. Aerts, D. (1998). The entity and modern physics: the creation discovery view of reality. In E. Castellani (Ed.), *Interpreting Bodies: Classical and Quantum Objects in Modern Physics (pp. 223–257)*. Princeton: Princeton Unversity Press.
2. Aerts, D. (2011). Quantum interference and superposition in cognition: Development of a theory for the disjunction of concepts. In Aerts, D., et al. (Eds.) *Worldviews, Science and Us* (pp. 169–211). Singapore: World Scientific.
3. Aerts, D. & Aerts, S. (1995). Applications of quantum statistics in psychological studies of decision processes. *Foundations of Science 1*, 85–97.
4. Aerts, D., Coecke, B., Durt, T. & Valckenborgh, F. (1997). Quantum, classical and intermediate I: a model on the Poincaré sphere. *Tatra Mountains Mathematical Publications 10*, 225.
5. Aerts D., Coecke B. & Smets S. (1999). On the Origin of Probabilities in Quantum Mechanics: Creative and Contextual Aspects. In: Cornelis G.C., Smets S., Van Bendegem J.P. (eds). Metadebates on Science. Einstein Meets Magritte: An Interdisciplinary Reflection on Science, Nature, Art, Human Action and Society, vol 6 (pp. 291–302). Springer, Dordrecht.

---

[f]This cannot be the case for classical stochastic processes, where the randomness only comes from a lack of knowledge about the state of the considered entity.

6. Aerts, D. & Sassoli de Bianchi M. (2014). The Extended Bloch Representation of Quantum Mechanics and the Hidden-Measurement Solution to the Measurement Problem. *Annals of Physics 351*, 975–1025.

7. Aerts, D. & Sassoli de Bianchi M. (2015a). The unreasonable success of quantum probability I. Quantum measurements as uniform fluctuations. *Journal Mathematical Psychology 67*, 51–75.

8. Aerts, D. & Sassoli de Bianchi M. (2015b). The unreasonable success of quantum probability II. *Journal Mathematical Psychology 67*, 76–90.

9. Aerts, D. and Sassoli de Bianchi, M. (2016): The GTR-model: a universal framework for quantum-like measurements. In: "Probing the Meaning of Quantum Mechanics. Superpositions, Dynamics, Semantics and Identity," 91–140. Eds. D. Aerts, C. De Ronde, H. Freytes and R. Giuntini, World Scientific Publishing Company, Singapore.

10. Aerts, D. & Sassoli de Bianchi, M. (2017a). Beyond-quantum modeling of question order effects and response replicability in psychological measurements. *Journal Mathematical Psychology 79*, 104–120.

11. Aerts, D. & Sassoli de Bianchi M. (2017b). Quantum measurements as weighted symmetry breaking processes: the hidden measurement perspective. *International Journal of Quantum Foundations 3*, 1–16.

12. Aerts D., Beltran L., Sassoli de Bianchi M., Sozzo S. & Veloz T. (2017b). Quantum Cognition Beyond Hilbert Space: Fundamentals and Applications. In: de Barros J., Coecke B., Pothos E. (eds) Quantum Interaction. QI 2016. *Lecture Notes in Computer Science 10106*, Springer, Cham, 81–98.

13. Aerts, D., Sassoli de Bianchi M. and Sozzo S. (2016). On the Foundations of the Brussels Operational-Realistic Approach to Cognition. *Frontiers in Physics 4*; doi: 10.3389/fphy.2016.00017.

14. Aerts, D. & Sozzo, S. (2012a). Entanglement of conceptual entities in Quantum Model Theory (QMod). *Quantum Interaction. Lecture Notes in Computer Science 7620*, 114–125.

15. Aerts, D. & Sozzo, S. (2012b). Quantum Model Theory (QMod): Modeling contextual emergent entangled interfering entities. *Quantum Interaction. Lecture Notes in Computer Science 7620*, 126–137.

16. Aerts, D., & Sozzo, S. (2014). Quantum entanglement in conceptual combinations. *International Journal of Theoretical Physics 53*, 3587–360.

17. Busemeyer, J. R., & Bruza, P. D. (2012). *Quantum Models of Cognition and Decision*. Cambridge: Cambridge University Press.

18. Busemeyer, J. R., & Diederich, A. (2009). *Cognitive Modeling*. London: Sage Publications.

19. Busemeyer, J. R., Pothos, E. M., Franco, R., & Trueblood, J. S. (2011). A quantum theoretical explanation for probability judgment errors. *Psychological review 118*, 193–218.

20. Coombs, C. H., Dawes, R. M. & Tversky, A. (1970). *Mathematical Psychology: An Elementary Introduction*. Prentice-Hall.

21. Khrennikov, A.Y., Basieva, I., Dzhafarov, E.N., Busemeyer, J.R. (2014). Quantum models for psychological measurements: An unsolved problem. *PlosOne 9*, e110909.

22. Moore, D. W. (2002). Measuring new types of question-order effects: additive and subtractive. *Public Opinion Quarterly 66*, 80–91.
23. Niestegge, G. (2008). An approach to quantum mechanics via conditional probabilities. *Foundations of Physics 38*, 241–256.
24. Rampin, M. (2005) *Al gusto di cioccolato. Come smascherare i trucchi della manipolazione linguistica.* Milano: Ponte alle Grazie.
25. Tentori, K., Crupi, V. & Russo, S. (2013). On the determinants of the conjunction fallacy: probability versus inductive confirmation. *Journal of Experimental Psychology: General 142*, 235-255.
26. Tversky, A. & Kahneman, D. (1983). Extensional versus intuitive reasoning: The conjunction fallacy in probability judgment. *Psychological Review 90*, 293–315.
27. Tversky, A. & Koehler, D. J. (1994). Support theory: A nonextensional representation of subjective probability. *Psychological Review 101*, 547–567.
28. Wang, Z. & Busemeyer, J. R. (2013). A quantum question order model supported by empirical tests of an a priori and precise prediction. *Topics in Cognitive Science 5*, 689–710.
29. Wang, Z., Busemeyer, J. R., Atmanspacher, H. & Pothos, Emmanuel M. (2013). The Potential of Using Quantum Theory to Build Models of Cognition. *Topics in Cognitive Science 5*, 672–688.
30. Wang, Z., Solloway, T., Shiffrin, R. M. & Busemeyer, J. R. (2014). Context effects produced by question orders reveal quantum nature of human judgments. *Proeedings of the National Academy of Sciences 111*, 9431–9436.
31. Zucchini, W. (2000). An introduction to model selection. *Journal of Mathematical Psychology 44*, 41–61.